SEISMIC WAVE PROPAGATION AND SCATTERING IN THE HETEROGENEOUS EARTH

AIP Series in
Modern Acoustics and Signal Processing

ROBERT T. BEYER, Series Editor-in-Chief
Physics Department, Brown University

EDITORIAL BOARD

BOOKS IN SERIES

SEISMIC WAVE PROPAGATION AND SCATTERING IN THE HETEROGENEOUS EARTH

Haruo Sato
Tohoku University, Japan

Michael C. Fehler
Los Alamos National Laboratory, U.S.A.

Springer

Haruo Sato
Department of Geophysics
Graduate School of Science
Tohoku University
980-77, Sendai, Japan

Michael C. Fehler
Geo Engineering
Earth and Environmental Sciences Division
Los Alamos National Laboratory
Los Alamos, NM 87545 USA

Series Editor
Robert T. Beyer
Physics Department, Brown University

Library of Congress Cataloging-in-Publication Data
Sato, Haruo.
Seismic wave propagation and scattering in the heterogeneous earth.
/ Haruo Sato, Michael C. Fehler.
   p. cm. — (Modern acoustics and signal processing)
   Includes bibliographical references (p.   –   ) and index.

   ISBN 978-1-4612-7457-5   ISBN 978-1-4612-2202-6 (eBook)
   DOI 10.1007/978-1-4612-2202-6

   1. Seismic waves.   2. Wave-motion, Theory of.   I. Fehler, Michael
C. II. Title. III. Series.
QE538.5.S267   1997   97-33260
551.21—dc21

Printed on acid free paper.

© 1998 Springer-Verlag New York, Inc.
Softcover reprint of the hardcover 1st edition 1998

Production managed by Anthony Battle; manufacturing supervised by Jacqui Ashri.
Photocomposed copy prepared from the authors' Microsoft Word files.

9 8 7 6 5 4 3 2 1

ISBN 978-1-4612-7457-5

Dedicated to
Yoko and Schön

SERIES PREFACE

". . . Soun is noght but air y-broke"
—*Geoffrey Chaucer*
end of the 14th century

Traditionally, acoustics has formed one of the fundamental branches of physics. In the twentieth century, the field has broadened considerably and become increasingly interdisciplinary. At the present time, specialists in modern acoustics can be encountered not only in Physics Departments, but also in Electrical and Mechanical Engineering Departments, as well as in Departments of Mathematics, Oceanography, and even Psychology. They work in areas spanning from musical instruments to architecture to problems related to speech perception. Today, six hundred years after Chaucer made his brilliant remark, we recognize that sound and acoustics is a discipline extremely broad in scope, literally covering waves and vibrations in all media at all frequencies and at all intensities.

This series of scientific literature, entitled *Modern Acoustics and Signal Processing (MASP)*, covers all areas of today's acoustics as an interdisciplinary field. It offers scientific monographs, graduate level textbooks, and reference materials in such areas as: architectural acoustics; structural sound and vibration; musical acoustics; noise; bioacoustics; physiological and psychological acoustics; speech; ocean acoustics; underwater sound; and acoustical signal processing.

Acoustics is primarily a matter of communication. Whether it be speech or music, listening spaces or hearing, signaling in sonar or in ultrasonography, we seek to maximize our ability to convey information and, at the same time, to minimize the effects of noise. Signaling has itself given birth to the field of signal processing, the analysis of all received acoustic information or, indeed, all information in any electronic form. With the extreme importance of acoustics for both modern science and industry in mind, AIP Press is initiating this series as a new and promising publishing venture. We hope that this venture will be beneficial to the entire international acoustical community, as represented by the Acoustical Society of America, a founding member of the American Institute of Physics, and other related societies and professional interest groups.

It is our hope that scientists and graduate students will find the books in this series useful in their research, teaching, and studies.

James Russell Lowell once wrote: "In creating, the only hard thing's to begin." This is such a beginning.

Robert T. Beyer
Series Editor-in-Chief

PREFACE

The structure of the earth has been extensively studied using seismic waves generated by natural earthquakes and man-made sources. In classical seismology, the earth is considered to consist of a sequence of horizontal layers having differing elastic properties, which are determined from travel-time readings of body waves and the dispersion of surface waves. More recently, three-dimensional inhomogeneity having scale larger than the predominant seismic wavelength has been characterized using travel-time data with velocity tomography. Forward and inverse waveform modeling methods for deterministic models have been developed that can model complicated structures allowing many features of complex waveforms to be successfully explained. Classical seismic methods are described in books like *Quantitative Seismology: Theory and Methods* by Aki and Richards [1980], *Seismic Waves and Sources* by Ben-Menahem and Singh [1981], *Theory and Application of Microearthquake Networks* by Lee and Stewart [1981], *Seismic Wave Propagation in Stratified Media* by Kennett [1985], and *Modern Global Seismology* by Lay and Wallace [1995]. High frequency (>1Hz) seismograms of local earthquakes, however, often contain continuous wave trains following the direct S-wave that cannot be explained by the deterministic structures developed from tomographic or other methods. Array observations have shown that these wave trains, known as "coda waves", are incoherent waves scattered by randomly distributed heterogeneities having random sizes and contrasts of physical properties. The characteristic scale of the heterogeneity that has the most influence on a given wave is not always much longer than but is sometimes the same order of the wavelength of the seismic wave. Strong random fluctuations in seismic velocity and density having short wavelengths superposed on a step-like structure are found in well-logs of boreholes drilled even in old crystalline rocks located in stable tectonic environments. These observations suggest a description of the earth as a random medium with a broad spectrum of spatial velocity fluctuations and the resulting importance of seismic wave scattering.

In the 1970s, geophysicists began to investigate the relationship between seismogram envelopes and the spectral structure of the random heterogeneity in the earth. Initial models were based on a phenomenological description of the scattering process. Later, in parallel with additional observational work, there have been theoretical studies using perturbation methods, the parabolic approximation, the phase screen method, and another empirical method known as the radiative transfer theory. These developments have gradually established a description of the scattering process of seismic waves in the inhomogeneous earth and have allowed a characterization of the statistical properties of the inhomogeneity.

This book focuses on developments over the last two decades in the areas of seismic wave propagation and scattering through the randomly heterogeneous structure of the earth with emphasis on the lithosphere. The characterization of the earth as a random medium is complementary to the classical stratified media characterization. We have tried to combine information from many sources to present a coherent introduction to the theory of scattering in acoustic and elastic materials that has been developed for the analysis of seismic data on various scales. Throughout the book, we include discussions of observational studies made using the various theoretical methods so the reader can see the practical use of the methods for characterizing the earth. The audience includes both undergraduate and graduate students in the fields of physics, geophysics, planetary sciences, civil engineering, and earth resources. In addition, scientists and engineers who are interested in the structure of the earth and wave propagation characteristics are included.

Many people have helped us. Keiiti Aki's encouragement and pioneering work in this field were major factors in getting this project started. Yoichi Ando kindly invited us to contribute to this book series. We benefited from careful reviewing of the manuscript by Keiiti Aki and Ru-Shan Wu. We thank Masakazu Ohtake, Ryosuke Sato, Alexei Nikolaev, Tania Rautian, Vitaly Khalturin, and Eystein Husebye for continuous encouragement. Many of our colleagues, friends, and graduate students have collaborated with us in the development of stochastic studies of seismic wave scattering, helping us to learn more than we knew: Shigeo Kinoshita, Frank Scherbaum, Leigh House, Peter Roberts, Rafael Benites, Steve Hildebrand, W. Scott Phillips, Hans Hartse, Kazushige Obara, Mitsuyuki Hoshiba, Anshu Jin, Bernard Chouet, Alexander Gusev, Yuri Kopnichev, Osamu Nishizawa, Satoshi Matsumoto, Kiyoshi Yomogida, Teruo Yamashita, Yasuto Kuwahara, Kinichiro Kusunose, Yanis Baskoutas, Kazuo Yoshimoto, Hisashi Nakahara, Ken Sakurai, Kazutoshi Watanabe, Katsuhiko Shiomi, Lee Steck, Lian-Jie Huang, Takeshi Nishimura, Fred Moreno, and Tong Fei. Michael Fehler gratefully acknowledges James Albright and C. Wes Myers for encouraging his work on this book. Ruth Bigio assisted in drafting some of the figures. We thank Maria Taylor of Springer-Verlag/AIP Press for her encouragement throughout this project, and Anthony Battle of Springer-Verlag for his cooperation.

Haruo Sato
Michael C. Fehler

CONTENTS

CHAPTER 1

Introduction

The region of the earth down to about 100 km is called the lithosphere. Rigorously speaking, lithosphere refers to the solid portion of the earth that overlays the low velocity zone or the asthenosphere, and the thickness varies from place to place depending on the tectonic setting; however, we will use this term loosely for the upper 100 km of the earth that consists of the crust and the uppermost mantle. The structure of the earth's crust has been investigated using layered models since the discovery of the Mohorovicic discontinuity or Moho at the base of the crust [Mohorovicic, 1909] and the Conrad discontinuity in the mid crust [Conrad, 1925]. The characterization of the earth as a random medium is complementary to the classical stratified medium characterization. Recent surveys using the reflection method, such as those conducted by the Consortium for Continental Reflection Profiling (COCORP) reveal that the Moho is not a simple discontinuity but a transition zone consisting of many segments of small reflectors, and the crust is heterogeneous on scales of a few kilometers to tens of kilometers [Schilt et al., 1979]. Well-log data collected in the shallow crust exhibit strong random heterogeneity with short wavelengths [Telford et al., 1976, p. 402]. Moreover, the development and application of regional velocity tomography [Aki et al., 1976; Husebye et al., 1976], which uses travel-time readings from seismograms of teleseismic waves, local earthquakes, or man-made sources such as explosions has allowed the delineation of the inhomogeneous velocity structure on scales from a few meters to a few tens of kilometers in various regions of the world.

Aki [1969] first focused interest on the appearance of continuous wave trains in the tail portion of individual seismograms of local earthquakes as direct evidence of the random heterogeneity of the lithosphere. Local earthquakes are those having magnitudes less than about 5, recorded at distances less than 100 km, and radiating frequencies ranging from 1–30 Hz. These wave trains, which are named "coda," look like random signals having an envelope whose amplitude gradually decreases with increasing time. Aki proposed that coda is composed of a superposition of incoherent waves scattered by distributed heterogeneities in the earth. Rautian and Khalturin [1978] showed that coda envelopes in central Asia decay stably irrespective of epicentral distance and the envelopes at all distances have a similar temporal dependence where time is measured from the initiation time of an earthquake, a time referred to as lapse time. This observed characteristic of seismic coda is strongly linked to the historic and continuing use of the duration of a recorded seismogram as a measure of the magnitude of a local earthquake [Solov'ev, 1965; Tsumura, 1967a].

Until recently, most seismograms of local and regional earthquakes, those recorded at distances to as long as 300 km, were recorded on regional networks whose primary function was to record first arrival times to be used for locating earthquakes. Since these networks relied on analog transmission of data from multiple stations over radio link or phone lines, the dynamic range of the recordings was limited. This necessitated that the stations be run at high gain to allow identification of the first arriving P-waves from small local earthquakes. High gains often clipped the early portions of seismograms. The first methods for analysis of coda waves were thus developed to be used on seismograms whose early portions were clipped. Coda analysis gained in popularity as a means for obtaining information about seismic source spectra and media properties from data collected by these high-gain networks. Coda waves with frequencies on the order of one Hz were thought to offer a useful seismological tool for the quantitative estimation of the strength of random heterogeneity [Aki and Chouet, 1975; Sato, 1978]. As seismometers were placed in boreholes where ambient seismic noise was greatly reduced compared to that on the surface and as the onset of digital recording allowed dynamic range to be increased, envelopes of entire seismograms including the coda portion were recorded and modeled to learn more about the heterogeneity of the earth's lithosphere using frequencies in the range of 1 to 30 Hz. Envelopes of seismograms of artificial sources and extremely small earthquakes having magnitudes less than zero that are recorded in boreholes have been useful for the study of scattering processes in the frequency range of the order of kHz revealing information about inhomogeneity on the scale of meters [Fehler, 1982].

Models for seismic wave propagation through inhomogeneous elastic media have been developed using deterministic approaches such as mode theory for layered structures or high-frequency approaches such as the eikonal approximation. However, array analysis has shown that coda waves are not regular plane waves coming from the epicenter, but are composed of scattered waves coming from all directions [Aki and Tsujiura, 1959]. Ray theoretical approaches are thus unsuitable for the study of coda. In the 1970s, S-coda waves were studied using the single scattering approximation to the wave equation as one end-member model and a model based on the diffusion equation as another end-member [Wesley, 1965; Aki and Chouet, 1975; Kopnichev, 1975; Sato, 1977a]. The single scattering theory based on the Born approximation for elastic media has been used to explain characteristics of observed three-component seismogram envelopes using a point shear-dislocation earthquake-like source [Sato, 1984a]. In parallel with the development of theoretical modeling, S-coda characteristics such as scattering coefficient and S-coda attenuation Q_C^{-1} have been measured throughout the world and compared with seismotectonic settings. There have been reports of temporal changes in these parameters in relation to the occurrence of large earthquakes [Gusev and Lemzikov, 1985; Jin and Aki, 1986].

Since the middle of the 1980s, the multiple scattering process has been modeled analytically using the radiative transfer theory, which has been extended from one based on a stationary solution [Wu, 1985] that was borrowed from astrophysics [Ishimaru, 1978] to the inclusion of explicit temporal dependence. The radiative transfer theory models the transfer of energy in the medium so the detailed mathematical treatment of phase can be ignored. This theory has been extended to include cases of importance to propagation in the earth such as nonisotropic scattering and nonspherical radiation from a point shear-dislocation earthquake-like source [Wu and Aki, 1988a; Shang and Gao, 1988; Zeng et al., 1991; Sato, 1994b]. In parallel, Monte-Carlo simulations have been developed for numerical syntheses of seismogram envelopes based on the radiative transfer approach in more complex structures [Gusev and Abubakirov, 1987; Hoshiba, 1994]. Although the radiative transfer theory has shown considerable utility for explaining observed seismic data, its formal basis in the wave equation lacks rigorous foundation. It has, however, provided tractable models for the analysis of multiple scattering, can be shown to be consistent with conservation of energy, and gives model predictions for single scattering cases that are consistent with those derived from the wave equation. The incoherent sum of scattered waves has been shown to be a suitable model not only for S-coda envelope but also for whole seismogram envelopes.

As a natural consequence of energy conservation, the excitation of coda waves in scattering media means that the direct wave that propagates along the minimum travel-time path from the source to the receiver loses energy with increasing propagation distance. Until the 1970s, however, there was little theoretical understanding of the contribution of scattering loss as a mechanism for attenuation of seismic waves. Intrinsic attenuation was considered dominant and frequency-independent. Using data from dense regional networks that were constructed in the U. S. A. and Japan for observations of microearthquakes, the frequency dependence of S-wave attenuation Q_S^{-1} was measured. Results showed that Q_S^{-1} decreases with increasing frequency for frequencies higher than 1 Hz. Combining attenuation measurements for lower frequencies made on surface waves, Aki [1980a] conjectured that Q_S^{-1} has a peak around 0.5 Hz. If scattering is the dominant mechanism of attenuation, the observed frequency dependence cannot be explained by the ordinary stochastic mean field theory for wave propagation through random media, which predicts that attenuation increases with frequency. To resolve the discrepancy, improvements were introduced to the stochastic theory to make it a more realistic model for the practical seismological measurement of amplitude attenuation. One improvement was to ignore the effects of forward scattering by integrating energy of scattered waves only for scattering angles larger than 90° [Wu, 1982a, b]; the other is to subtract the travel-time fluctuation caused by the slowly varying velocity perturbation before using the stochastic averaging procedure [Sato, 1982a]. The spectral structure of the random inhomogeneity has been quantitatively studied using these models and measurements of attenuation.

As the concept of scattering loss was accepted in the seismological community, Wu [1985] introduced the seismic albedo as a phenomenological measure of the contribution of scattering attenuation to the total attenuation. Fehler et al. [1992] proposed a method to estimate the seismic albedo by analysis of whole S-seismogram envelopes. The seismic albedo measurement is based on the solution of the radiative transfer theory for the multiple isotropic scattering process for an impulsive spherically symmetric source radiation [Hoshiba et al. 1991; Zeng et al., 1991]. Since this method, known as the multiple lapse-time window analysis method, was developed, we have been able to measure the ratio of scattering attenuation and intrinsic attenuation quantitatively and many measurements have been reported worldwide.

In general, when waves propagate through random media, the interaction excites coda waves and, in addition, changes the characteristics of the early portion of the seismogram envelope. Although the source duration of small earthquakes is often less than 1 s, the duration of S-wave envelopes are found to be much longer than 1 s at hypocentral distances greater than 100 km [Sato, 1989]. There is an observed time delay between the S-wave onset and the occurrence of the maximum of the S-wave envelope. Both the envelope broadening and the peak delay were quantitatively studied using the stochastic averaging method for the parabolic approximation, which was originally developed for optical waves or acoustic waves through media with random refractive index. The theory gives a good explanation for the observed characteristics of the early portions of S-wave envelopes at large travel distances as a result of strong diffraction.

Measurements of amplitude fluctuation of direct waves arriving from artificial explosions using data from arrays up to a few hundred km long were interpreted with a model based on a stochastic description of random inhomogeneity to quantify the statistical properties of the crust [Nikolaev, 1975]. Phase and amplitude correlation measurements of teleseismic P-waves on a seismic array revealed the scale of inhomogeneity [Aki, 1973]. These observations suggest a description of the earth as a random medium with a broad spectrum of spatial velocity

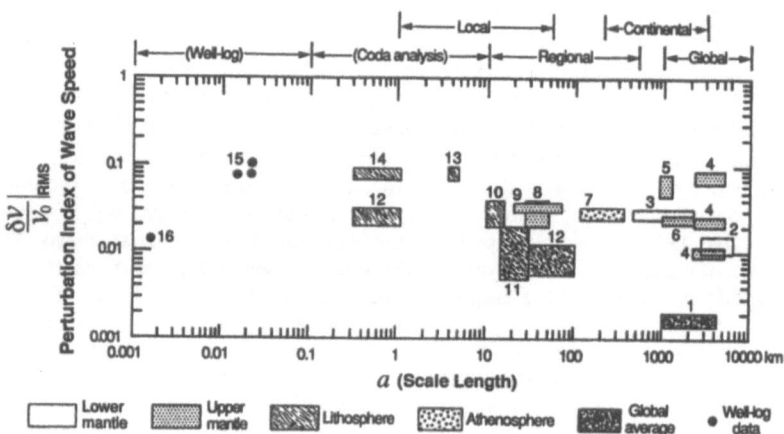

FIGURE 1.1. Fractional velocity fluctuations for different scales estimated by various methods: 1, global average, mode splitting of free oscillations; 2–3, lower mantle, velocity tomography; 4–9, upper mantle, surface wave inversion and velocity tomography; 10–12, lithosphere, transmission fluctuation; 13–14, lithosphere, coda-wave excitation; 15–16, upper crust, acoustic well-log. [From Wu and Aki, 1988b, with permission from Birkhäuser Verlag AG, Switzerland.]

fluctuations and the resulting importance of the scattering processes.

Wu and Aki [1988b] compiled measurements of fractional velocity fluctuations made in the 1980s. Figure 1.1 shows the RMS fractional velocity fluctuation plotted against scale length of the heterogeneity of the earth. This figure illustrates the wide range of scales over which heterogeneities are spread within the earth. Various methods for the estimation of the heterogeneity strength are used depending on the scale length. Well-log data are reliable for short scale lengths of 1–100 m in the shallow crust; coda excitation measurements of local earthquakes are useful for scale lengths of 0.1–10 km in the lithosphere; the resolution of velocity tomography analysis using travel-time readings of seismic waves is a few to a few tens of km for the crust to upper mantle; surface-wave inversion and free oscillation analysis become useful for scale lengths longer than a few tens of km. Several special issues of journals which focus on these subjects have been published [Husebye, 1981; Wu and Aki, 1988c, 1989, 1990; Sato, 1991b; Korn et al., 1997].

This book focuses on developments over the last two decades in the areas of seismic wave propagation and scattering characteristics through the randomly inhomogeneous earth structure, especially in the lithosphere. In Chapter 2, we will briefly review data and observations that support the view that the lithosphere is heterogeneous. Chapter 3 is an introduction to coda-wave excitation using phenomenological modeling, which forms the basis of S-coda analysis and the coda-normalization method. We discuss the Born approximation for inhomogeneous elastic media in Chapter 4. Scalar wave theory is introduced as a mathematical introduction. In Chapter 5, we first review the frequency dependence of observed Q_S^{-1} and discuss several of the proposed mechanisms of intrinsic absorption. Then we introduce an improved stochastic averaging method that is consistent with observational methods, and we will derive a formula describing the frequency dependence of scattering attenuation in random media. In Chapter 6, a method for synthesizing three-component seismogram envelopes of a local earthquake based on the summation of incoherent waves scattered by random elastic inhomogeneities will be developed. This model includes the effects of nonspherical radiation from a realistic point shear-dislocation source in addition to nonisotropic scattering from the inhomogeneities. Chapter 7 contains an introduction to approaches for seismogram envelope synthesis based on the radiative transfer theory for the case of spherical source radiation and multiple isotropic scattering. After the theory is developed, we present the multiple lapse-time window analysis for the estimation of seismic albedo. We will also introduce extensions of the theory to cases of non-spherical source radiation, nonisotropic scattering, and wave-type conversions. In Chapter 8, we will review the parabolic approximation for scalar waves and its stochastic treatment. Correlation measurements of teleseismic P-waves by an array and the broadening of S-wave seismogram envelopes will be discussed. Finally, in Chapter 9, we will summarize the results developed in the last two decades and discuss issues for future research.

CHAPTER 2

Heterogeneity in the Lithosphere

Geologists and geophysicists have numerous ways to investigate and characterize heterogeneity in the earth. Geophysical characterization includes measurement of physical properties such as seismic velocities and density of rocks. Geological characterization includes mineralogical composition and grain size distribution that are both controlled by the processes by which the rock evolved. Geologists observe the surface of the earth and analyze rocks that originated from within the earth for signs of heterogeneity. The wide variation of rocks erupted from volcanoes provides geochemical and geological evidence of heterogeneity within the earth. Tectonic processes such as folding, faulting, and large scale crustal movements associated with plate tectonics contribute to making the lithosphere heterogeneous. Rocks recovered from boreholes show wide variation and rapid changes in chemical composition with depth. Geophysical measurements in wells show correlation and lack of correlation with chemical composition of the rocks, indicating that mineral composition alone is not the only factor that controls the physical properties of rocks. Deterministic seismic studies reveal a wide spatial variation in elastic properties within the earth's lithosphere. Scattering of high-frequency seismic waves shows the existence of small scale heterogeneities in the lithosphere. In this chapter, we will review some of the methods by which the heterogeneous structure of the earth's crust can be evaluated.

2.1 GEOLOGICAL EVIDENCE

The earth has heterogeneities on many scales. Rocks have crystals that range in size from fractions of mm to a few cm in scale. Properties of minerals that make up the bulk of rocks in the earth's crust vary a great deal [Simmons and Wang, 1971]. For example, the bulk modulus of quartz, one of the major constituents of crustal rocks, is about 0.39×10^{12} dyn/cm^2 whereas that of the mineral plagioclase, another major constituent, is about 0.65×10^{12} dyn/cm^2 [Simmons and Wang, 1971]. Thus, the relative abundance of these two minerals in a rock can greatly influence its elastic properties. In addition to mineralogy, fractures influence the elastic properties of a rock [Simmons and Nur, 1968]. Fractures range in size from submicroscopic to many tens of meters. Since fractures are more compliant than intact minerals, the spatial variations in fracture content and size can have a larger influence on elastic properties of crustal rocks than mineral composition. Table 2.1 lists the P- and S-wave velocities of some common rocks that compose the earth's crust. Figure 2.1 shows laboratory measurements of velocity variation with pressure for granite from Westerly, Rhode Island, U. S. A. The variation with pressure is due to the closure of fractures having lengths ranging from 0.01 mm to 1 cm and is typical of most crustal rocks. The P-wave velocity is more sensitive to the presence of fluids in the fractures than the S-wave velocity since fluids transmit compressional waves but not shear waves.

The earth's crust contains a wide variation of rock types. The variations in rock composition can range on scales of a few mm to many km. Holliger and Levander [1992] examined geological maps and properties of rocks for a region of northern Italy that is believed to be an exposed section of the earth's lower crust and concluded that the spatial variation of this region has a characteristic scale of 200–800 m. Intrusions of magma into preexisting country rock can result in dikes and sills that have different composition from the country rock. These dikes and sills can be as small as a few mm wide resulting in a rapid spatial variation in rock properties. Variations in rock properties in volcanic regions can occur on scales of a few m to a few km due to variations in composition of magmas erupted at differing stages of a volcano's life. The variation in tectonic provinces occurs over tens to hundreds of km. For example, the Cascade range in the western U. S. A. is largely made up of young volcanic rocks whose elastic properties are dramatically different from those of the old Precambrian rocks of the central U. S. A.

The earth's crust has largely been formed through volcanic processes. Large silicic batholiths like the Sierra Nevada, U. S. A. are the intrusive remains of volcanic complexes that have been eroded away. Geochemists argue that silicic rocks that intrude into the shallow crust and erupt at volcanoes were formed by either fractionation of iron-rich rocks that intrude into the lower crust from the mantle or by the transfer of heat from intruded iron-rich mantle rocks to silicic rocks in the deep crust [Perry et al., 1990]. In either case, there will be high-velocity material remaining within the silicic crust. The velocity of the high-density material may be as high as 7.5 km/s [Fountain and Christensen, 1989]. If heat is transferred from mantle-derived magmas, the resulting magmas may have velocities of about 7.0

Table 2.1. Velocities of rocks in the earth's crust. Data are for near-surface rocks [Press, 1966].

Rock Type	Location	P-Wave Velocity	S-Wave Velocity
Granite	Westerly Rhode Island U. S. A.	5.76 km/s	3.23 km/s
Quartz Monzonite	Westerly Rhode Island U. S. A.	5.26	2.89
Andesite	Colorado U. S. A.	5.23	2.73
Basalt	Germany	5.0–6.4	2.7–3.2

km/s. The intrusion process thus results in considerable heterogeneity in the earth's crust.

Extensive study of most volcanoes reveals that erupted magmas show considerable geochemical variation [Perry et al., 1990]. This variation may result from differences in the origin of the mantle-derived magmas, from the evolution of a single crustal magma system with time, or due to mixing of various sources of magma. In any case, the result of magmatic processes is a wide spatial variation in chemical composition of the earth's crust and a wide variation in mechanical properties of the rocks. Figure 2.2 shows a schematic cross section through the subsurface of the Mount Taylor, New Mexico, U. S. A. volcanic field for the period 1.5–2.5 million years ago developed largely from geochemical data by Perry et al. [1990]. Note that the types of magmas erupted require that several different types of intrusions were present in the earth's crust. These magmas all have differing mechanical properties and would eventually crystallize into rocks with differing seismic velocities.

Other geological processes that contribute to heterogeneity in the lithosphere include erosion and metamorphism that act to transport rocks or change their character in place. Tectonic processes, such as faulting and folding, move rocks relative to one another and result in heterogeneity. Large scale movements of lithospheric plates distribute rocks having a common origin over a wide range. The collision of tectonic plates at plate boundaries, such as subduction zones or collision zones, causes rocks of differing types to come into contact.

The direct relationship between material elastic parameters and seismic veloci-

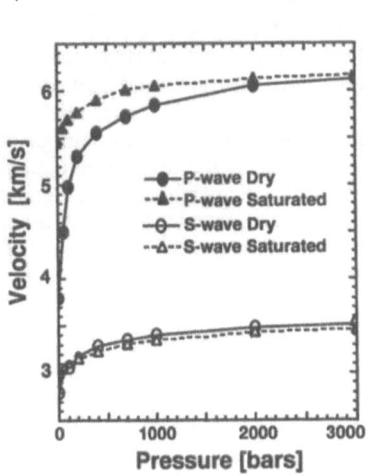

FIGURE 2.1. Laboratory measurements of the variation of wave velocity in dry and water saturated Westerly granite. Data from Nur and Simmons [1969].

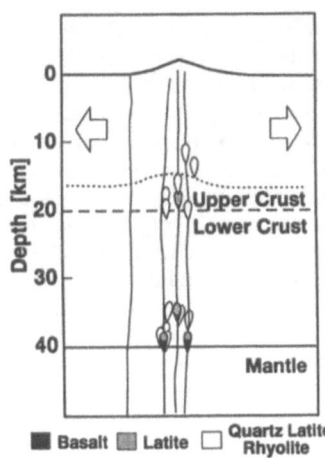

FIGURE 2.2. Model for the subsurface of the Mount Taylor volcanic field based on geochemical and geological evidence. Note crustal heterogeneity in the model. [From Perry, et al., 1990, copyright by the American Geophysical Union.]

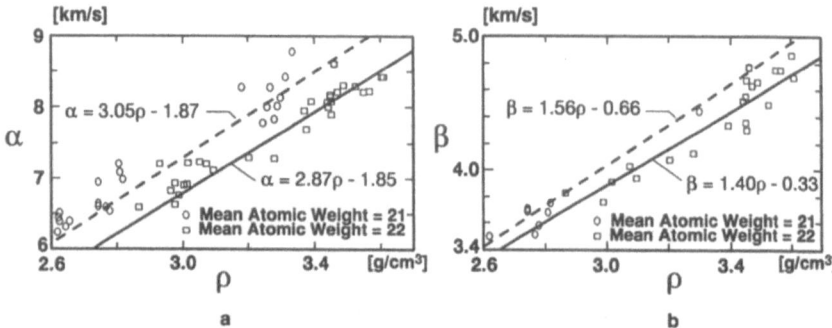

FIGURE 2.3. (a) P-wave velocity against mass density and (b) S-wave velocity against mass density for common lithospheric igneous and metamorphic rocks measured at 10 kbar. Dashed lines show fits to data for rocks having mean atomic weights between 20.5 and 21.5 by Birch [1961] for P-waves and Manghnani et al. [1974] for S-waves. Solid lines show fits for rocks having mean atomic weights between 21.5 and 22.5 by Manghnani et al. [1974] for both P- and S-waves. Data from Manghnani et al. [1974] and Birch [1960, 1961].

ties are given by

$$\alpha = \sqrt{\frac{\lambda + 2\mu}{\rho}} \quad \text{and} \quad \beta = \sqrt{\frac{\mu}{\rho}} \tag{2.1}$$

where $\alpha, \beta, \lambda, \mu, \rho$ are P-wave velocity, S-wave velocity, Lamé coefficients, and mass density, respectively. These relations give the impression that velocities decrease with increasing density. However, velocity is generally observed to increase with increasing density since λ and μ are influenced by mass density. From experimental data on rocks of many types, Birch [1960, 1961] found that seismic velocity increases roughly linearly with mass density for rocks having the same mean atomic weight, which is the atomic weight of the minerals that comprise the rocks averaged in proportion to the mass they contribute to the rock. Mean atomic weight for most crustal rocks ranges from about 21 for silica-rich rocks like granite to 22 for iron-rich igneous rocks. Figure 2.3 shows velocities measured at 10 kbar pressure on common lithospheric rocks having mean atomic weights between 20.5 and 22.5 plotted vs. mass density and the relationship between seismic velocities and mass densities derived from the data. For P-waves measured at 10 kbar pressure, Birch [1961] found $\alpha[\text{km/s}] = 3.05\,\rho[\text{g/cm}^3] - 1.87$ for rocks having mean atomic weight ~21. Kanamori and Mizutani [1965] found $\alpha = 2.8\,\rho - 1.3$ at 6 kbar for dunite, peridotite and eclogite in Japan. Christensen [1968] made laboratory measurements on rocks typical of those suspected to compose the upper mantle and found that S-wave velocity varies as $\beta = 1.63\,\rho - 0.88$ at 10 kbar for mean atomic weight ~22. Manghnani et al. [1974] measured both P- and S-wave velocities for granulite facies rocks and eclogite and found $\alpha = 2.87\,\rho - 1.85$ and $\beta = 1.40\,\rho - 0.33$ at 10 kbar where mean atomic weight ~22.

Christensen and Mooney [1995] reported on laboratory P-wave velocity measurements of many rocks that compose the earth's crust. They grouped the rocks into a total of 29 categories by common rock type. They made measurements using a common laboratory technique on all the rocks at various pressures corresponding to depths of 5 to 50 km. They give an extensive analysis of these data including the changes in density with depth and the changes in velocity and mass density with temperature. They give a relationship between P-wave velocity and mass density that is appropriate for rocks at 10, 20, 30, 40, and 50 km depths. For 20 km depth, they find $\alpha[\text{km/s}] = 2.41 \rho[\text{g/cm}^3] - 0.454$. For rocks typical of the crust and upper mantle, they propose that a better fit to the data is obtained using a relationship of the form $(\alpha[\text{km/s}])^{-1} = -2.3691 \cdot 10^{-3} (\rho[\text{g/cm}^3])^3 + 0.2110$. Christensen [1996] reported on laboratory measurements of P- and S-wave velocities of 678 crustal rocks. He investigated the average ratio of P- to S-wave velocity for crustal rocks by comparing his data with average crustal composition obtained from seismic refraction studies of the crust that are summarized by Christensen and Mooney [1995]. He found that the average ratio is 1.768 for the continental crust. He estimates that Poisson's ratio varies from 0.253 in the upper crust to 0.283 at a depth of 30 km and down to 0.279 in the lower crust.

2.2 WELL-LOGS

2.2.1 Velocity Inhomogeneity Revealed by Well-Logs

Direct evidence for the existence of random inhomogeneities can be found in log data from wells drilled in the earth. Figure 2.4a shows velocity and density log data from well YT2 drilled through lava, tuff, and volcanic breccia in Kyushu, Japan [Shiomi et al., 1996]. The velocity structure was determined from the travel times of ultrasonic waves having frequencies of a few tens of kHz. Rock density is measured from the intensity of gamma rays received at a borehole detector when an artificial source of gamma rays is located about 0.4 m below the receiver in the borehole. The intensity of received gamma rays can be shown to be a function of the formation density [Telford et al., 1976]. Wave propagation velocity usually increases with increasing depth in the earth; however, considerable spatial variation of velocity is evident in the logs. We find that the P-wave velocity has a clear correlation with S-wave velocity as shown in scattergram in Figure 2.4b. As predicted by Birch's law [Birch, 1961], the P-wave velocity has a positive correlation with mass density.

Making a scattergram from bandpass-filtered trace pairs, we can estimate the correlation coefficient. The correlation coefficient generally decreases as the pass band center-wavelength becomes shorter. The correlation coefficient between P- and S-wave velocities is as large as 0.7 even when the center-wavelength is as small as a few meters as shown in Figure 2.5. Those between the P- and S-wave velocities and the mass density drop to less than 0.7 for center-wavelengths less

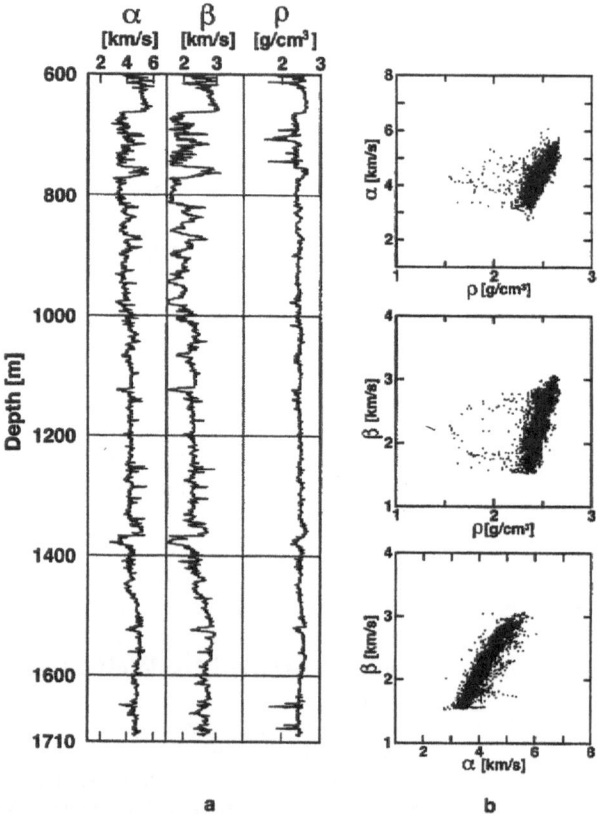

FIGURE 2.4. (a) Well-logs showing P- and S-wave velocities and mass density vs. depth for well YT2 in Kyushu, Japan. (b) Scattergrams showing correlation among the physical properties measured at the same depth. Data from New Energy and Industrial Technology Development Organization, Japan [1992a, b]. [Courtesy of K. Shiomi.]

than 30 m. We note that the transducer-receiver separation for the velocity logging tool of 0.61 m works as a high-wavenumber-cut filter.

2.2.2 Autocorrelation Function and Power Spectral Density Function

Spectral characteristics of inhomogeneities, such as those found in well-log data and elsewhere in the earth, are conveniently described by using mathematical tools such as the autocorrelation function and the power spectral density function. These concepts will be described, and several types of power spectral density functions

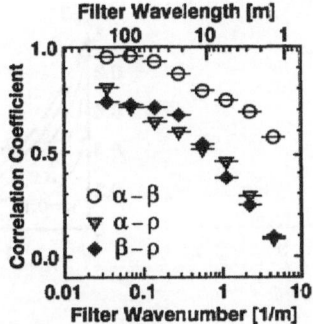

FIGURE 2.5. Correlation coefficients between different physical properties calculated from bandpass-filtered traces of the logs shown in Figure 2.4a for well YT2 in Kyushu, Japan for depth range of 600–1700 m. [Courtesy of K. Shiomi.]

that are used for characterizing inhomogeneities in the earth's lithosphere will be introduced.

Ensemble of Homogeneous and Isotropic Random Media

We suppose wave velocity V is not a constant but depends on the location $\mathbf{x}$. It is decomposed into a sum of the mean velocity V_0 and the perturbed velocity δV:

$$V(\mathbf{x}) \equiv V_0 + \delta V(\mathbf{x}) = V_0 \left[1 + \xi(\mathbf{x}) \right] \qquad (2.2)$$

where we call $\xi(\mathbf{x})$ the fractional fluctuation of wave velocity. We introduce an ensemble of inhomogeneous media $\{\xi(\mathbf{x})\}$, and denote the average over this ensemble by $\langle \cdots \rangle$, where V_0 is chosen so that

$$V_0 = \langle V(\mathbf{x}) \rangle \quad \text{and} \quad \langle \xi(\mathbf{x}) \rangle = 0 \qquad (2.3)$$

In addition, we suppose that $\xi(\mathbf{x})$ is a homogeneous (stationary) and isotropic random function of coordinate $\mathbf{x}$ [Uscinski, 1977, p. 3].

First, we define the autocorrelation function (ACF) as an ensemble average by

$$R(\mathbf{x}) \equiv \langle \xi(\mathbf{y}) \xi(\mathbf{y} + \mathbf{x}) \rangle \qquad (2.4)$$

The ACF is a statistical measure of the spatial scale and the magnitude of irregularity in the medium. We note that "homogeneous" or "stationary" means that the ACF is a function of lag-distance $\mathbf{x}$ irrespective of $\mathbf{y}$, and "isotropic" means that the ACF is a function of $r \equiv |\mathbf{x}|$. The magnitude of the fractional fluctuation is given by the mean square (MS) fractional fluctuation:

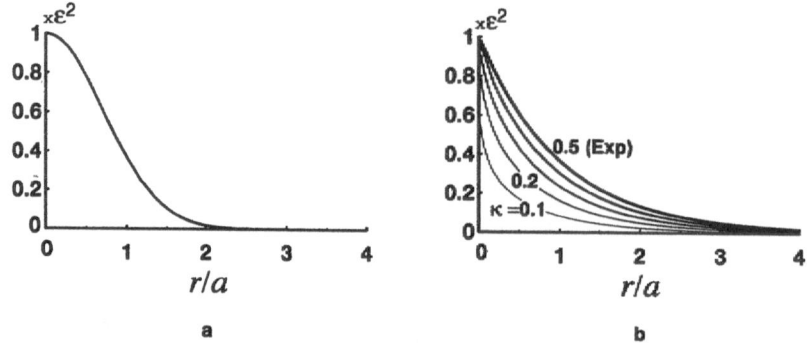

FIGURE 2.6. Plots of (a) Gaussian ACF and (b) von Kármán ACF in 3-D. The von Kármán ACF becomes an exponential ACF when κ=0.5.

$$\varepsilon^2 \equiv R(0) = \left\langle \xi(\mathbf{x})^2 \right\rangle \tag{2.5}$$

We call $R(\mathbf{x})/\varepsilon^2$ the normalized ACF. The spatial variation of randomness is characterized by another parameter, correlation distance (or length) a. The Fourier transform of the autocorrelation function in 3-D space is the power spectral density function (PSDF) where $\mathbf{m}$ is the wavenumber vector:

$$P(\mathbf{m}) \equiv \int_{-\infty}^{\infty}\int_{-\infty}^{\infty}\int_{-\infty}^{\infty} R(\mathbf{x})\ e^{-i\mathbf{m}\mathbf{x}} d\mathbf{x} \tag{2.6}$$

We may explicitly write arguments of R and P as $r \equiv |\mathbf{x}|$ and $m \equiv |\mathbf{m}|$, respectively, because of isotropy.

Gaussian ACF

The most popular form for the ACF is the Gaussian ACF (see Figure 2.6a):

$$R(\mathbf{x}) = R(r) = \varepsilon^2\, e^{-r^2/a^2} \tag{2.7}$$

where $r \equiv |\mathbf{x}|$ and a is the correlation distance. The PSDF is also Gaussian:

$$P(\mathbf{m}) = P(m) = \varepsilon^2 \sqrt{\pi^3}\, a^3\, e^{-m^2 a^2/4} \tag{2.8}$$

The Gaussian ACF is used to describe media that are poor in short wavelength components since the form of the PSDF goes rapidly to zero for large m.

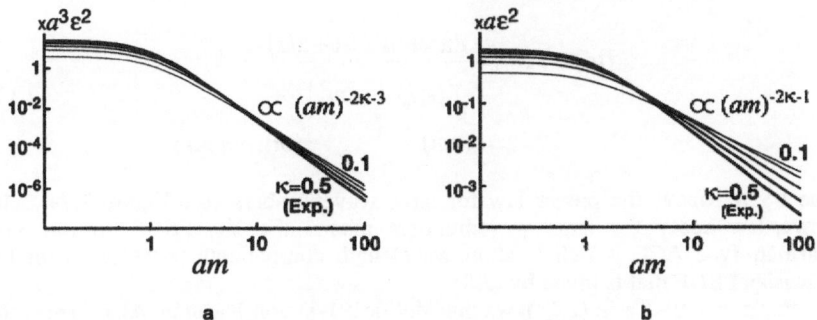

FIGURE 2.7. PSDF for von Kármán type in (a) 3-D and (b) 1-D, where m is wavenumber.

Exponential ACF

The next example is the exponential form (see Figure 2.6b):

$$R(\mathbf{x}) = R(r) = \varepsilon^2 e^{-r/a} \tag{2.9}$$

We note that the PSDF obeys the power law for large wavenumbers (see Figure 2.7a):

$$P(\mathbf{m}) = P(m) = \frac{8\pi\varepsilon^2 a^3}{\left(1 + a^2 m^2\right)^2} \tag{2.10}$$

$$\propto (am)^{-4} \quad \text{for} \quad am \gg 1$$

von Kármán ACF

This is an extension of the exponential ACF appropriate for turbid media [Uscinski, 1977, p. 5] (see Figure 2.6b):

$$R(\mathbf{x}) = R(r) = \frac{\varepsilon^2 2^{1-\kappa}}{\Gamma(\kappa)} \left(\frac{r}{a}\right)^{\kappa} K_{\kappa}\left(\frac{r}{a}\right) \quad \text{for } \kappa = 0 \sim 0.5 \tag{2.11}$$

where Γ is the gamma function and K_{κ} is the modified Bessel function of the second kind of order κ. We note that $\lim_{z \to 0} z^{\kappa} K_{\kappa}(z) = 2^{-1+\kappa}\Gamma(\kappa)$ for $\kappa > 0$ [Abramowitz and Stegun, 1970, p. 375]. The von Kármán ACF coincides exactly with the exponential ACF when $\kappa = 0.5$. The PSDF corresponding to the von Kármán ACF is

$$P(\mathbf{m}) = P(m) = \frac{8\pi^{3/2}\varepsilon^2 a^3 \Gamma(\kappa + 3/2)}{\Gamma(\kappa)(1 + a^2 m^2)^{\kappa + \frac{3}{2}}}$$

$$\propto (am)^{-2\kappa - 3} \qquad \text{for} \quad am \gg 1$$

(2.12)

The PSDF obeys the power law for large wavenumbers (see Figure 2.7a). The power-law decay for large wavenumbers means that the PSDF for the von Kármán type ACF is rich in short wavelength components compared with the Gaussian PSDF that is given by (2.8).

Replacing $\mathbf{x} \to z$ in (2.11), we may define a 1-D von Kármán ACF. Then, the corresponding PSDF (see Figure 2.7b) is

$$P(m) = \frac{2\sqrt{\pi}\varepsilon^2 a \Gamma(\kappa + 1/2)}{\Gamma(\kappa)(1 + a^2 m^2)^{\kappa + \frac{1}{2}}}$$

$$\propto (am)^{-2\kappa - 1} \qquad \text{for} \quad am \gg 1$$

(2.13)

If we consider that well-log data are derived from a stationary random process, we can calculate the ACF from the spatial average instead of the ensemble average. Figure 2.8a shows normalized ACFs of log data obtained from well YT2 (see Figure 2.4) for velocities and densities where the same depth range has been used for each. Correlation lengths scatter over a few tens of meters for these sampled data;

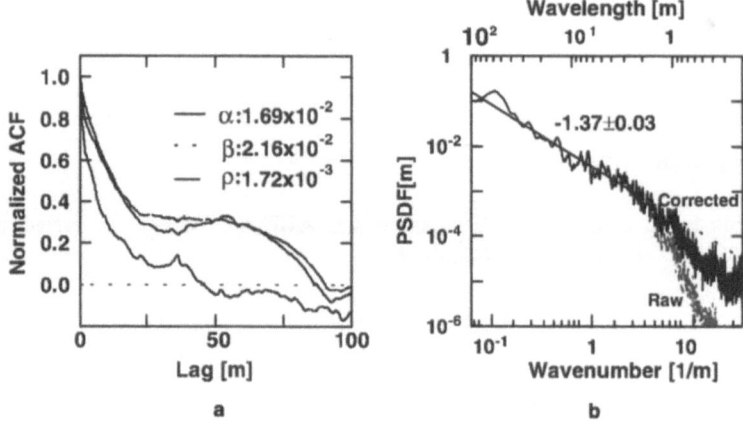

FIGURE 2.8. Statistical characteristics of well-log data from well YT2 in Kyushu, Japan for depth range 600–1700 m (see Figure 2.4a): (a) normalized ACFs for fractional fluctuations of three physical properties. Numerals are the MS fractional fluctuation for each property; (b) PSDF of the P-wave velocity fractional fluctuation, where gray shows raw data and dark the data after logging tool correction. Numeral is the power index. [Courtesy of K. Shiomi.]

however, the shape is not consistent with a Gaussian ACF but more closely follows the exponential or von Kármán type. Suzuki et al. [1981] reported that the exponential type ACF fits the P-wave velocity fluctuations measured in logs of pretertiary basement rock in Kanto, Japan.

Figure 2.8b shows the PSDF of the P-wave velocity in YT2. The PSDF decreases according to the −1.37th power of the wavenumber for a wide range of wavelengths from one meter to a few hundred meters, where the high-wavenumber-cut filtering effect due to the logging tool has been corrected [Shiomi et al., 1996, 1997]. A similar power-law characteristic was found for the KTB deep wells in Germany [Wu et al., 1994], where the power is reported to be −1.1. The power of the wavenumber may be correlated with tectonic activity. As the total length of log data increases, the power law has been found to hold well for an increased range of wavenumbers indicating self-affinity. But we will restrict ourselves to the case that the statistical characteristics are independent of the sample size. We will use the forms of ACFs and PSDFs presented in this section throughout the book. We may expect that the correlation distance and the MS fractional fluctuation vary from place to place, with depth, and in relation to seismotectonic setting.

2.3 DETERMINISTIC IMAGING USING SEISMOLOGICAL METHODS

One of the main focuses of seismology is the deterministic characterization of the spatial heterogeneity of the earth's lithosphere. These efforts are undertaken for economic as well as purely scientific goals. Characterizing the spatial heterogeneity of the lithosphere has enabled investigators to better understand the mechanism by which the earth's crust is formed, volcanic processes, and the nature of active seismic zones. In petroleum exploration, the shallow crust is investigated and subsurface structures that are considered to be good petroleum reservoirs are identified. For each study of subsurface structure, the spatial resolution of the desired information about structure must be determined prior to data collection. A brief discussion of the role that various seismic techniques play in the determination of crustal structure can be found in Braile et al. [1995].

Deterministic characterization has been undertaken using both travel times of seismic phases and waveform characteristics. Figure 2.9 shows travel-time data from teleseismic P-wave arrivals recorded on a seismic network in the Jemez volcanic field in northern New Mexico, U. S. A. Teleseismic arrivals are those from earthquakes occurring at distances larger than about 1000 km. The relative arrival times measured at 49 stations spread across the volcanic field with spacing of about 3 km are contoured for various wave arrival directions. The variations in arrival times at the stations as a function of incidence direction are a direct indication of spatial heterogeneity in the crust beneath the volcanic field.

The reflection and refraction methods are the oldest deterministic methods used to characterize the crust. Refractions from the crust-mantle boundary were discovered in 1909 by Mohorovicic and provided the first direct evidence of the large contrast in seismic velocity between the crust and mantle. In current applications,

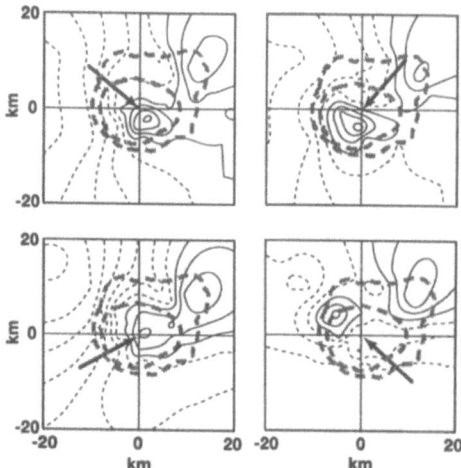

FIGURE 2.9. Relative travel-time anomalies observed for teleseisms recorded by an array of seismometers located in the Jemez volcanic field of northern New Mexico, U. S. A. The inner thick dashed line shows the location of the ring fracture associated with the volcanic caldera. Average anomalies are shown for wavefronts arriving from directions as indicated by arrows. The contour interval is 0.1 s, where the solid lines are for positive and the dashed lines are for negative. Positive anomalies indicate late arrivals. [Courtesy of L. Steck.]

both amplitudes and arrival times are used in the reflection and refraction methods. Refraction studies are used mostly for studying regional seismic structure to depths as great as the crust-mantle boundary. Sizes of regions studied are generally on the order of a few tens to hundreds of km. Reflection studies are generally used on a more local scale for studies of a few to a few tens of km and to depths of about 10 km.

Seismic tomography was introduced in the 1970s as an extension of the methods developed in materials testing and the medical community. Initially seismic tomography was applied only to arrival times but later methods were introduced to take account of waveform characteristics. Seismic tomography can be conducted on scales ranging from the whole earth to the laboratory scale.

2.3.1 Velocity Tomography

Velocity tomography was introduced into seismology by Aki et al. [1976, 1977], who showed how to use travel times from distant earthquakes recorded on a closely spaced network of seismometers to determine the 3-D seismic structure of the region beneath the network. Subsequently, Aki and Lee [1976] and Crosson [1976] showed how to use travel times of local earthquakes recorded on a regional seismic array to simultaneously determine the 3-D velocity structure and earthquake locations. Since that time, velocity tomography has been widely used on regional and global scales with travel times from local and distant earthquakes to determine the P- and S-wave velocity structure of the lithosphere.

The basis of the tomography method is that travel time through a structure can be written as

$$t_{ij} = \int_{c_{ij}} p(\mathbf{x})\,dc \qquad (2.14)$$

where t_{ij} is the travel time for source i to receiver j, c_{ij} is the travel path from source i to receiver j, $p(\mathbf{x})$ is the slowness (inverse velocity) at location $\mathbf{x}$ in the earth and dc is the infinitesimal line element. Since the predicted ray path is not known until the velocity structure is known, inversion of (2.14) for $p(\mathbf{x})$ using known t_{ij} is a nonlinear problem. Generally, solution proceeds by the following approach. An initial velocity structure is assumed and the travel times through that structure are calculated from

$$t_{ij}^0 = \int_{c_{ij}^0} p^0(\mathbf{x})\,dc \qquad (2.15)$$

where $p^0(\mathbf{x})$ is the initial slowness structure, c_{ij}^0 is the travel path between source and receiver for this initial model, and t_{ij}^0 is the travel time along this initial path. The ray path and travel times can be determined using ray-tracing methods, such as initial value approaches [Cerveny, 1987], ray bending [Um and Thurber, 1987], or solutions of the eikonal equation [Vidale, 1988; Fei et al., 1995]. Defining the difference between the measured travel time and t_{ij}^0 as the delay time δt_{ij}, we find that

$$\delta t_{ij} = t_{ij} - t_{ij}^0 = \int_{c_{ij}} p(\mathbf{x})\,dc - \int_{c_{ij}^0} p^0(\mathbf{x})\,dc \approx \int_{c_{ij}^0} (p(\mathbf{x}) - p^0(\mathbf{x}))\,dc = \int_{c_{ij}^0} \delta p(\mathbf{x})\,dc \qquad (2.16)$$

We may solve (2.16) for perturbations in slowness that give a better fit to the observed data than the initial model. The basic assumption in deriving (2.16) is that the changes in the slowness model do not result in significant changes in the ray path. It can be shown that this assumption is good so long as the slowness perturbations are small. Equations of the form of (2.16) are known as Radon transforms, named after Radon who showed a formal inverse transform to find the slowness $\delta p(\mathbf{x})$ in the case that the rays follow straight lines [Radon, 1917]. As pointed out by Chapman [1987], (2.16) shows that the data δt_{ij} are obtained as an integral over the slowness and are thus smoother than the slowness. Inverting (2.16) for slowness involves computing derivatives of data, which are numerically unstable with seismic data. Thus, seismologists usually include smoothness constraints in the solution for slowness. These constraints limit the resolution of the spatial variation in structure that can be found with travel-time tomography.

Deans [1983] discusses the problem of inverting Radon transforms using real data. In general, the inversion is nonunique due to the fact that data are not available over all paths c_{ij} in (2.16). The projection-slice theorem illustrates how available data may help to constrain a slowness model. For the 2-D case in which slowness variations are small enough that rays follow straight paths, consider that we have travel-time anomaly measurements $\delta t(r, \theta)$ along coordinate r at angle θ to the x-axis, where coordinate r is perpendicular to ray-paths that travel through the medium at angle $\theta + \pi/2$. In 2-D, the theorem can be written as $\delta \tilde{t}(k_r, \theta) = \delta \tilde{p}(k_r \cos \theta, k_r \cos \theta)$, where $\delta \tilde{t}$ stands for the 1-D Fourier transform of δt with respect to r and $\delta \tilde{p}$ for the 2-D Fourier transform of the slowness in x-y space [see Menke, 1984a, p. 178]. The projection slice theorem states that the Fourier transform of the travel-time anomaly data along a line perpendicular to the ray-paths traveling through the structure is equal to a slice of the Fourier transform taken through the slowness model. We can thus think of the image as built up in the Fourier domain from slices of travel-time data for rays propagating at various angles through the structure. Since observations of seismic travel time can usually be made for only a limited range of propagation angles, we have limited constraints on the structure of the earth available from travel-time tomography.

When studying the earth, (2.16) is valid only if the ray path does not depart from the path calculated for the original, unperturbed slowness structure. Thus, an iterative approach is used to solve (2.16), and slowness is constrained to vary slowly during each iteration. In addition, when (2.16) is to be solved using data collected in the earth, the inversion for slowness is not a well posed problem. Parts of the velocity structure are overdetermined and, due to noise in the data, generally overdetermined with inconsistent data. Portions of the structure are also underdetermined due to the limited angle that waves propagate through the region of study. Various methods for solving equations like (2.16) can be found in Menke [1984a] or Tarantola [1987].

When interpreting results of tomography, we must consider the resolution limits. In deriving (2.16) we have assumed that seismic information propagates along rays that are infinitely narrow. Ray theory is a high-frequency approximation to the wave equation, and we know that it is not appropriate for finite wavelength waves. One approximate measure of the spatial resolution of tomography is given by a measure of ray width [Nolet, 1987]: $\delta r = \sqrt{Z \lambda_w / 8}$, where δr is the minimum separation of two objects to be resolved, Z is the travel distance between source and receiver, and λ_w is the wavelength. When the spatial structure reaches a certain level of complexity, scattering effects become important and travel-time tomography cannot be applied. Devaney [1982] introduced the concept of diffraction tomography or inverse scattering, which uses both the amplitude and phase of incident energy. Diffraction tomography is based on a comparison of the observed wavefield with a scattered wavefield that is calculated using the Born (or Rytov) approximation for the perturbed media. Devaney [1982] showed that resolution of features as small as a half wavelength is possible using diffraction tomography. Wu and Toksöz [1987] examined the application of diffraction tomography to geophysical data sets. Williamson [1991] investigated the effects of scattering on travel-time tomography by calculating the scale at which images calculated with travel time and diffraction formulations begin to differ. He found that the images begin to depart significantly when structure has variations on a scale smaller than $\delta r = \sqrt{Z \lambda_w}$, which should be viewed as the limiting resolution of travel-time tomography imposed by scattering effects.

Tomography has been applied to study the velocity structure of the mantle [Vasco et al., 1994], regional lithospheric structures using teleseismic data [Weiland et al., 1995], regional structures using local earthquakes and explosions as sources [Thurber, 1993; Pujol, 1996], the small-scale structure of a man-made geothermal reservoir [Block et al., 1994], the small-scale structure of the region between two boreholes, and the structure near a borehole using surface seismic sources and borehole receivers in a configuration known as Vertical Seismic Profiling (VSP). We show results from three studies. Figure 2.10 shows a horizontal slice through a 3-D S-wave velocity tomogram calculated by Block et al. [1994] using data collected at a geothermal site in New Mexico, U. S. A. Data from microearthquakes induced by hydraulic fracturing and small explosions were used in the analysis. Figure 2.11 shows a P-wave velocity tomogram constructed using arrival-time anomalies like those shown in Figure 2.9 from teleseisms recorded on an array of receivers spread across the Jemez volcanic field in New Mexico. The low-velocity anomaly located in the center of the image is thought to be due to the presence of molten magma beneath the center of the volcanic caldera. Beneath island arcs, strong velocity inhomogeneities are caused by the subduction of oceanic slabs and the related volcanic activities. Figure 2.12 shows an EW section of the

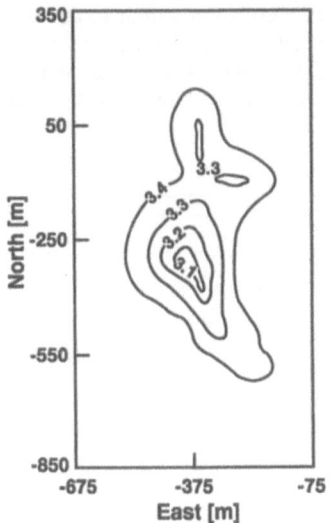

FIGURE 2.10. Horizontal slice through tomogram of S-wave velocity structure at depth 3500 m for the Fenton Hill hot dry rock geothermal energy site in New Mexico, U. S. A. Contours show velocity in km/s. Tomogram was calculated using travel times from microearthquakes induced by hydraulic fracturing in crystalline rock by Block et al. [1994].

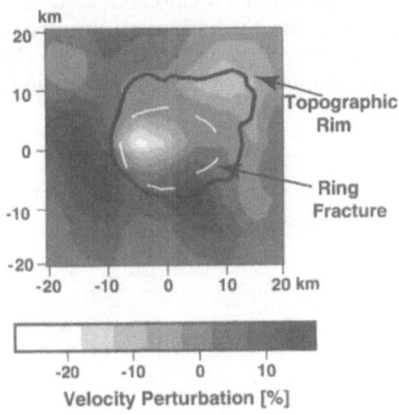

FIGURE 2.11. Horizontal slice through P-wave velocity tomogram of the Jemez volcanic field in New Mexico, U. S. A. at depth 10 km beneath the surface. Locations of the ring fracture where collapse occurred during 2 major eruptions that occurred about 1 million years ago and highest elevation on edge of Valles caldera (topographic rim) are shown. Percent variation in velocity from average at 10 km depth are shown. Results were derived from measurements of travel-time anomalies from teleseismic events. [Courtesy of L. Steck.]

perturbation in P-wave velocity structure beneath northeastern Japan. The subducting Pacific plate appears as a high-velocity zone; low-velocity bodies are found in the crust beneath the double volcanic zone that runs from north to south [Zhao et al., 1992].

Autocorrelation Function of Velocity Tomograms

In Section 2.2.2 we introduced the concept of the autocorrelation function as a means of characterizing heterogeneity. Here, we investigate the autocorrelation of velocity tomograms. This should be done with caution since artifacts introduced by the construction of the tomogram may influence the autocorrelation. For example, tomograms are often constructed using constraints, such as a smoothing of the derived velocity model, to regularize the system of equations that must be solved to find the velocity model. In addition, the poor resolution of some portions of the model will lead to artifacts that influence the autocorrelation function.

The normalized ACF in three directions of the velocity tomogram for the Jemez volcanic field is shown in Figure 2.13. A 2-D slice through the velocity model derived from analysis of travel times from teleseisms is shown in Figure 2.11. The calculated autocorrelation function shows a shorter correlation length in the vertical direction than the horizontal direction. The shorter characteristic length in the vertical direction is probably due to the change in velocity with depth, which is large compared to the change in velocity in the horizontal direction caused by the velocity

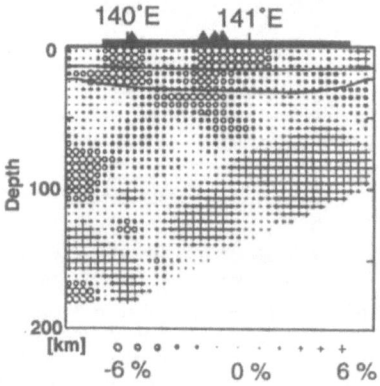

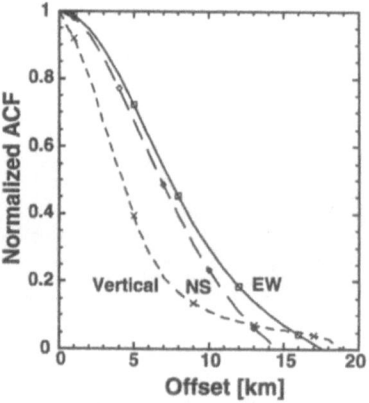

FIGURE 2.12. EW section showing the perturbation of P-wave velocity beneath northeastern Japan, where the Pacific plate is subducting from east to west. Large open circles show regions of low velocity; large plusses show regions of high velocity. [From Zhao et al., 1992, copyright by the American Geophysical Union.]

FIGURE 2.13. Normalized ACF of P-wave velocity tomogram for the Jemez volcanic field constructed from teleseismic data. The ACF is calculated in three directions: vertical, north-south (NS) and east-west (EW). A slice through the tomogram is shown in Figure 2.11.

anomaly beneath the center of the volcanic field.

The ratio of the scale length in the horizontal direction to that in the vertical direction was estimated by analysis of log data from closely spaced wells at the KTB deep boreholes in Germany and found to be 1.8 [Wu et al., 1994], which is similar to that shown in Figure 2.13 for the Jemez volcanic field. The difference in correlation length scales measured for borehole data may be due to the characteristic thickness of rock layers in the region under study. If the boreholes cross many rock layers in the study region, as was the case for the KTB site, the characteristic length in the vertical direction may be smaller than in the horizontal direction. If no layers are encountered and the rock mass is geologically homogeneous, the elastic heterogeneity may be dominated by microfractures, which may be isotropically oriented, leading to similar autocorrelation functions in the vertical and horizontal directions.

Digitizing the lithological map of the Ivrea zone in northern Italy, which is considered a typical exposure of the lower continental crust, Holliger and Levander [1992] estimated the ACF in two orthogonal directions. They found that the ACF is a von Kármán type having nonisotropic randomness: the shorter correlation distance is 150–180 m and the longer is 550–750 m, that is, the aspect ratio is 3–5.

It is desirable to compute the autocorrelation function directly from arrival-time data rather than from velocity models derived from travel-time data. In Section 8.1, we will discuss the relationship between amplitude and phase fluctuations and media parameters. Müller et al. [1992] developed a method for estimating fractional fluctuation of medium slowness and correlation length from measurements of the autocorrelation function of travel times measured parallel to a wavefront. Using 2-D finite difference simulations, they tested the validity of their method and show that it is reliable when the ratio of seismic wavelength to correlation distance is less than about 0.5. Roth [1997] describes an extension of the method proposed by Müller et al. [1992] to the case that measurements are not made parallel to the wavefront. He tested the method using simulations and shows that it is reliable for cases where the propagation distances is less than about ten times the correlation distance. He applied the method to active seismic data collected using air guns in the ocean off Sweden and found that the correlation distance is 330–600 m.

2.3.2 Refraction Surveys

The refraction technique is probably the oldest method used to characterize the earth's crust. Although refraction studies conducted today generally provide less resolution of crustal structure than reflection studies used for petroleum exploration, refraction studies do provide information about larger regions of the crust. As conducted today, refraction surveys can be considered transmission surveys or refraction combined with wide-angle reflection surveys since waves other than refractions are identified in the data collected and provide additional constraints on the models derived.

Processing methods for refraction seismic studies include forward modeling of travel times, forward modeling of waveforms using methods such as finite difference or reflectivity [Mooney, 1989], and travel-time tomography [Zelt and Smith, 1992]. Each method has limitations: forward modeling of travel times may provide

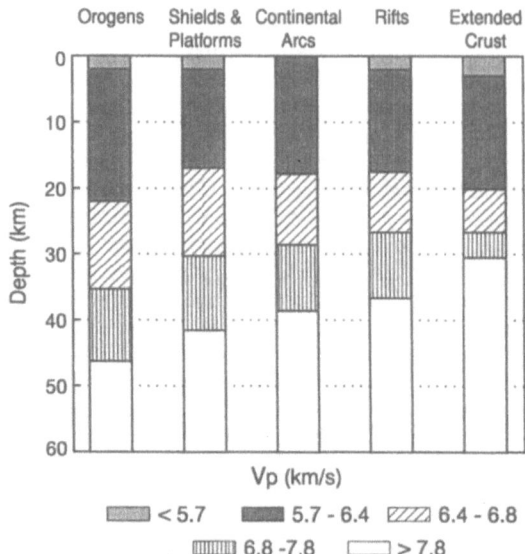

FIGURE 2.14. Compilation of average layered P-wave velocity structures for five tectonic provinces based on results of measurements made worldwide. [From Christensen and Mooney, 1995, copyright by the American Geophysical Union.]

some information about crustal heterogeneity but provides little information about uniqueness of the derived model. Forward modeling of waveforms provides constraints from data of both amplitudes and phase and can result in more reliable models; however, due to the difficulty of matching data and the large amount of computation involved in forward modeling, usually only simple models can be tested. Inversion of travel times provides information about uniqueness of the model and usually provides a good fit between predicted arrival times and measured times, but the limited angles at which rays propagate in refraction surveys limit the resolving power of this method.

Christensen and Mooney [1995] summarize results of refraction surveys made worldwide and discuss the implications of these surveys for our understanding of crustal composition. They divided the results into those obtained in five tectonic provinces and generated average P-wave velocity structural models as illustrated in Figure 2.14. The layered velocity structures are the first information that is usually derived from refraction surveys and represent the most basic information about inhomogeneity in the crust. Using information about velocities of various types of rocks under in situ conditions, Christensen and Mooney [1995] developed models for the average composition of the tectonic provinces. They argue that the average P-wave velocity of the crust is 6.45 km/s and the average for the upper mantle is 8.09 km/s.

2.3.3 Reflection Surveys

Due to the widespread use by the petroleum industry, data from reflection surveys make up a majority of the data collected for imaging the earth's crust. In typical industry surveys, each source location is recorded by as many as 500 receivers. Source locations at sea can be as closely spaced as 30 m, and surface areas of the scale of 25 km^2 can be surveyed. Generally, such reflection data are processed using a method known as migration [Schneider, 1978], which is an approach to backpropagate the wavefield measured at the surface to develop an image of the reflectivity of the earth's subsurface. Migration is based on the representation theorem for scalar waves [see Schneider, 1978] by using reciprocity, which shows how to relate known values of the wavefield on a surface S bounding a medium to the wavefield at any point interior to the surface:

$$u(\mathbf{x},t) = \int_{-\infty}^{\infty} dt' \oint_{S} dS(\mathbf{x}')\big[-G(\mathbf{x},t;\mathbf{x}',t')\mathbf{n}\nabla' u(\mathbf{x}',t')+u(\mathbf{x}',t')\mathbf{n}\nabla' G(\mathbf{x},t;\mathbf{x}',t')\big] \quad (2.17)$$

where $u(\mathbf{x},t)$ is the wavefield at location $\mathbf{x}$ interior to S and at time t, $u(\mathbf{x}',t')$ is the wavefield at point $\mathbf{x}'$ on the surface S, $\mathbf{n}$ is the outward pointing unit normal vector to the surface, and ∇' is the derivative with respect to $\mathbf{x}'$ on the surface. $G(\mathbf{x},t;\mathbf{x}',t')$ is the Green function for a source located at $\mathbf{x}'$ at time t' with a receiver located at $\mathbf{x}$ at time t. Since we know the wavefield near the earth's surface from measurements, we choose a Green function that vanishes at the surface, and we obtain the migration integral for reflection data

$$u(\mathbf{x},t) = \int_{-\infty}^{\infty} dt' \oint_{S} dS(\mathbf{x}')u(\mathbf{x}',t')\mathbf{n}\nabla' G(\mathbf{x},t;\mathbf{x}',t') \quad (2.18)$$

Eq. (2.18) needs boundary and initial conditions. The boundary condition at the

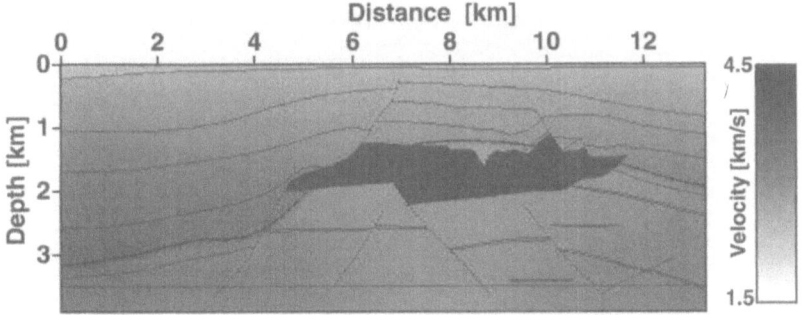

FIGURE 2.15. Slice through 3-D model of region containing a salt body thought to be typical of the Gulf of Mexico. The velocity through the salt is as much as a factor of three times that of the surrounding strata. Solid lines represent layer interfaces. [Courtesy of T. Fei.]

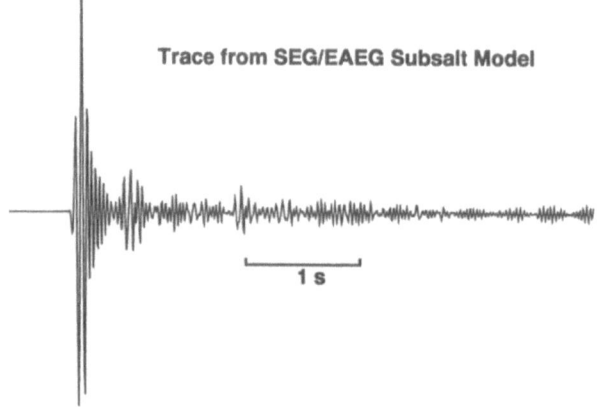

Trace from SEG/EAEG Subsalt Model

FIGURE 2.16. Synthetic seismogram calculated using 3-D finite difference solution of the scalar wave equation (constant density) for the salt model whose cross section is shown in Figure 2.15. Shot and receiver are located on the earth's surface (in water) above the location of the salt body. Data are calculated as in Aminzadeh et al. [1994].

surface is the observed seismic data. At other boundaries, we use a radiation condition that states that the wavefield goes to zero at infinity. The initial condition is the causality condition that limits the time integral from time zero to some finite time. Physically, (2.18) allows us to predict the scattered wavefield below the surface of the earth. To obtain a reflection coefficient for a layer, we use a simple imaging concept. We take the scattered wavefield just above a virtual reflector at the time it takes to propagate energy from the source to the reflector and back to the receiver.

In general, the Green function for (2.18) has no analytic representation. Two popular numerical techniques to compute the Green function consist of solving a finite difference representation of the scalar wave equation or solving the eikonal

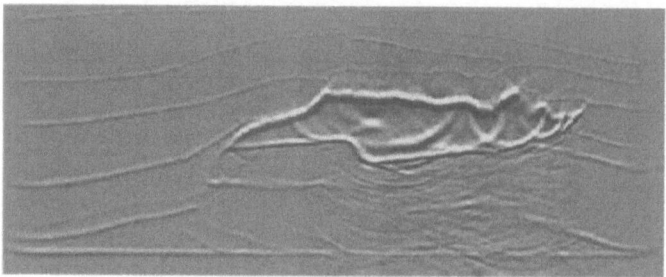

FIGURE 2.17. Results of migrating synthetic zero-offset data for the salt structure shown in Figure 2.15. Migration adequately accounts for complexity of seismograms like those shown in Figure 2.16 and the structure is nearly recovered by the deterministic treatment of the data. [Courtesy of T. Fei and the Gulf of Mexico Imaging Consortium.]

equation to obtain an asymptotic ray equation solution. The finite difference solution can be used to propagate the wavefield at the surface backward in time and into the earth; this is called reverse time migration. The formal limit of resolution of seismic migration for wavelength λ_w is given by $\delta r = \lambda_w / 4$ [Claerbout, 1985].

Figure 2.15 shows a 2-D cross section through a synthetic model developed to represent the earth's structure in a typical region of the Gulf of Mexico where salt features are common. In the Gulf region, the velocity through salt is dramatically higher than that of the surrounding strata, and the interface between the salt and the strata is irregular and often steeply dipping. The scattering by the salt results in extremely complicated seismograms during reflection profiling. Figure 2.16 shows a synthetic seismogram calculated using 3-D finite differencing of the scalar wave equation (constant density) for the 3-D model whose cross section is shown in Figure 2.15 [Aminzadeh et al., 1994]. Note the complexity of the seismogram calculated for this structure.

Figure 2.17 shows the result of migrating zero-offset data calculated for the 3-D salt model. Zero offset data are calculated by assuming that the source and the receiver are located at the same location on the earth's surface. The migration was calculated using almost 300,000 seismograms over the 3-D structure. Surprisingly, complicated seismograms like those shown in Figure 2.16 can be properly processed to yield an image of the earth's subsurface that is close to the actual model shown in Figure 2.15 [Fei et al., 1996].

On a larger scale, observations of earthquake waveforms have provided evidence for reflected phases in the earth's crust. One notable observation of reflections, interpreted as coming from a mid-crustal reflector near Socorro, New Mexico, U. S. A. was made by Sanford and Long [1965] and later refined by Hartse et al. [1992]. A waveform from an earthquake showing where reflected phases are observed is shown in Figure 2.18. Extensive investigation of the reflected phases has led to the conclusion that the reflections are caused by a mid-crustal magma intrusion that is perhaps as thin as 60 m, located at a depth of 19 km, and covering an area as large as 1,700 km^2 [Hartse et al., 1992]. The existence of S-wave reflectors in the mid-crust was also reported to be associated with the Nikko–Shirane volcano in northern Kanto, Japan [Matsumoto and Hasegawa, 1996].

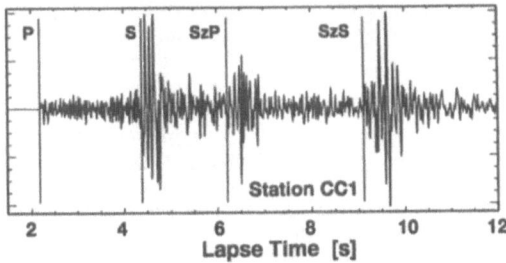

FIGURE 2.18. Seismogram from an earthquake located near Socorro, New Mexico, U. S. A., where lapse time is measured from the earthquake's origin time. Phases SzP and SzS are phases reflected from a magma body located in the mid-crust. [Data courtesy of H. Hartse, A. Sanford, and J. Knapp.]

In many continental regions, sequences of reflections are often observed from the lower crust and the vicinity of the crust–mantle boundary. The surveys show that the lower continental crust is heterogeneous compared with upper crust and the upper mantle as shown in Figure 2.19 [Warner, 1990b]. Warner [1990a] discusses the amplitudes of these reflections and indicates that some data show as many as 40 strong spatially consistent reflectors in 200 km-long 2-D seismic lines. Such spatially-consistent and strong reflectors are not observed at shallow or mid-crustal depths. Warner [1990b] argues that these reflectors are due to the intrusion of iron-rich mantle material into the silica-rich crust that flattens out into layers as it reaches a depth where the density contrast between the intrusion and the surrounding material no longer allows it to continue its ascent to the surface. To further investigate the sequence of reflections observed from the lower crust, Holliger et al. [1993] examined geological maps of a region of northern Italy that is considered an outcrop of material that used to be in the lower crust. They estimated the spatial variations of material properties by comparing geological units with seismic velocities measured on rocks from the region and developed model sections for the lower crust. They used 2-D finite difference calculations to generate synthetic seismograms for the near and far offset seismic response for a region that contains a lower crust similar to the one developed from geological data. They found that the layered appearance of seismic data from the lower crust may be explained by small scale

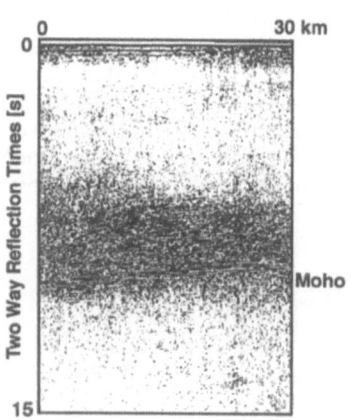

FIGURE 2.19. Record section showing bright subhorizontal layered reflections in the lower crust. The Moho is located at the base of the layering. [From Warner, 1990b, with permission from Elsevier Science - NL, Sara Burgerhartsraat 25, 1055 KV, Amsterdam, The Netherlands.]

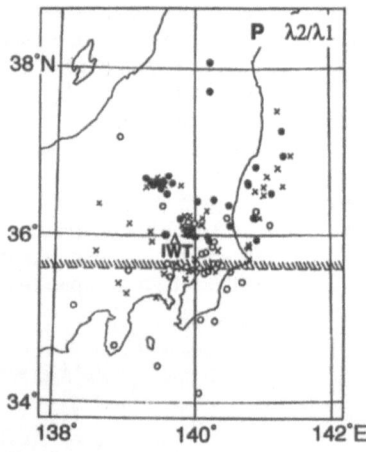

FIGURE 2.20. Distribution of aspect ratio of P-wave particle motion for earthquakes in Kanto, Japan observed at borehole seismic station IWT. The aspect ratio is plotted at the position of each epicenter using open circles (0.–0.48), crosses (0.48–0.69) and closed circles (0.69–1.0). [From Matsumura, 1981, with permission from Center for Academic Publications Japan, Bunkyo-ku, Tokyo, Japan.]

(500–1000 m) spatial variation in material properties with maximum velocity variations of about 0.55 km/s against a flat background velocity of 6.3 km/s.

2.4 SCATTERING OF HIGH-FREQUENCY SEISMIC WAVES

High-frequency seismograms contain features that reflect the random inhomogeneities in the earth. For local earthquakes, those recorded at distances of less than about 100–200 km, "high frequency" generally means higher than 1 Hz in this book. Traditionally, networks record waveforms of local earthquakes with a bandwidth of about 1–30 Hz although recent instrumentation records higher frequencies. Recording of frequencies higher than about 30 Hz requires that the seismic sensor be placed in a borehole at depths below the highly attenuating surface layers. When active sources such as explosions are used, frequencies as high as many kHz can be recorded, especially when both source and receiver are in boreholes.

When we examine the particle motion around the direct-wave arrival, we find evidence of scattering along the propagation path from the source to the receiver. The 3-D particle motion trajectory, which gives some information about the types of seismic waves and their directions of travel, can be analyzed using the 3-D covariance matrix. In a simple medium, the P-wave should be linearly polarized along the direction of travel and the S-wave is polarized in the plane perpendicular to the direction of travel. In most cases, the P-wave particle motion is observed to be elliptical, which indicates scattering. The aspect ratio of the ellipsoid, given by the square root of the ratio of the middle eigen value to the maximum eigen value of the covariance matrix composed of three-component data for a short interval of time around the P-wave, indicates the strength of scattering. If the ratio is zero, the particle motion is needle-like indicating no scattering. On the other hand, if the ratio is close to one, the particle motion is spherical representing strong scattering.

Matsumura [1981] measured the ratios of eigen values for P-waves from seismograms of local earthquakes recorded at station IWT in the Kanto district, Japan, which is run by the National Research Institute for Earth Science and Disaster Prevention, Japan (NIED). At station IWT, a three-component velocity-type seismograph is installed in a borehole in pretertiary formation rock at a depth of 3510 m. The predominant frequency of the data is about 10 Hz, and no phases reflected from the free surface are included in the 0.8 s time window used for the analysis. He found differences in scattering strength for earthquakes occurring in different regions, as shown in Figure 2.20, where the scattering is stronger for earthquakes in the north than in the south. He found a similar pattern in scattering strength for S-waves. Nishizawa et al. [1983] examined the particle motion of 10 kHz-band P-wave seismograms of microearthquakes induced by water injection at Fenton Hill in New Mexico. They found stronger scattering for waves which traverse through the known location of a fracture zone.

In this section we will describe some of the observed characteristics of seismograms of local and regional earthquakes that can be interpreted using scattering models. We will briefly describe some of the approaches used in the modeling, which will be further developed in Chapters 3–8.

2.4.1 S-Coda Waves

The most prominent evidence for the short-wavelength random heterogeneity of the earth is the appearance of coda waves in seismograms. On typical seismograms of local earthquakes, like those illustrated in Figure 2.21, the direct S-wave is followed by wave trains whose amplitude decreases smoothly with increasing lapse

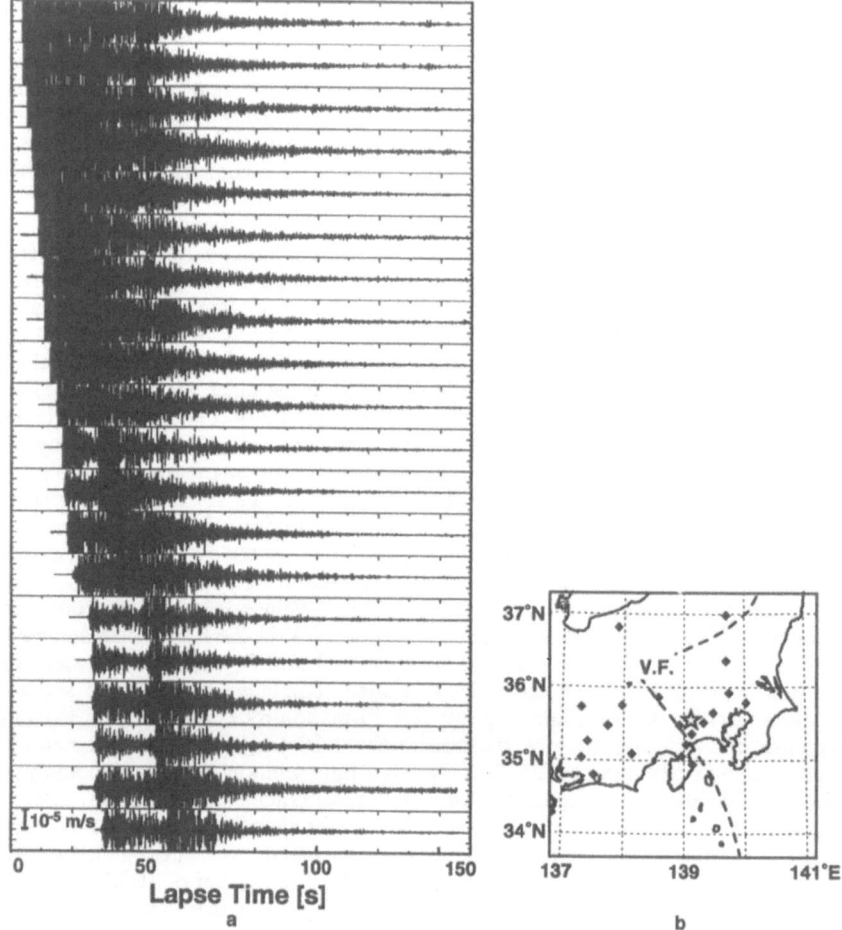

FIGURE 2.21. (a) Horizontal-component velocity seismograms of an earthquake whose magnitude is M_L=4.6 and focal depth is 19.3 km. Seismograms are arranged from top to bottom by increasing distance from the earthquake epicenter. (b) Distribution of the NIED network stations (diamonds) in Kanto-Tokai, Japan whose seismograms are shown. Star indicates the epicenter. [Courtesy of K. Obara.]

time. These wave trains are called "S-coda waves", or simply "S-coda" or "coda". Initially the word "coda" was used to refer to the oscillations of the ground continuing after the passage of surface waves or the tail portion of a seismogram. Recently, this word has been used to refer to all wave trains except direct waves: "P-coda" for waves between direct P- and S-waves and "S-coda" for waves following the direct S-waves. Direct S-wave amplitude decreases with increasing epicentral distance; however, average S-coda amplitudes, for example, those at lapse time of 100 s, have nearly equal amplitudes irrespective of epicentral distance.

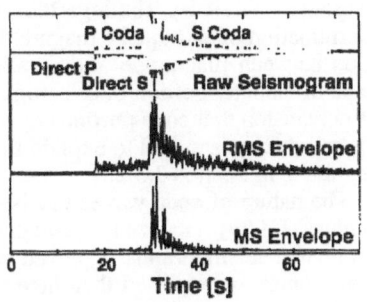

FIGURE 2.22. Example of a velocity seismogram of a local earthquake.

To characterize the S-coda envelopes we often calculate the smoothed trace of the square of the seismogram, which is called the MS seismogram envelope. Sometimes taking the square root of the MS trace, we make the RMS seismogram envelope. In Figure 2.22, we show a velocity seismogram of a local earthquake (top), the corresponding RMS seismogram envelope (middle), and MS seismogram envelope (bottom). The MS envelope, whose amplitude is linearly proportional to energy density, is appropriate for comparison with the synthesis based on the radiative transfer theory. On the other hand, the RMS envelopes reflect the visual image of the seismograms themselves.

Rautian and Khalturin [1978] studied coda amplitude for a wide range of lapse times and frequency bands. They found that early portions of the coda are different from station to station; however, the coda of bandpass-filtered seismograms have a common shape at all stations after about two times and always after three times the S-wave travel-time from the source to the receiver. Figure 2.23a shows the RMS envelopes measured from bandpass-filtered seismograms at two stations for a local earthquake in Central Asia, where the station separation is about 45 km. The figure shows the similarity of the shape of the coda portion of the envelope at the two stations. Figure 2.23b shows coda amplitude vs. lapse time for a suite of small earthquakes in Kanto, Japan recorded at a single station [Tsujiura, 1978]. The similarity of the curve shape for all the earthquakes is clear. We also note that there is little difference between vertical and horizontal components of S-coda envelopes of small local earthquakes as shown in Figure 2.24.

The magnitude of local earthquakes can be determined from the average of the direct wave amplitudes measured at many stations after a distance correction is applied to data from each station. The magnitude calculated from amplitudes has been found to be proportional to the logarithm of the duration of a local seismogram, which is the length of time measured from the P-wave arrival to the time when the S-coda amplitude decreases to the level of microseisms or noise [Solov'ev, 1965]. The proportionality is shown in Figure 2.25, where earthquake local magnitudes determined from measurements of amplitude by the Japan Meteorological Agency (JMA) are plotted against the duration time at a station in Wakayama, Japan

[Tsumura, 1967a, b]. The logarithm of duration time has been used for the quick determination of earthquake magnitudes in many regions of the world. This correlation between magnitude and duration time is consistent with the similarity in shape of the later portion of seismograms observed at regional seismic stations and the conclusion that coda portions of seismograms are composed of scattered waves. We will discuss a model to explain the correlation of coda duration and earthquake magnitude in Section 3.3.2.

The nature of coda waves has been studied using array observations. Aki and Tsujiura [1959] analyzed correlations of seismograms among six vertical-component seismographs deployed on granitic rock at the foot of Mt. Tsukuba in Kanto, Japan and reported that there was little energy at the receiver array that had left the epicenter region of the earthquake as plane waves. One way to find the propagation direction of component waves using array observations is to use a frequency–wavenumber power spectrum. For a stationary time series $u(\mathbf{x}, t)$, we de-

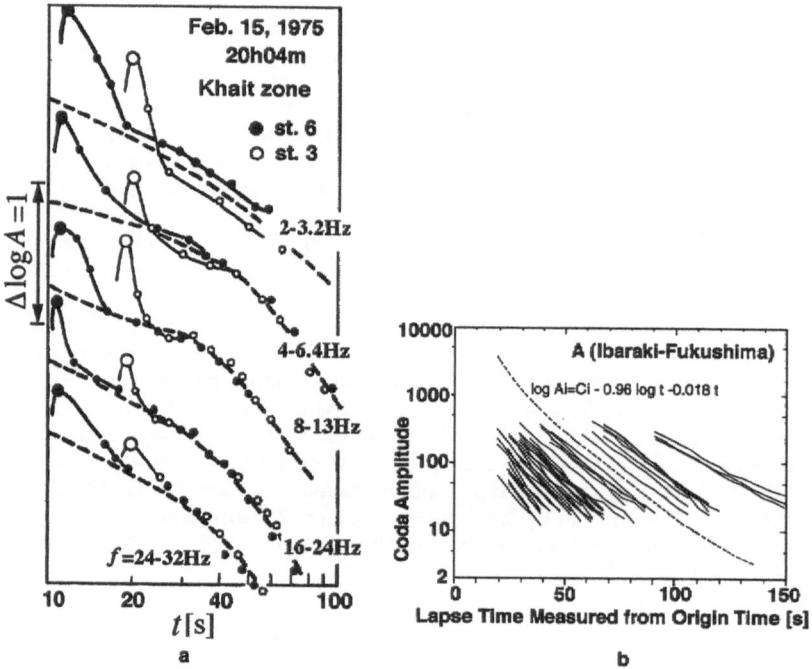

a   b

FIGURE 2.23. (a) RMS seismogram envelopes in different frequency bands beginning at the S-wave arrival-time (large symbols) for an event recorded at two stations having different epicentral distances in Central Asia. [From Rautian and Khalturin, 1978, copyright by the Seismological Society of America.] (b) RMS coda-amplitude decay with lapse time for local earthquakes of the 6 Hz-band recorded at Tsukuba in Kanto, Japan, where the broken curve is the average decay curve. [From Tsujiura, 1978, copyright by Earthquake Research Institute, University of Tokyo.]

fine the frequency-wavenumber (*f–k*) power spectral density as the Fourier transform of the autocorrelation function $\langle u(\mathbf{x},t)u(\mathbf{x}+\mathbf{x}',t+t')\rangle$ [Lacoss et al., 1969]:

$$P_{f_w}(\mathbf{k},f) = \int\limits_{-\infty}^{\infty}\int\limits_{-\infty}^{\infty}\int\limits_{-\infty}^{\infty} \langle u(\mathbf{x},t)u(\mathbf{x}+\mathbf{x}',t+t')\rangle\, e^{-i(\mathbf{k}\mathbf{x}'+2\pi ft')}\, d\mathbf{x}'\, dt' \qquad (2.19)$$

For wavefield data having a finite duration in a given frequency band having center frequency *f*, we usually make a contour plot of the estimated *f–k* power spectral density P_{f_w} in the k_x - k_y plane. A peak in the plot indicates the direction of approach and the apparent propagation velocity of the plane wave that crosses the array. Figure 2.26 shows results of an *f–k* analysis of data recorded by an eight-element seismic array located SE of the center of the Valles Caldera of northern New Mexico, U. S. A. (see Figure 2.11). The analysis was performed on narrow-band filtered data from three 2 s time windows, one surrounding the direct P-arrival, one surrounding the direct S-arrival, and one beginning 20 s after the direct S-arrival. The *f–k* analysis of waveform data that includes the direct waves (Figure

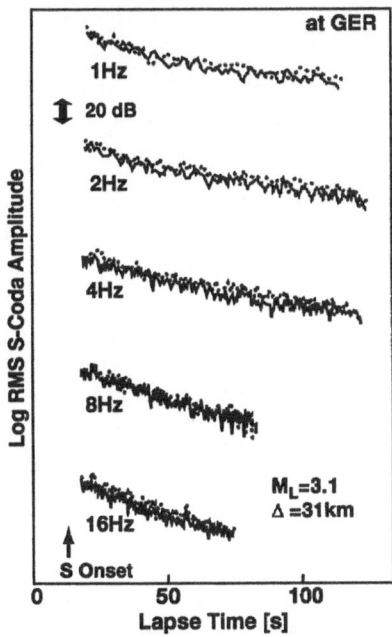

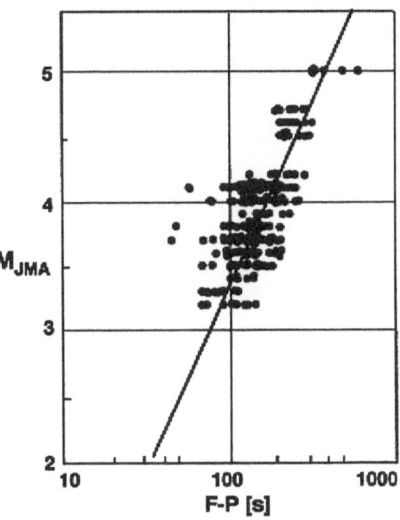

FIGURE 2.24. Octave bandwidth RMS envelopes of coda of seismograms from a local earthquake recorded at hard rock site GER in central Japan. Solid curve is for UD component and dotted curve is for NS component.

FIGURE 2.25. Relationship between the magnitude M_{JMA} determined using amplitude data by JMA and the F-P duration time at a station of the Wakayama Observatory, Japan. [From Tsumura, 1967a, with permission from the Seismological Society of Japan.]

2.26b and c) shows that the direct waves are dominated by energy arriving from the direction of the event indicated by the peaks in the SSE portion of the plots. The S-coda, on the other hand, shows no consistent arrival direction, as shown by Figure 2.26d where high amplitude contours appear in all quadrants of the plot.

Spudich and Bostwick [1987] proposed to use seismograms recorded at a single station for a cluster of earthquakes as a virtual seismic array. Source and receiver positions are exchangeable because of the reciprocity of the Green function of elastodynamics, so the earthquake cluster can be considered an array of seismic stations within the earth that records seismograms from a single source located at the position of the real seismic receiver that recorded the earthquakes. Making the

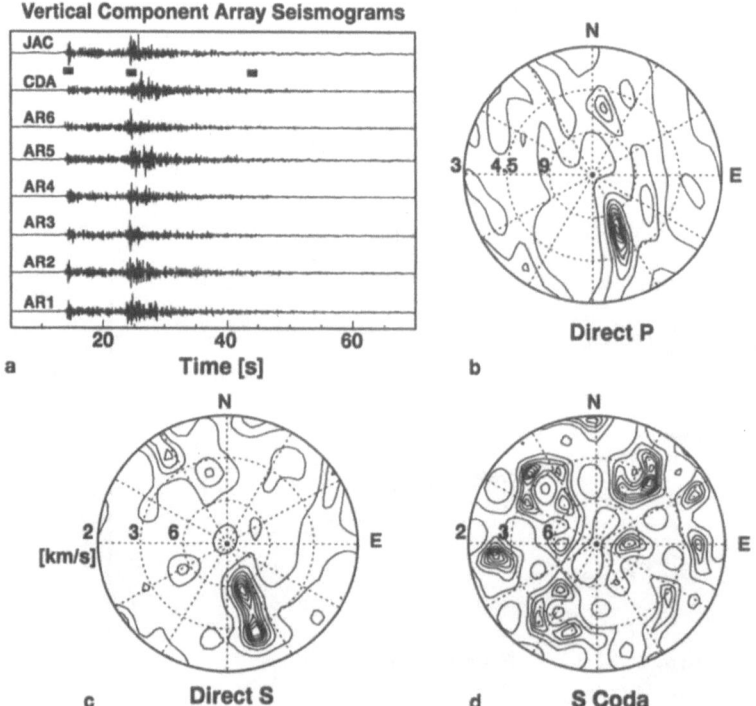

FIGURE 2.26. (a) Vertical component seismograms recorded by an eight element array located in the Valles Caldera, New Mexico, U. S. A. of a small local earthquake located at a distance approximately 80 km SSE of the array, where bold bars indicate 2 s long time windows used for *f–k* analyses. (b) Results of *f–k* analysis of vertical component data surrounding the direct P-arrival. Data were filtered into the 2–5 Hz frequency band prior to analysis. Numbers inside circle refer to velocity of waves crossing array. (c) Results of *f–k* analysis of EW component data surrounding the direct S-arrival. Frequency band is 1.3–3 Hz. (d) Results of *f–k* analysis of EW component S-coda data beginning 20 s after the direct S-arrival. Frequency band is 1.3–3 Hz.

f–k analysis of a set of aftershocks of the 1984 Morgan Hill, California earthquake, they measured the propagation directions and slownesses of the component waves as they travel through the earthquake focal region. The early S-coda, starting immediately after the direct S-wave and ending at twice the S-wave travel-time, was dominated by waves that are multiply scattered near the station since the propagation direction is upward and almost the same as the direct S-wave. Using the same method, Scherbaum et al. [1991] analyzed microearthquake clusters in northern Switzerland. All events were relocated using a master event technique and inter-event arrival-times estimated by a cross-correlation method. This relocation method greatly reduces the errors in relative event locations making the array analysis more reliable. The *f–k* analysis plots show two different patterns: early coda immediately following the direct S-wave was composed of wavelets leaving the source region with the same slowness vector as the direct S-waves; however, latter S-coda waves are composed of wavelets leaving the source region in a variety of directions. The transition between the two types often takes place at 1.5–2 times the S-wave travel-time from source to receiver.

Semblance is another measure of the coherency of waves that is sensitive to amplitude of waves. Semblance coefficient is defined as [Neidell and Taner, 1971]

$$S_C(t,\mathbf{p}) = \frac{\displaystyle\sum_{j=K(i)}^{K(i)+N}\left[\sum_{i=1}^{M} u\left(\mathbf{x}_i, t_j\right)\right]^2}{M\displaystyle\sum_{j=K(i)}^{K(i)+N}\sum_{i=1}^{M} u\left(\mathbf{x}_i, t_j\right)^2} \tag{2.20}$$

where M is the number of stations and N the number of samples. When the aperture of the array is small and we may consider the waves as plane waves, we may set the starting time of the window as $K(i) = t + \mathbf{p}\mathbf{x}_i$, where $\mathbf{x}_i$ is the coordinate of the *i*th station, $\mathbf{p}$ the apparent slowness in 2-D, and t the arrival time at the center of array. Semblance may be viewed as the ratio of the power of the stacked beam to the product of the total power in the traces and the number of channels M. The time resolution of semblance measurements decreases as the number of time samples used in the estimation increases but the resolution increases with increasing number of time samples.

Kuwahara et al. [1990, 1991, 1997] calculated semblance coefficients using array observations of microearthquakes in Kanto, Japan to analyze propagation characteristics of waves. Seismograms from 13 seismometers having average spacing of 50 m were used. Figure 2.27 shows an example of the temporal variation of arrival azimuth, slowness, and the semblance coefficient for vertical components of motion. Results for the vertical component show that the P-coda has almost the same propagation direction and apparent velocity as those of the direct P-wave that arrives from the direction of the epicenter. The semblance coefficient for the P-coda is quite high, but it rapidly drops in the S-coda. The S-coda is composed of waves with widely distributed propagation directions and low semblance coefficient.

The above observations strongly suggest an incoherent nature for high-frequency coda waves. We cannot expect phases other than direct P- and S-waves if the propagation medium is transparent. To explain the observed smooth temporal decay of coda amplitude that is independent of hypocentral distance, Aki and Chouet [1975] proposed a model in which S-coda is composed of S-waves that have been scattered by heterogeneities distributed in a large region outside the zone containing the direct wave path from the source to the receiver. Reverberations in soft layers or the trapping and release of seismic energy by lakes or ponds cannot explain the observed characteristics. In Chapters 3 and 7, we will introduce recently developed models of S-coda that focus on S-wave scattering by distributed heterogeneities.

If we consider that S-coda wave excitation is dominated by scattering of S-waves from heterogeneities in the earth, conservation of energy says that the energy is supplied from the direct S-wave as schematically illustrated in Figure 2.28. Measured S-wave attenuation per distance in the lithosphere is of the order of 0.01 km^{-1} [Aki, 1980a, b], which is almost the same order as the reciprocal of the mean free path of S-waves measured from the S-coda excitation [Sato, 1978]. This coincidence supports the idea of scattering as a mechanism of attenuation. Quantitative laboratory measurements of ultrasonic (~0.5 MHz) elastic wave propagation in granitic blocks were made by Nishizawa et al. [1997]. They reported a larger attenuation of direct wave amplitude and stronger excitation of coda waves for Inada

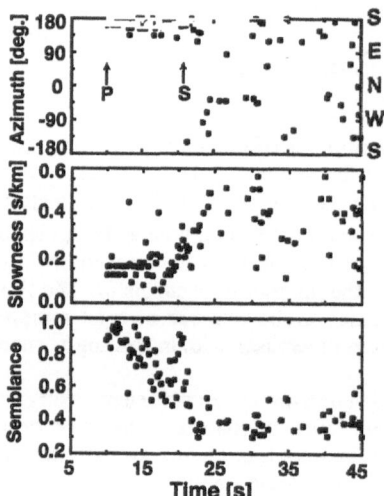

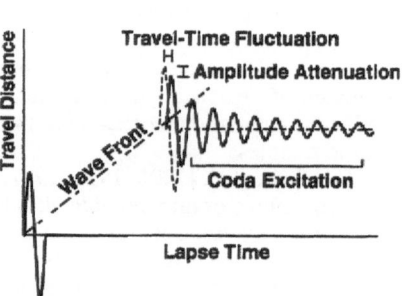

FIGURE 2.27. Arrival azimuth (top), slowness (middle) and semblance coefficient value (bottom) vs. time for a local earthquake in Kanto, Japan determined from analysis of vertical component data recorded by a seismograph array. [From Kuwahara et al., 1991, with permission from the Seismological Society of Japan.]

FIGURE 2.28. Concept of scattering attenuation and coda excitation. The source waveform at the origin (solid line) is unaltered if the medium is transparent (dashed line), but heterogeneity causes scattering that produces coda, fluctuations in travel time, and amplitude attenuation.

granite, which has coarse grain sizes (> 10 mm), compared with those in Westerly granite, which has fine grain sizes (~1 mm). Assuming that the P-wave velocities of the laboratory samples are 5 km/s and spectra peaked at 0.25 MHz, the P-wave wavelength is 20 mm. Thus, the wavelength is closer to the grain size for the Inada granite than for the fine-grained Westerly granite. Their experiments clearly show that the energy transfer from direct waves to coda waves is controlled by the scattering characteristics that depend on the grain sizes of rock media.

Scattering attenuation was left out of early models for the mechanism of attenuation in the earth [Jackson and Anderson, 1970]; however, the importance was pointed out by Aki [1980a] and theoretical models for scattering attenuation were proposed by Wu [1982a, b] and Sato [1982a, b]. Since then, there have been mathematical investigations of scattering attenuation caused by random heterogeneity in the lithosphere. Amplitude and travel-time fluctuations of seismic rays traveling through structures composed of random heterogeneities superposed on a layered velocity structure have been studied numerically [Mereu and Ojo, 1981; Levander and Holliger, 1992]. A model for the attenuation caused by scattering in randomly inhomogeneous elastic media based on the Born approximation will be given in Chapter 5.

2.4.2 Three-Component Seismogram Envelopes

The whole seismogram starting from the P-wave onset until the end of the S-coda reflects not only the source process characterized by the fault-plane geometry and the source-time function but also the scattering characteristics of the heterogeneous earth. Figure 2.29 shows typical three-component seismograms recorded at five stations having epicentral distances between 10 and 60 km located in the vicinity of the Izu Peninsula, Japan. The earthquake source is a strike slip type. The P-wave first motions are shown on the lower hemisphere projection in the figure. We find considerable spatial variation in the amplitudes of the P and S-codas that are functions of both source-receiver azimuths and hypocentral distances.

At station NRY, located near the P-wave nodal line, the direct P-phase is unclear on the vertical component and the S-phase is a large pulse having a period of 0.2 s on the NS component. The P-coda amplitude gradually increases with time on all three components, and the maximum peak amplitude of the S-phase on the vertical component occurs a little later than those on the horizontal components. At stations YMK, SMD and JIZ located near the maximum P-wave radiation directions, the direct P-wave is dominated by one pulse having a period of 0.2 s on the vertical component. P-coda envelopes are concave between the P- and S-phases, decreasing with increasing time after the direct P-arrival. As a result of scattering, the durations of the S-phases on the horizontal components are much larger than the source duration time of 0.2 s. The maximum peak arrival of S-phase on the vertical component appears to occur a little later than those on the horizontal components at each station. Modeling the wave trains of P- and S-coda as incoherent scattered waves, we may sum up the incoherent singly scattered waves and synthesize the whole seismogram envelope. We will introduce a method in Chapter 6 for synthesizing three-component seismogram envelopes by using scattering amplitudes from the Born approximation for elastic waves.

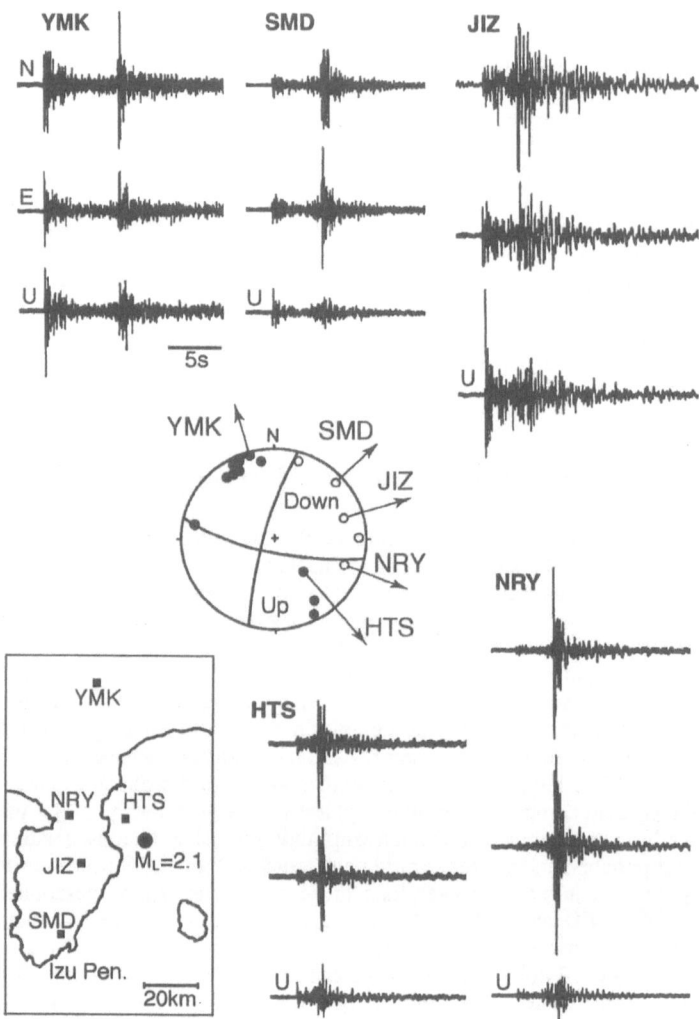

FIGURE 2.29. Three-component seismograms of a strike slip microearthquake (closed circle on the map) recorded near the Izu Peninsula, Japan at five stations (solid squares), where the amplification gains are the same for three components but different for different stations. Initial P-motions are plotted on the lower hemisphere. [From Sato, 1991a, with permission from Elsevier Science - NL, Sara Burgerhartsraat 25, 1055 KV, Amsterdam, The Netherlands.]

2.4.3 Broadening of S-Wave Seismogram Envelopes

Another piece of evidence supporting the existence of heterogeneities in the lithosphere is the observed broadening of the envelopes of the S-wave seismograms of earthquakes recorded at distances between 100 and 300 km. The source duration of earthquakes having local magnitude less than 5 is shorter than 1 s as estimated from the empirical relationship describing the fault rupture process [Kikuchi and Ishida, 1993]; however, we find that the duration of observed S-wave first arrival packets at long distances is much longer than 1 s. In Figure 2.30 we show typical seismogram envelopes, recorded at NIED station ASO in Kanto, Japan, of two earthquakes having different hypocentral distances. The maximum peak, indicated by an open circle on each bandpass-filtered RMS seismogram, occurs several seconds after the S-wave onset that is indicated by a vertical bar. The seismometer is installed on hard rock; therefore, it is difficult to explain the broadening by reverberation in shallow soft deposits. The delay in arrival time of the maximum amplitude is not caused by source effects since we find that the average delay time increases with increasing travel distance. We also find a delay in the time of arrival of the half-maximum amplitude as indicated by a closed circle in Figure 2.30, that is, the wave that is initially an impulse at the source collapses and broad-

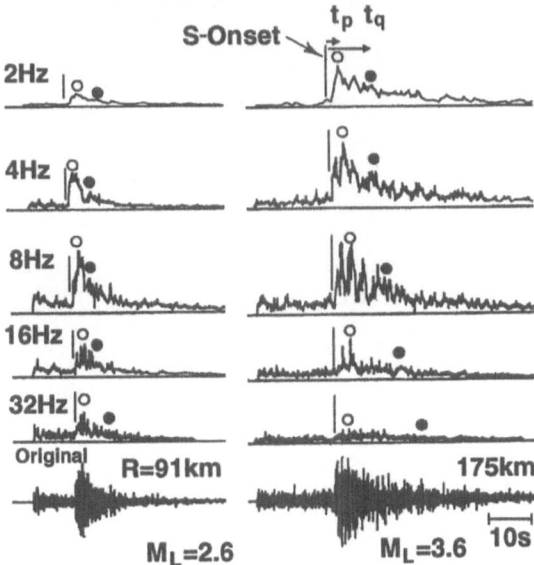

FIGURE 2.30. RMS envelopes of bandpass-filtered seismograms (NS component, octave-width frequency) of two earthquakes in southeast Honshu, Japan, recorded at station ASO. S-wave onset (vertical bar), arrival of the maximum peak (open circle), and the arrival of half-maximum amplitude (closed circle) are shown. [From Sato, 1989, copyright by the American Geophysical Union.]

ens with increasing travel distance. Increasing packet duration is called envelope broadening. It was initially proposed that S-wave seismogram broadening is due to strong diffraction and multiple forward scattering caused by slowly varying velocity structure, which was modeled by employing a stochastic treatment of the parabolic approximation of the wave equation [Sato, 1989]. Later, Obara and Sato [1995] used the model to investigate the regional difference in the S-wave envelope broadening relative to the location of the volcanic front in Japan. This work will be discussed in Chapter 8.

CHAPTER 3

Phenomenological Modeling of Coda-Wave Excitation

As discussed in Section 2.4.1, the excitation of S-coda waves is one of the most compelling pieces of evidence supporting the existence of random heterogeneity in the lithosphere. Aki and Chouet [1975] summarized the characteristics of high-frequency S-coda waves of local earthquakes, like those shown in Figure 2.21, as follows: (1) the spectral contents of the later portions of the S-coda observed at different stations are nearly the same; (2) the total duration of a seismogram, defined as the length of time between the P-wave onset and the time when the coda amplitude equals the level of microseisms, is a reliable measure of earthquake magnitude; (3) bandpass-filtered S-coda traces of different local earthquakes recorded within a given region have a common envelope shape whose time dependence is independent of epicentral distance; (4) the temporal decay of S-coda amplitudes are independent of earthquake magnitude at least for $M_L < 6$; (5) the S-coda amplitude depends on the local geology of the recording site; (6) array measurements show that S-coda waves are not regular plane waves coming directly from the epicenter [Aki and Tsujiura, 1959]. S-coda waves have the same site amplification factor as that of direct S-waves confirming that coda-waves are composed primarily of S-waves [Tsujiura, 1978]. Clear S-coda waves have even been identified on seismograms recorded at the bottom of deep boreholes drilled in hard rock beneath soft deposits [Sato, 1978; Leary and Abercrombie, 1994].

We will present a phenomenological model for coda-wave generation based on a view of the earth's lithosphere as composed of a random and uniform distribution of point-like scatterers in a homogeneous background medium having a constant propagation velocity as schematically illustrated in Figure 3.1. We neglect diffraction effects caused by gradual changes in velocity. This model was originally proposed by Aki and Chouet [1975] for the case of collocated source and receiver. An extension to the case of source-receiver separation was done by Sato [1977a] for body waves and by Kopnichev [1975] for surface waves.

Alternative phenomenological models for coda-wave generation have been proposed. Prior to the work of Aki and Chouet [1975], Wesley [1965] proposed a diffusion-like process as an explanation for seismogram envelopes. The observed long duration of the coda of lunar seismograms was studied using the diffusion model [Dainty and Toksöz, 1981]. Due to the extremely low intrinsic attenuation of the lunar crust and large amount of scattering, the diffusion approach works well to explain wave propagation in the lunar crust. Frankel and Wennerberg [1987] developed a model called the energy-flux model based on the uniform distribution of scattered-wave energy that was found in finite difference simulations of wave propagation in 2-D random media.

The parameters that control the shape and amplitude of the coda envelopes predicted by the phenomenological models are the total scattering coefficient and the coda-wave attenuation. Applying these models to observed seismogram envelopes has resulted in measurements of these two parameters throughout the world [Herraiz and Espinosa, 1987; Rautian et al., 1981; Kopnichev, 1985]. There have been many reports of temporal change in coda characteristics as precursors to earthquakes and volcanic eruptions [Jin and Aki, 1986; Sato, 1988c; Fehler et al., 1988]. Temporal change in coda has been studied using earthquake doublets, which are earthquakes that are thought to have identical locations but occur at differing times [Got et al., 1990; Aster et al., 1996]. The most practical tool that has originated from the study of coda-waves is the coda-normalization method, which is based on the assumption of a uniform spatial distribution of coda energy for long lapse times. The coda-normalization method allows us to estimate the difference of site amplification factors as a function of frequency, to distinguish differences in source spectral characteristics, and to measure attenuation using data from only a single station [Aki, 1969; Phillips and Aki, 1986; Aki, 1980a].

In this chapter we first introduce the mathematics for the phenomenological modeling of S-coda-wave excitation. Then, we describe measurements of coda characteristics and introduce the coda-normalization method.

Scattering Characteristics

Randomly inhomogeneous media will be modeled as homogeneous background media having propagation velocity V_0 that are filled with distributed point-

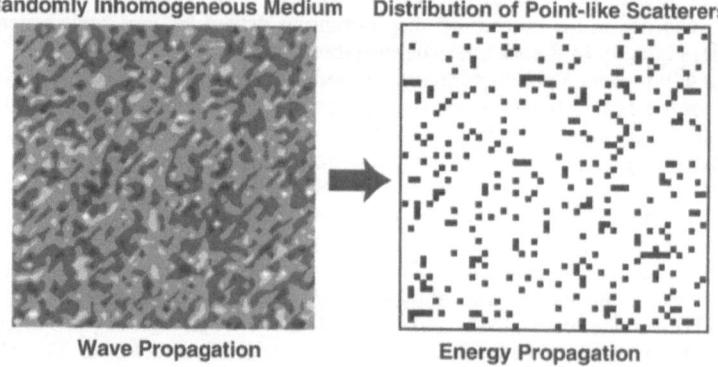

Randomly Inhomogeneous Medium　　**Distribution of Point-like Scatterers**

Wave Propagation　　　　　　　　Energy Propagation

FIGURE 3.1. Modeling a random medium as a distribution of point-like scatterers.

like scatterers with number density n (see Figure 3.1). The distribution is assumed to be randomly homogenous and isotropic, and the scattering is characterized by the differential scattering cross section $d\sigma / d\Omega$. As illustrated in Figure 3.2, we imagine a stationary process in which the incident wave with energy-flux density J^0 interacts with a scatterer and generates spherically outgoing waves with energy-flux density J^1. The energy-flux density is defined as the amount of energy passing through a unit area perpendicular to the propagation direction per unit time. Then, the amount of energy scattered per unit time into a given solid angle element $d\Omega$ is $J^1 r^2 d\Omega$, where $r^2 d\Omega$ is the corresponding surface element. We define the differential scattering cross section as the ratio

$$\frac{d\sigma}{d\Omega} = \frac{J^1 r^2}{J^0} \tag{3.1}$$

For a medium filled with such scatterers, the scattering power per unit volume is given by the product of the number density and the differential scattering cross section, which is called the scattering coefficient [Aki and Chouet, 1975]:

$$g \equiv 4\pi n \frac{d\sigma}{d\Omega} \tag{3.2}$$

Quantity g has dimension of reciprocal length. We may characterize the scattering power using only the scattering coefficient. In this formulation, we do not distinguish between a small number distribution of strong scatterers and a large number of weak scatterers. The total scattering coefficient is defined as the average over all directions:

$$g_0 \equiv \frac{1}{4\pi} \oint g \, d\Omega = n \oint \frac{d\sigma}{d\Omega} d\Omega = n\sigma_0 = \ell^{-1} = {}^{Sc}Q^{-1} k \tag{3.3}$$

where σ_0 is the total scattering cross section which is the integral of the differential scattering cross section over a solid angle. The reciprocal of the total scattering coefficient is the mean free path ℓ. The incident wave energy decreases with increasing travel distance due to scattering, where we define scattering attenuation ${}^{Sc}Q^{-1}$ for

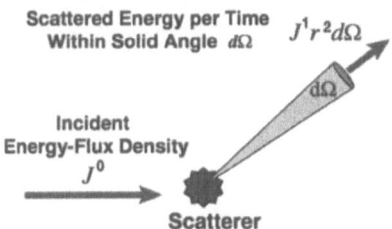

FIGURE 3.2. Concept of the differential scattering cross section of a single scatterer.

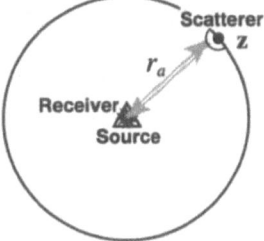

FIGURE 3.3. Geometry of the single backscattering model.

waves of wavenumber k. For plane-wave incidence, the energy-flux density at travel distance x decays as $\exp(-g_0 x) = \exp(-^{Sc}Q^{-1}kx)$.

The simplest model is isotropic scattering:

$$\frac{d\sigma}{d\Omega} = \frac{\sigma_0}{4\pi} \quad \text{and} \quad g = g_0 \tag{3.4}$$

Since the scatterers are considered randomly distributed, the scattered waves are incoherent, so that phase may be neglected, and the scattered wave power may be obtained as a sum of power from individual scattered waves.

3.1 SINGLE SCATTERING MODELS

3.1.1 Single Backscattering Model for a Common Source and Receiver Location

Aki and Chouet [1975] proposed the single backscattering model to explain the time dependence of the scattered energy density at the source location in 3-D space. They considered the case of impulsive spherical radiation of total energy W from the source located at the origin as illustrated in Figure 3.3. They did not consider polarization and partition of energy into three components. The incident energy-flux density upon a scatterer located at $\mathbf{z}$ is given by

$$\frac{W}{4\pi r_a^2}\delta\left(t - \frac{r_a}{V_0}\right) \tag{3.5}$$

where $r_a = |\mathbf{z}|$. Single backscattered energy-flux density at the origin from a single scatterer is given by

$$\frac{W}{4\pi r_a^2}\delta\left(t - \frac{2r_a}{V_0}\right)\frac{1}{r_a^2}\left.\frac{d\sigma}{d\Omega}\right|_\pi \tag{3.6}$$

The delta function includes the time delay for the round trip between source and receiver, and subscript π indicates backward scattering. Dividing the energy-flux density by wave velocity V_0, we get the energy density. Summing up the energy density for all scatterers,

$$E^1(\mathbf{x}=0,t) = \sum_{Scatterers} \frac{W}{4\pi r_a^2}\delta\left(t - \frac{2r_a}{V_0}\right)\frac{1}{r_a^2}\left.\frac{d\sigma}{d\Omega}\right|_\pi \frac{1}{V_0} \tag{3.7}$$

where the superscript 1 means single scattering. Multiplying by the number density of scatterers n, we replace the summation with an integral over the whole space. The energy density is given by

$$E^1(\mathbf{x}=0,t) = \int_{-\infty}^{\infty}\int_{-\infty}^{\infty}\int_{-\infty}^{\infty}\frac{W}{4\pi r_a^2}\delta\left(t-\frac{2r_a}{V_0}\right)\frac{n}{r_a^2}\frac{d\sigma}{d\Omega}\bigg|_\pi\frac{1}{V_0}d\mathbf{z}\qquad(3.8)$$

Substituting (3.2) and integrating the above in spherical coordinates,

$$\begin{aligned}E^1(\mathbf{x}=0,t) &= \int_0^\infty \frac{W}{4\pi r_a^2}\delta\left(r_a-\frac{V_0 t}{2}\right)\frac{V_0}{2}\frac{1}{r_a^2}\frac{g_\pi}{4\pi}\frac{1}{V_0}4\pi r_a^2 dr_a\\ &= \frac{Wg_\pi}{8\pi\left(\dfrac{V_0 t}{2}\right)^2}H(t) = \frac{Wg_\pi}{2\pi V_0^2 t^2}H(t) \propto t^{-2}\end{aligned}\qquad(3.9)$$

where H is the step function. The single scattered energy density at the source location decreases with the inverse square of lapse time, that is, the RMS coda amplitude decreases with the inverse of lapse time. The single scattered energy is proportional to the backscattering coefficient $g_\pi = 4\pi n(d\sigma/d\Omega)\big|_\pi$.

For practical analysis, we introduce phenomenological attenuation by multiplication with an exponential damping factor at angular frequency ω:

$$E^1(\mathbf{x}=0,t) = \frac{Wg_\pi H(t)}{2\pi V_0^2 t^2}e^{-Q_c^{-1}\omega t}\qquad(3.10)$$

where Q_c^{-1} is called the coda-attenuation factor. The interpretation of this factor will be discussed in Sections 3.3.2 and 7.1.1.

3.1.2 Single Isotropic Scattering Model for General Source and Receiver Locations

Three-Dimensional Case

We calculate the spatiotemporal change in energy density for a receiver located at a distance r from a source located at the origin. As for the single backscattering model, we do not consider polarization and the partition of energy into three components of motion. We suppose that the scattering is isotropic as in (3.4). When energy W is impulsively radiated spherically from the source, the energy-flux density at a scatterer located at $\mathbf{z}$ is given by (3.5). The energy-flux density at the receiver at $\mathbf{x}$ is given by

$$\frac{W}{4\pi r_a^2}\delta\left(t-\frac{r_a+r_b}{V_0}\right)\frac{1}{r_b^2}\frac{\sigma_0}{4\pi}\qquad(3.11)$$

where $r_b = |\mathbf{x} - \mathbf{z}|$. Dividing by V_0 and multiplying by the number density of scatterers n, the energy density is given by

$$E'(\mathbf{x},t) = \sum_{Scatterers} \frac{W}{4\pi r_a^2} \delta\left(t - \frac{r_a + r_b}{V_0}\right) \frac{1}{r_b^2} \frac{\sigma_0}{4\pi} \frac{1}{V_0}$$

$$\rightarrow \frac{W g_0}{(4\pi)^2} \int_{-\infty}^{\infty}\int_{-\infty}^{\infty}\int_{-\infty}^{\infty} \frac{\delta(r_a + r_b - V_0 t)}{r_a^2 r_b^2} d\mathbf{z} \tag{3.12}$$

where $g_0 = n\sigma_0$, and r_a and r_b are distances from the source to a scatterer and the scatterer to the receiver, respectively. The delta function in (3.12) means that the integral in 3-D is reduced to an integral over the surface of a prolate spheroid with foci at the source and the receiver. Integration of (3.12) is accomplished by the introduction of prolate spheroidal coordinates (w, v, ϕ) [Morse and Feshbach, 1953, p. 661] as

$$z_1 = \frac{r}{2}\sqrt{(v^2 - 1)(1 - w^2)}\cos\phi$$

$$z_2 = \frac{r}{2}\sqrt{(v^2 - 1)(1 - w^2)}\sin\phi \tag{3.13}$$

$$z_3 = \frac{r}{2}(1 + vw)$$

where $r \equiv |\mathbf{x}|$. The ranges of coordinates are $-1 \le w \le 1, 1 \le v < \infty$ and $0 \le \phi < 2\pi$. The coordinates of the source at the origin and the receiver at r in the prolate spheroidal coordinate system are $v = 1$ and $w = -1$ and $v = 1$ and $w = 1$, respectively as illustrated in Figure 3.4. We note that

$$r_a \equiv \sqrt{z_1^2 + z_2^2 + z_3^2} = \frac{r}{2}(v + w) \text{ and } r_b \equiv \sqrt{z_1^2 + z_2^2 + (z_3 - r)^2} = \frac{r}{2}(v - w) \tag{3.14}$$

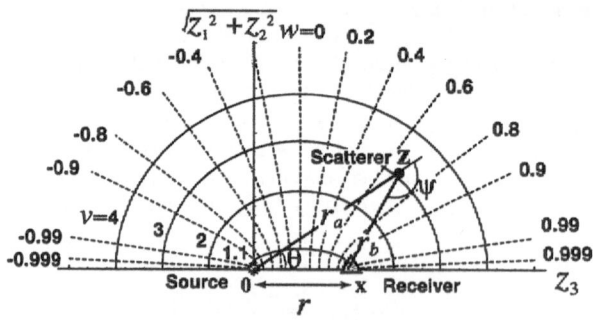

FIGURE 3.4. Geometry of the single scattering process. The source and the receiver are located at the foci of the prolate spherical coordinate system.

The infinitesimal volume element in (3.12) is given by

$$dz = \left| \frac{\partial(z_1, z_2, z_3)}{\partial(w, v, \phi)} \right| dv\,dw\,d\phi = \left(\frac{r}{2}\right)^3 (v^2 - w^2) dv\,dw\,d\phi = \frac{r r_a r_b}{2} dv\,dw\,d\phi \quad (3.15)$$

A prolate spheroidal shell is defined by coordinate v:

$$r_a + r_b = rv \qquad (3.16)$$

Thus the delta function in (3.12) defines a spheroidal shell given by spheroidal co-ordinate $v = V_0 t / r$. We refer to the single scattering isochron defined by constant v as the "Isochronal scattering shell" corresponding to a given lapse time t. Scattering angle ψ and radiation angle θ (see Figure 3.4) are given by

$$\psi = \cos^{-1}\left(\frac{2 - v^2 - w^2}{v^2 - w^2}\right) \quad \text{and} \quad \theta = \cos^{-1}\frac{1 + v\,w}{v + w} \qquad (3.17)$$

As shown in Figure 3.5a, scattering angle ψ approaches π as lapse time increases, that is, as v increases. This means that the source and receiver can be considered collocated at large lapse times.

Transforming integral (3.12) into an integral over a prolate spheroidal surface of constant v, we can integrate analytically as follows [Sato, 1977a]:

$$E^1(\mathbf{x}, t) = \frac{W g_0}{(4\pi)^2} \int_0^{2\pi} d\phi \int_1^\infty dv \int_{-1}^1 dw \left(\frac{r}{2}\right)^3 (v^2 - w^2) \frac{\frac{1}{r}\delta\left(v - \frac{V_0 t}{r}\right)}{\left(\frac{r}{2}\right)^4 (v^2 - w^2)^2}$$

$$= \frac{W g_0}{4\pi r^2} \int_1^\infty dv\,\delta\left(v - \frac{V_0 t}{r}\right) \int_{-1}^1 dw \frac{1}{v^2 - w^2} \qquad (3.18)$$

$$= \frac{W g_0}{4\pi r^2} K\left(\frac{V_0 t}{r}\right) H(V_0 t - r)$$

where

$$K(v) \equiv \int_{-1}^1 dw \frac{1}{v^2 - w^2} = \frac{2}{v}\tanh^{-1}\frac{1}{v} = \frac{1}{v}\ln\frac{v+1}{v-1} \quad \text{for } v > 1 \qquad (3.19a)$$

Function $K(v)$ logarithmically diverges as $v \to 1_+$ and has asymptotic behavior given by

$$K(v) \approx \frac{2}{v^2} \quad \text{for } v \gg 1 \qquad (3.19b)$$

Figure 3.5b is a plot of K, where the asymptote is plotted using a dotted curve. The asymptotic time dependence of scattered energy density is given by

$$E^1(\mathbf{x},t) \approx \frac{Wg_0}{2\pi V_0^2 t^2} \quad \text{for} \quad V_0 t \gg r \tag{3.20}$$

which coincides with the single backscattering model (3.9) for $g_x = g_0$.

We may scale physical quantities using the total scattering coefficient, the propagation velocity, and the radiated energy as

$$\overline{\mathbf{x}} = g_0 \mathbf{x}, \quad \overline{t} = g_0 V_0 t, \quad \text{and} \quad \overline{E} = E / (g_0^3 W) \tag{3.21}$$

where the overbar denotes the nondimensional normalized quantity. Then, the normalized energy density is given by

$$\overline{E}^1(\overline{\mathbf{x}},\overline{t}) = \frac{1}{4\pi \overline{r}^2} K(\frac{\overline{t}}{\overline{r}}) H(\overline{t} - \overline{r}) \tag{3.22}$$

where $\overline{r} = |\overline{\mathbf{x}}|$, and the asymptotic behavior of (3.20) is given by

$$\overline{E}^1(\overline{\mathbf{x}},\overline{t}) \approx \frac{1}{2\pi \overline{t}^2} \quad \text{for} \quad \overline{t} \gg \overline{r} \tag{3.23}$$

The spatiotemporal change in the normalized energy density is shown in Figure 3.6. Each temporal trace shown in Figure 3.6a asymptotically approaches the broken curve which is independent of distance from the source. The difference between (3.22) and (3.23) is less than 10% for lapse times greater than twice the travel time. The spatial sections show the uniform distribution of coda energy around the source located at the origin.

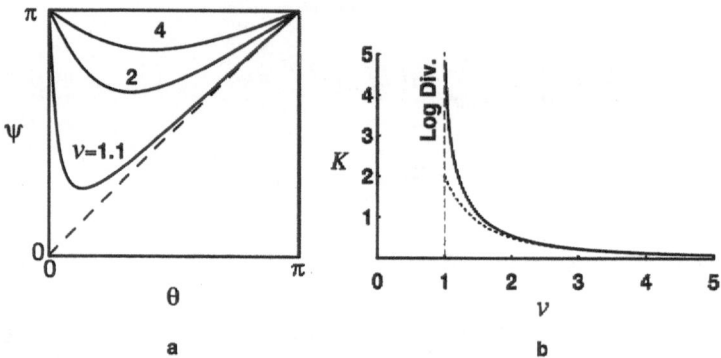

FIGURE 3.5. (a) Plot of ψ vs. θ for (3.17) for different ν corresponding to normalized lapse time. (b) Plot of $K(\nu)$ from (3.19a), where the dotted curve shows the asymptote (3.19b).

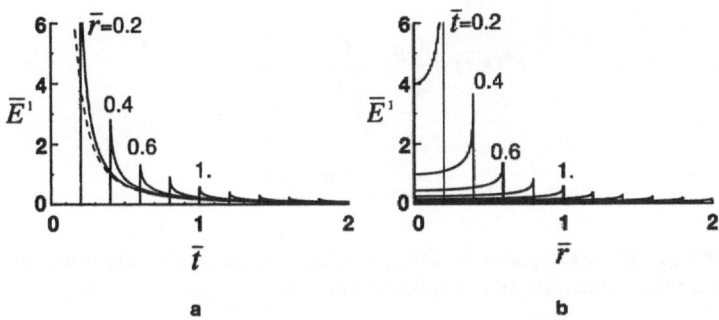

FIGURE 3.6. (a) Time trace and (b) spatial distribution of normalized energy density (3.22) for the single isotropic scattering model, where the broken curve shows the asymptote (3.23).

The spatial integral of the energy density gives the total energy:

$$\int_0^\infty E^1(\mathbf{x},t)d\mathbf{x} = \int_0^{V_t} E^1(\mathbf{x},t)4\pi r^2 dr = W g_0 V_0 t \qquad (3.24)$$

The total energy increases monotonously with increasing lapse time. To satisfy the conservation of total energy, we multiply by the exponential scattering attenuation term $e^{-g_0 V_0 t}$ to account for energy lost due to scattering by the direct energy propagation term (3.5). Later, in the study of the multiple scattering process in Chapter 7, we correctly introduce the scattering attenuation term into the formulation. As discussed in Section 3.1.1, we introduce a phenomenological attenuation factor, coda attenuation Q_c^{-1} at angular frequency ω

$$E^1(\mathbf{x},t) = \frac{W g_0}{4\pi r^2} K\left(\frac{V_0 t}{r}\right) H\left(V_0 t - r\right) e^{-Q_c^{-1}\omega t}$$

$$\approx \frac{W g_0}{2\pi V_0^2 t^2} e^{-Q_c^{-1}\omega t} \qquad V_0 t \gg r \qquad (3.25)$$

The resultant formulas (3.10) and (3.25) have been used worldwide for measurements of coda characteristics as will be summarized in Section 3.3.

Two-Dimensional Case

For the study of surface wave scattering, Kopnichev [1975] derived a formula for isotropic scattering in 2-D space. The single scattering energy density at a receiver located at $\mathbf{x}$ for impulsive circular radiation from a source at the origin is given by

$$E^1(\mathbf{x},t) = \frac{Wg_0}{2\pi r} \frac{H(V_0 t - r)}{\sqrt{\left(\frac{V_0 t}{r}\right)^2 - 1}} \tag{3.26}$$

$$\approx \frac{Wg_0}{2\pi V_0 t} \quad \text{for } V_0 t \gg r$$

where $r = |\mathbf{x}|$. The energy density diverges at the arrival time of the direct wave and decreases with the inverse power of lapse time near the source location.

3.2 MULTIPLE SCATTERING MODELS

3.2.1 Diffusion Model

As lapse time increases, we expect that multiple scattering will dominate compared to single scattering. For large lapse times, it is reasonable to assume that direct energy is small and that multiple scattering produces a smooth spatial distribution of energy density. Consider a medium having a randomly homogeneous and isotropic distribution of isotropic scatterers in which energy W is spherically radiated from a source located at the origin and the source time function is a delta function in time. A strong multiple scattering process can be described by the diffusion equation [Morse and Feshbach, 1953, p.191]

$$(\partial_t - D_C \Delta) E(\mathbf{x},t) = W\delta(\mathbf{x})\delta(t) \tag{3.27}$$

where the diffusivity for isotropic scattering $D_C = V_0/(3g_0) = V_0 \ell / 3$. The analytical solution for (3.27) is known as the diffusion solution

$$E^D(\mathbf{x},t) = \frac{W}{(4\pi D_C t)^{3/2}} e^{-\frac{r^2}{4D_C t}} H(t) \tag{3.28}$$

$$\propto t^{-\frac{3}{2}} \qquad \text{at} \quad r = 0$$

Scaling by using (3.21) to put (3.28) into nondimensional form

$$\bar{E}^D(\bar{\mathbf{x}},\bar{t}) = \left(\frac{3}{4\pi\bar{t}}\right)^{\frac{3}{2}} e^{-\frac{3\bar{r}^2}{4\bar{t}}} H(\bar{t}) \tag{3.29}$$

Energy decreases with the −1.5th power of lapse time near the source location, which is slower than that of the single scattering model. The diffusion solution is the continuous limit of a discretized random walk. Figure 3.7 shows the spatial

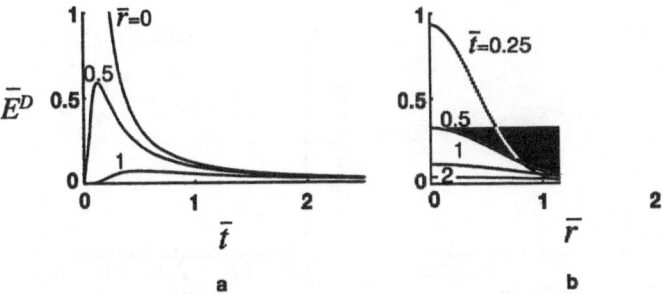

FIGURE 3.7. Diffusion model: (a) time traces at different scaled distances and (b) spatial distribution of normalized energy density (3.29) at different scaled times.

distribution of the energy density for various scaled times and the time traces for various scaled distances. For the diffusion equation, energy spreads in front of the wavefront violating causality. In Chapter 7, we will show that the energy density calculated using the radiative transfer theory for the multiple isotropic scattering process converges to the diffusion solution for large lapse times.

The total energy, given by the spatial integral of the energy density in (3.29), is conserved. Introducing intrinsic attenuation $^IQ^{-1}$ at angular frequency ω, we may write (3.28) for practical analysis as

$$E^D(\mathbf{x},t) = \frac{W}{(4\pi D_c t)^{3/2}} e^{-\frac{r^2}{4D_c t} - {}^IQ^{-1}\omega t} H(t) \tag{3.30}$$

The diffusion model solution (3.30) was used for the analysis of coda recorded near the hypocenter of earthquakes [Wesley, 1965; Aki and Chouet, 1975] and the coda of lunar-quakes [Nakamura, 1977; Dainty and Toksöz, 1981] as will be discussed in Section 3.3.

3.2.2 Energy-Flux Model

From analysis of 2-D finite difference simulations of wave propagation in inhomogeneous media, Frankel and Clayton [1986] observed the excitation of coda-waves. The results of the numerical simulations, as illustrated in Figure 3.8, led Frankel and Wennerberg [1987] to the conclusion that waves scattered from the direct wave rapidly spread over the spherical volume behind the direct wavefront. They proposed a phenomenological model for the spatiotemporal distribution of energy density that is consistent with observations that seismogram envelopes recorded at different distances asymptotically approach a common decay curve and with the similarity of coda amplitude in the region behind the S-wavefront for large lapse times.

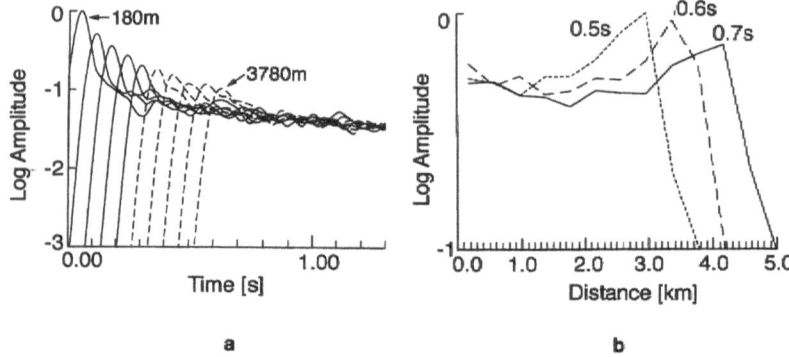

FIGURE 3.8. Average log-amplitude envelopes of synthetic seismograms from 2-D finite difference calculations at 25–35 Hz: (a) time traces at different distances and (b) spatial sections at different lapse times for radiation from a source at the origin, where the random acoustic media are characterized by the exponential ACF with ε=10% and a=40 m. [From Frankel and Wennerberg, 1987, copyright by the Seismological Society of America.]

For total energy W radiated spherically from a source at the origin, Frankel and Wennerberg [1987] a priori assume that scattering leads to a spatially uniform distribution of coda energy density $E^{EF}(\mathbf{x},t)$ within a sphere of radius $V_0 t$. The direct energy decreases due to scattering attenuation at an exponential rate given by $^{Sc}Q^{-1}$ with increasing travel distance. This model strictly discriminates between direct and scattered waves. When there is no intrinsic absorption, the conservation of total energy at angular frequency ω can be written as

$$W e^{-^{Sc}Q^{-1}\omega t} + \frac{4\pi}{3}(V_0 t)^3 E^{EF}(\mathbf{x},t) = W \quad \text{for} \quad V_0 t > r \qquad (3.31)$$

where $r = |\mathbf{x}|$, the first term on the left-hand side corresponds to the direct wave energy, and the second term is the energy scattered within a volume behind the direct wavefront, Then,

$$E^{EF}(\mathbf{x},t) = \frac{3W\left(1 - e^{-^{Sc}Q^{-1}\omega t}\right)}{4\pi(V_0 t)^3} H(t - r/V_0)$$

$$\approx \frac{3W\omega\,^{Sc}Q^{-1}}{4\pi V_0^3 t^2} H(t - r/V_0) \quad \text{for} \quad ^{Sc}Q^{-1}\omega t << 1 \qquad (3.32)$$

For small lapse times, the energy density decreases with the inverse second power of lapse time in agreement with (3.9). The amplitude of direct-wave energy density at the wavefront depends on the source duration T_0.

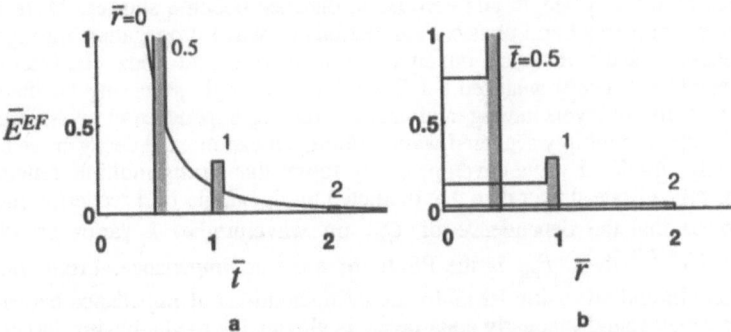

FIGURE 3.9. Energy-flux model: (a) time traces at different scaled distances and (b) spatial distribution of normalized energy density (3.33) at different scaled times. Shaded area corresponds to the direct-wave energy density.

Using the relationship between the total scattering coefficient and the direct-wave scattering attenuation described by (3.3) to replace $^{Sc}Q^{-1}$ with $g_0 V_0/\omega$ and scaling quantities using (3.21), we may write (3.32) as

$$\bar{E}^{EF}(\bar{\mathbf{x}}, \bar{t}) = \frac{3(1 - e^{-\bar{t}})}{4\pi \bar{t}^3} H(\bar{t} - \bar{r}) \tag{3.33}$$

As illustrated in Figure 3.9, the spatial distribution of energy density is uniform within the sphere behind the wavefront, and the temporal decay is common irrespective of distance except near the direct wave. This model has no clear mechanism to explain how scattered energy is spread over space but incorporates the effects of both multiple scattering and causality.

When we introduce intrinsic absorption $^I Q^{-1}$, we may modify (3.32) to

$$E^{EF}(\mathbf{x}, t) = \frac{3W(1 - e^{-^{Sc}Q^{-1}\omega t}) e^{-^I Q^{-1}\omega t}}{4\pi(V_0 t)^3} H\left(t - \frac{r}{V_0}\right) \tag{3.34}$$

Frankel and Wennerberg [1987] point out that the phenomenological exponential decay factor of coda amplitude Q_c^{-1} is not a simple combination of scattering and intrinsic attenuation, but is far more sensitive to $^I Q^{-1}$ than $^{Sc}Q^{-1}$.

3.2.3 Simulations of Coda-Wave Excitation

Frankel and Clayton [1986] measured the effects of wavelength and correlation distance on the spatial coherence of coda-waves in a random medium from their

2 - D numerical simulations. They observed a decrease in correlation for a fixed separation as the wavelength and correlation distance become shorter. There have been several numerical and physical simulations of wave propagation through inhomogeneous media focusing on envelope formation and coda characteristics. Menke and Chen [1984] analyzed a 1-D model of strongly scattering media composed of a series of layers having randomly fluctuating impedance to investigate the coda-envelope of multiply scattered waves. Fitting an exponential decay curve to the numerically simulated wave envelopes, they found that strong multiple reflections make the fall-off rate slower than that predicted by the single backscattering model. They found that the dependence of Q_C^{-1} on wavenumber k varies as $Q_C^{-1} \propto k^{-2} P_{\text{Imp}}(2k)^{-1/2}$, where P_{Imp} is the PSDF of acoustic impedance. From various numerical simulations using RMS fractional fluctuations of impedance between 1 and 20%, they found that early coda decay is slower for media having larger impedance fluctuations. Numerical simulations of 2-D SH-wave propagation through a medium containing 50 cavities using the boundary integral method were made by Yomogida and Benites [1995] who studied the relation between the seismogram envelopes, the wavelength of the SH wave, and the diameter of the cavities. They found a coincidence between Q_C^{-1} and the apparent attenuation of direct wave amplitude Q_S^{-1}. Their method will be described in more detail in Section 5.4 along with a discussion of their scattering attenuation study.

Hestholm et al. [1994] numerically simulated wave propagation through a complex heterogeneous 2-D medium consisting of a random velocity structure characterized by a nonisotropic von Kármán type ACF superimposed on a layered velocity structure. Their medium had an irregular surface topography and irregular Moho boundary along with a low-velocity layer near the surface. They reported the importance for coda formation of scattering by the irregular interfaces and surface including conversion from body waves to surface waves. Semblance analysis of synthesized array seismograms showed the dominance of scattered S-waves in S-coda as predicted but showed the dominance of P- and Rg-waves in P-coda.

Physical model simulations of wave propagation through cracked media have also been conducted. Matsunami [1991] measured ultrasonic wave propagation through a plate with many holes. Changing the number of holes and frequency of incident waves, he found a strong correlation between the strength of scattering attenuation and the excitation level of coda in 2-D. He also concluded that there is a large contribution of intrinsic attenuation to coda attenuation. Vinogradov et al. [1992] experimentally studied the excitation of scattered waves through a thin Plexiglas sheet containing many parallel cracks. They reported not only the excitation of coda but a delay of the peak arrival in highly cracked media.

It has been well established that there is a strong link between the medium heterogeneity and the characteristics of seismogram envelopes; however, we note that there is not yet a wave theoretical method to derive coda attenuation Q_C^{-1} from the stochastic characterization of random media. A relation between coda attenuation and media properties based on the phenomenological radiative transfer theory will be discussed in Section 7.1.1.

3.3 CODA ANALYSIS

The phenomenological models for high-frequency seismograms of local earthquakes described in Sections 3.1 and 3.2 can be used to estimate the scattering characteristics of the lithosphere from seismic data. The two most commonly measured parameters are the total scattering coefficient and the coda attenuation. The total scattering coefficient was introduced in (3.3) and is the parameter that governs the strength of S-coda excitation. The coda attenuation Q_c^{-1} empirically characterizes the exponential decay of the coda amplitude envelope with increasing lapse time. Temporal change of coda attenuation has been proposed as a precursory indicator for the occurrence of large earthquakes.

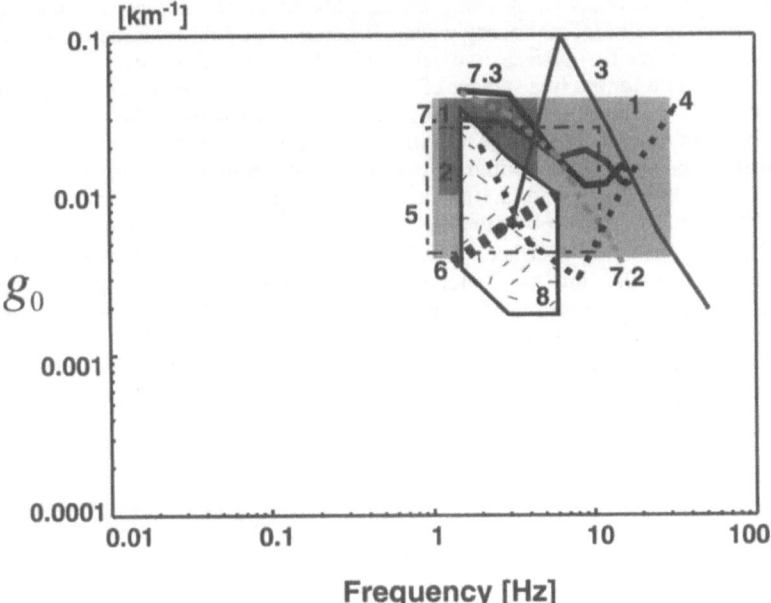

FIGURE 3.10. Total scattering coefficient g_0 for S-to-S wave scattering vs. frequency from regional measurements made throughout the world. **Based on the single scattering model** (Plots include backscattering coefficient g_π): 1, Kanto, Japan [Sato, 1978]; 2, Kanto, Japan [Aki, 1980b]; 3, New Brunswick, Canada [Dainty et al., 1987]; 4, western Nagano, Japan [Kosuga, 1992]; 5, central Greece [Baskoutas, 1996]. **Based on the multiple lapse-time window analysis** (Isotropic scattering is assumed. See Section 7.2): 6, Kanto-Tokai, Japan [Fehler et al., 1992]; 7.1, central California; 7.2, Hawaii; 7.3, Long Valley in California [Mayeda et al. 1992]; 8, 16 measurements in Japan [Hoshiba, 1993].

3.3.1 Coda-Excitation Measurements

The energy density of coda waves in a frequency band having central frequency f is written as a sum of three components of the mean square particle velocity of S coda $\dot{u}_i^{S\,Coda}(t;f)$ as

$$E^{S\,Coda}(t;f) = \left\langle \sum_{i=1}^{3} \frac{\rho_0}{2} \left| \dot{u}_i^{S\,Coda}(t;f) \right|^2 + \text{Elastic Energy} \right\rangle_T \qquad (3.35)$$

$$\approx \sum_{i=1}^{3} \rho_0 \left\langle \left| \dot{u}_i^{S\,Coda}(t;f) \right|^2 \right\rangle_T$$

where ρ_0 is the mass density, elastic energy is the potential energy stored in the media, and $\langle \cdots \rangle_T$ is a moving time average over a few cycles around time t. The last expression is obtained by using the equality of kinetic energy and elastic energy, which is valid for stationary waves. When we use seismograms recorded at a hard rock site on the surface, we take half the observed velocity amplitude to roughly account for the free surface effect. When only one component of motion is available, we often multiply the energy density calculated from the single component by three to account for the missing data since, as illustrated in Figure 2.24, vertical component coda amplitudes are nearly equal to those of horizontal components. Substituting (3.35) in (3.10), (3.25), or (3.30), we can estimate total scattering coefficient g_0 for S-to-S scattering.

The total scattering coefficient has been measured in many regions throughout the world. Roughly, this parameter is the ratio of S-coda energy to radiated S-energy. Some investigators used an empirical relationship between radiated energy and local earthquake magnitude to estimate W. Others used joint analysis of direct S-wave and S-coda envelopes which will be discussed in detail in Section 7.2. The

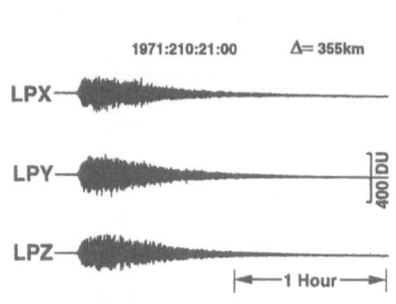

1971:210:21:00   Δ= 355km

LPX

LPY

LPZ

400 IDU

|—— 1 Hour ——|

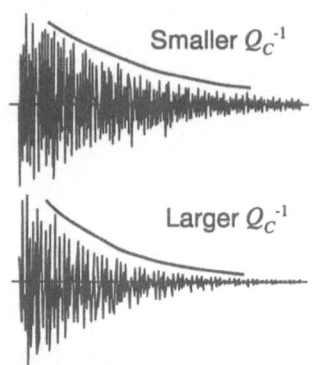

Smaller Q_C^{-1}

Larger Q_C^{-1}

FIGURE 3.11. Seismogram of a lunar-quake showing long coda duration. [From Nakamura, Fig. 1, 1977, copyright by Springer-Verlag GmbH & Co. KG, Germany.]

FIGURE 3.12. Seismograms showing large and small coda attenuation.

scatter is a factor of two for individual measurements of g_0. We compile reported values of g_0 in Figure 3.10, where we include reported backscattering coefficient g_π as total scattering coefficient g_0 by assuming isotropic scattering. We find that total scattering coefficient g_0 for S-to-S wave scattering is of the order of 10^{-2} km^{-1} for frequencies 1–30Hz; however, regional differences related to tectonic activity have not yet been quantitatively clarified.

Compared to seismograms recorded on the Earth, lunar seismograms, as shown in Figure 3.11, contain evidence for strong scattering and low intrinsic attenuation. As noted by Nakamura [1977] and Dainty and Toksöz [1981], lunar seismograms often have coda durations exceeding one hour. Applying the diffusion model of Section 3.2.1 for the explanation of the spindle-like envelopes, Dainty and Toksöz [1981] estimated that g_0 is as large as 0.05–0.5 km^{-1} at 0.45 Hz for long-range data (70–150 km).

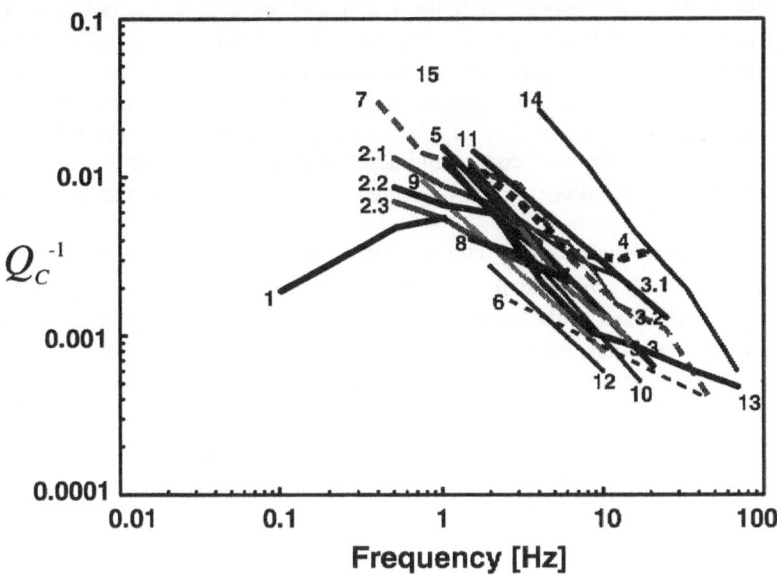

FIGURE 3.13. Coda attenuation Q_C^{-1} against frequency for various regions: 1, central and south central Alaska [Biswas and Aki, 1984]; 2.1, (Iceland), 2.2, (Galapagos), 2.3, (Guam), oceanic lithosphere [Jin et al., 1985]; 3.1, Central California; 3.2, Hawaii; 3.3, Long Valley in California [Mayeda et al. 1992]; 4, Campi Flegrei volcano, southern Italy [Del Pezzo et al., 1985]; 5, Dead sea [Eck, 1988]; 6, Garm, Central Asia [Rautian and Khalturin., 1978]; 7, Hindu-Kush [Roecker et al., 1982]; 8, Kanto-Tokai, Japan [Fehler et al., 1992]; 9, New England, U. S. A. [Pulli, 1984]; 10, southern Norway [Kvamme and Havskov, 1989]; 11, Petatlan, Gurrero, Mexico [Rodriguez et al., 1983]; 12, South Carolina, USA [Reha, 1984]; 13, western Nagano, Japan [Kosuga, 1992]; 14, shallow crust at Ashio, Kanto, Japan [Baskoutas and Sato, 1989]; 15, shallow crust at Western Nagano, Japan [Kosuga, 1992].

3.3.2 Coda-Attenuation Measurements

As reported by Rautian and Khalturin [1978], for a given region, the S coda has a common amplitude decay curve for lapse time greater than the twice the S-wave travel-time. The shape of this decay curve is quantified using a parameter known as coda attenuation. For practical analysis of data from a single station, we use (3.10), (3.25), or (3.30) in (3.35) to write the MS velocity amplitude of coda in a frequency band with central frequency f vs. time as the product of a power of lapse-time and an exponential decay factor as

$$\left\langle \left| \dot{u}^{S\,Coda}(t;f) \right|^2 \right\rangle_{T} \propto \frac{1}{t^n} \exp\left[-Q_C^{-1}(f)2\pi f t\right] \tag{3.36}$$

where power n is 1–2 depending on the dominance of surface, diffusive, or body waves. Recently, most investigators fix the power at two for the geometrical decay in the single scattering model. The exponential decay term, characterized by coda attenuation Q_C^{-1}, is independent of the source and station location but depends on the frequency band. It is possible to measure Q_c^{-1} from analysis of records obtained at a single station, which allows measurements to be made even in regions of sparse station coverage. Coda attenuation Q_c^{-1} characterizes the seismogram coda amplitude decay with lapse time as schematically illustrated in Figure 3.12. Larger Q_c^{-1} means rapid decay of coda amplitude. While (3.36) is valid for a single frequency, coda attenuation is often measured from octave-width, bandpass-filtered seismograms [Tsujiura, 1978]. If we make the band-width too small, the filtered coda envelope changes rapidly and a stable estimation of Q_c^{-1} becomes difficult. Estimations of Q_c^{-1} are typically made from plots of the logarithm of the product of the lapse time raised to the correct power and the MS coda amplitude measured over a few cycles against lapse time. We can then estimate Q_c^{-1} directly from the decay gradient against lapse time using the least square method. Takahara and Yo-

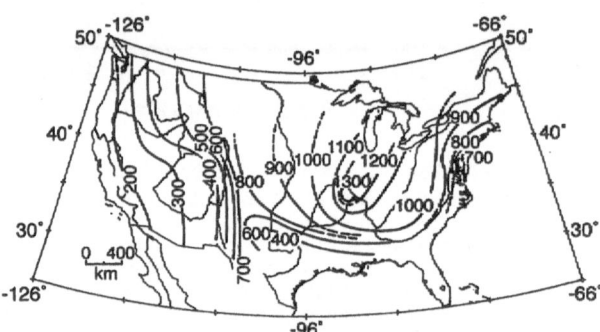

FIGURE 3.14. Contour curves of coda attenuation Q_c at 1 Hz in United States. [From Singh and Herrmann, 1983, copyright by the American Geophysical Union.]

mogida [1992] proposed a way to estimate Q_c^{-1} based on the maximum likelihood method. In some studies, the estimate is stabilized by using data from many seismograms and simultaneously finding the decay gradient that best fits all the data [Fehler et al., 1988].

Regional measurements of Q_c^{-1} made throughout the world have been compared with seismotectonic activity and review papers on S-coda measurements have been published [Herraiz and Espinosa, 1987; Matsumoto, 1995]. Figure 3.13 is a compilation of reported Q_c^{-1}. The variation from region to region is more than a factor of 10. In general, Q_c^{-1} is about 10^{-2} at 1 Hz and decreases to about 10^{-3} at 20 Hz. Q_c^{-1} is generally larger in volcanic regions and in the shallow crust.

The variation of Q_c^{-1} with frequency also has some relationship to tectonic activity. The frequency dependence can be written in the form of a power of frequency f as $Q_c^{-1} \propto f^{-n}$ for $f > 1$ Hz. The power n ranges between 0.5 and 1. Figure 3.14 shows contour curves of Q_c at 1 Hz developed by Singh and Herrmann [1983] from analysis of short period WWSSN seismograms of local earthquakes in the U. S. A. Note that these measurements are Q_c and not Q_c^{-1}. Q_c is highest in the central U. S. A. where exposed rocks are oldest. To the west, it decreases to 400–500 in Utah, and to 200–300 in Nevada. Q_c decreases to 800–700 in the Appalachian mountains and eastern U. S. A. This figure clearly shows that Q_c is higher (Q_c^{-1} is smaller) in tectonically stable areas and lower in active areas. Jin and Aki [1988] measured Q_c in China. Figure 3.15 shows their contours of Q_c at 1 Hz along with the locations of large historical earthquakes. They found that large historical earthquakes took place in low Q_c regions. Note that Q_c is as low as 100 in Tibet where active continental collision has caused rapid and large uplift.

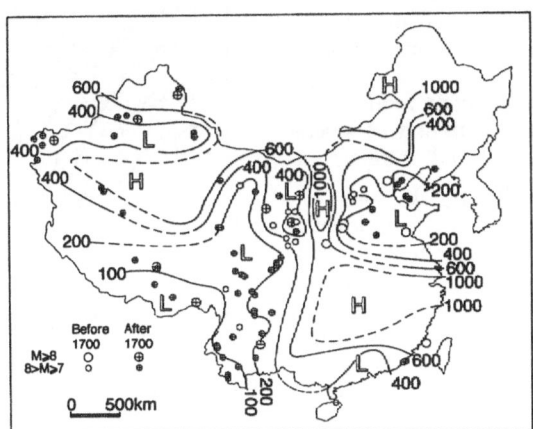

FIGURE 3.15. Contour curves of coda attenuation Q_c at 1 Hz in China. Circles indicate locations of historical earthquakes of magnitude greater than 7. [From Jin and Aki, 1988, copyright by the Seismological Society of America.]

Matsumoto and Hasegawa [1989] reported large Q_C^{-1} near active volcanoes and the coast of the Japan sea where the S-wave velocity is relatively low.

Coda attenuation Q_C^{-1} has been widely measured in the world, mostly for a frequency range between 1 to 30 Hz, as a parameter for characterizing the heterogeneity of the lithosphere. There have been some attempts to reveal shorter-wavelength heterogeneity: seismogram envelopes containing frequencies of more than several hundred Hz have been analyzed as a part of surveys for fracture detection and characterization in gold mines [Cichowicz and Green, 1989; Gibowicz and Kijko, 1994].

Lapse-Time Dependence of Coda Attenuation

Examination of data over a wide range of lapse times led Rautian and Khalturin [1978] to conclude that coda envelopes cannot be described by a single Q_C^{-1} value. Figure 3.16 is a log-log plot of bandpass-filtered coda envelopes of seismograms from regional earthquakes in Central Asia vs. lapse time. The coda envelopes show a systematic change in decay rate with lapse time. For example, for the 1.25 Hz band, we find three segments in the S-coda envelope decay curve, which correspond to three different values of Q_C^{-1}. Decreasing Q_C^{-1} with increasing lapse time was observed in the Hindu–Kush [Roecker et al., 1982], in France [Gagnepain–Beyneix, 1987], in southern Norway [Kvamme and Havskov, 1989], in Antarctica [Akamatsu, 1991], and in the shallow crust in central Japan [Kosuga, 1992].

Single scattering coda models that are based on the assumption of spatial homogeneity of g_0 and intrinsic attenuation predict that Q_C^{-1} is independent of lapse time. Most of the investigators who found a lapse-time dependence of the coda decay rate suggested that the later portion of the coda is dominated by energy that has propagated in zones with attenuation lower than energy in the early coda. Gusev [1995a] investigated the single isotropic scattering model for the case where g_0 decreases according to an inverse power of depth and found that Q_C^{-1} de-

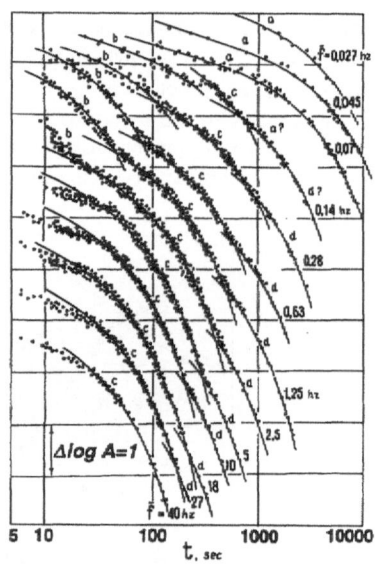

FIGURE 3.16. Bandpass-filtered RMS coda amplitude vs. lapse time for a local earthquake in Central Asia, where the numeral is the center frequency in Hz of the passband. [From Rautian and Khalturin, 1978, copyright by the Seismological Society of America.]

creases with increasing lapse time and that such a model is consistent with many observations of lapse-time dependence of coda attenuation. Peng [1989] plotted the spatial autocorrelation function of Q_C^{-1} measured in southern California in different frequency bands and different time windows. He found that the autocorrelations are similar among different frequencies, but depend on the time window selected to measure Q_C^{-1}: longer and later time windows give slower decays in the autocorrelation with increasing lag distance. Lapse-time dependence of coda decay is still an unresolved issue and will be briefly discussed again in Section 7.1.

Duration Magnitude

The most widely used coda measurement is the determination of earthquake magnitude from the S-coda duration. As shown in Figure 2.25, there is a good relationship between magnitudes determined from direct-arrival-wave-packet amplitude and the coda length of seismograms. This relationship has lead to the widespread use of coda duration to estimate magnitude since coda duration can be more reliably measured from poorly calibrated instruments than amplitude. For the single scattering model, the end of the coda wave t_F is the time when the energy density of the scattered waves given by (3.25) is equal to that of noise E_{Noise}:

$$\log W + \log g_0 = 2\log t_F + \log 2\pi V_0^2 + \left(2\pi f Q_C^{-1} \log e\right) t_F + \log E_{Noise} \qquad (3.37)$$

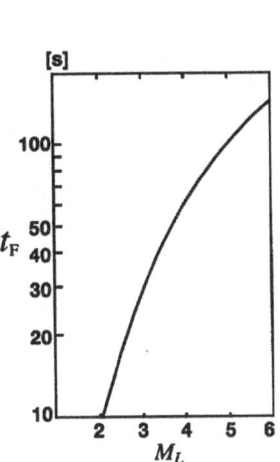

FIGURE 3.17. Plot of coda duration against local magnitude given by (3.39).

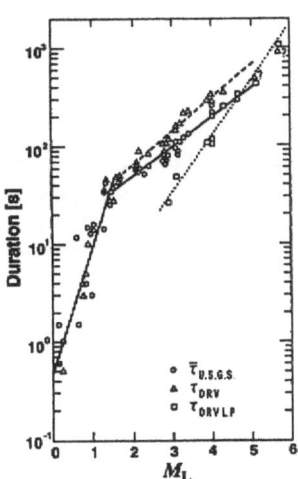

FIGURE 3.18. Plot of coda duration against local magnitude in California. [From Bakun and Lindh, 1977, copyright by the Seismological Society of America.]

We may estimate source energy W [erg] by using the Gutenberg–Richter formula [Gutenberg, 1956]:

$$\log W = 1.5M_L + 11.8 \qquad (3.38)$$

Putting the observed relationship $fQ_c^{-1} = 10^{-2}\,\text{s}^{-1}$, as shown in Figure 3.13, into (3.37),

$$M_L = 1.33\log t_F + 0.018\,t_F + \text{Const.}(E_{Noise}, g_0) \qquad (3.39)$$

The second term makes the relation between magnitude and coda duration convex as illustrated in Figure 3.17, where the constant is chosen so that $M_L=5$ at $t_F=100$ s. The constant usually depends on the noise level and the site amplification factor at the observation site. We may replace t_F with the lapse time measured from the P-wave onset t_{F-P} for local earthquakes located near the receiver. If we approximate (3.39) by

$$M_L = C_0 + C_1 \log t_{F-P} \qquad (3.40)$$

for lapse time between 10 s and 200 s, coefficient C_1 is 2.5 for 10–60 s, and increases to 4 for 10–200 s. These values agree well with the observed values for small local earthquakes: 2.85 in Wakayama, Japan as shown in Figure 2.25 [Tsumura, 1967a] and 2.92–3.32 for $t_{F-P} > 40$ s in California as shown in Figure 3.18 [Bakun and Lindh, 1977].

3.3.3 Temporal Change in Coda Characteristics

The generally observed correlation of Q_c^{-1} with the level of tectonic activity or the amount of lithospheric fracturing that was described in Section 3.3.2 suggests that seismic coda monitoring could provide information about the temporal change in fractures and attenuation caused by changes in tectonic stress during the earthquake cycle. Since coda waves sample a volume of rock within a region, measurements of coda characteristics may be more sensitive to small temporal changes than measurements of velocity or attenuation using direct waves which sample a 1-D ray path between the source and the receiver [Aki, 1985].

Chouet [1979] was the first to report a temporal change in Q_c^{-1} from observations made in Stone Canyon, California. He found a significant increase in Q_c^{-1} in the 1.5 to 24 Hz frequency range during an observational period of about one year; however, the temporal change could not be correlated with any seismic activity. Jin and Aki [1986] reported a temporal change in Q_c^{-1} associated with the occurrence of the Tangshan earthquake ($M_S=7.8$, July 27, 1976) in China. They measured the average shape of the coda envelope vs. lapse time using data from many earthquakes by first correcting for source size and then plotting a parameter they called reduced $B(t)$ from coda-wave amplitude $u^{\text{S Coda}}$ as a function of lapse time based on the single isotropic scattering model:

$$B(t) \equiv \frac{\left|u^{\text{S Direct}}(t_S)\right|^2}{\left\langle\left|u^{\text{S Coda}}(t)\right|^2\right\rangle_{\text{T}}} K\left(\frac{t}{t_S}\right) \propto e^{2\pi f Q_C^{-1}(t-t_S)} \qquad (3.41)$$

where $u^{\text{S Direct}}(t_S)$ is the maximum amplitude of the direct S-wave, t_S is the S-wave travel time, and K is given by (3.19a). Figure 3.19 shows that there was a change in Q_C^{-1} in the frequency range 1.6 to 2.9 Hz measured at a station about 120 km from the mainshock epicenter during three time periods. The trend of $\ln B(t)$ vs. $t - t_S$ was a straight line for 1969–1972. However, $\ln B(t)$ vs. $t - t_S$ has a bend during 1973–1976, where early coda shows anomalously strong attenuation that was about 3 times stronger than before. $\ln B(t)$ vs. $t - t_S$ became linear after the mainshock that occurred in July, 1976; however, Q_C^{-1} was higher than during the first time period. Jin and Aki [1986] also reported a difference in Q_C^{-1} between the time periods before and after the Haicheng, China earthquake (M_S=7.3, Feb. 4, 1975).

Gusev and Lemzikov [1985] reported a precursor-like decrease in the coda decay parameter, corresponding to an increase in Q_C^{-1}, before the Ust–Kamchatsk earthquake (M_S=7.8, 1971). The anomaly started at the end of 1970 and took the minimum value a half year before the mainshock. They also observed an increase in the scatter of measurements for one year before the main shock (see Figure 3.20). Gusev [1995b] summarized coda observations made during the 24-year period from 1967 to 1992 in Kamchatka by plotting coda magnitude residuals ΔK_C at each station as illustrated in Figure 3.21a. The coda magnitude residual ΔK_C is equal to the log of the ratio of the coda amplitude at 100 s lapse time at the station to the network average of the coda amplitude at 100 s lapse time. Coda amplitudes were read from photographic recordings of the ground motion measured by 1.2 s period displacement seismometers filtered between 1 and 10Hz. The plots in Figure 3.21b show moderate but statistically significant oscillations around a constant level. He reported two prominent anomalies. One anomaly lasted for three years at station KBG and was followed by two

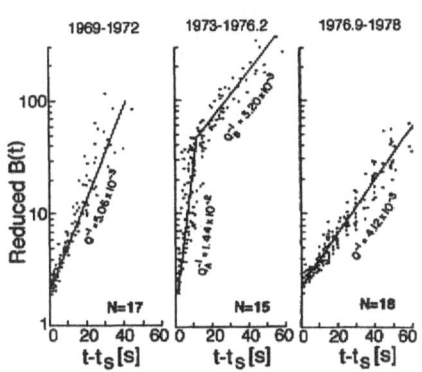

FIGURE 3.19. Coda decay measurements in the 1.6 to 2.9 Hz band made before and after the Tangshan earthquake (July 27, 1976) in China. N is the number of earthquakes used. [From Jin and Aki, 1986, copyright by the American Geophysical Union.]

M 8 earthquakes (E69, E71) within 100km distance. Another anomaly that lasted for 1.5 year at station APH preceded a major (volume of 2.5 km³) fissure volcanic eruption (V75) within 70 km of the station. These changes can be interpreted as an increase of 30% in S-wave attenuation in the lithosphere near the station.

An increase in scatter among individual measurements of Q_C^{-1} was reported before an earthquake with M_L=5.2 (Feb. 26, 1983) in Central Asia [Sato et al., 1988]. A change in the relationship between coda duration and local magnitude determined from the network average of maximum amplitudes was reported by Sato [1987] for the western Nagano earthquake (M_s=6.8, September 14, 1984) in Japan. Data from February 1982 to December 1984 were analyzed. As shown in Figure 3.22, during the period May 1983 – September 1984 mostly preceding the

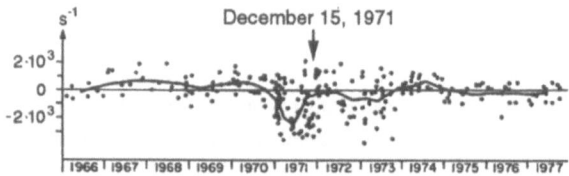

FIGURE 3.20. Measure of coda decay parameter corresponding to Q_c vs. time before and after the Ust–Kamchatsk earthquake of Ms=7.8. The solid curve is a running mean over eight data points. [From Gusev and Lemzikov, 1985, with permission from Elsevier Science - NL, Sara Burgerhartsraat 25, 1055 KV, Amsterdam, The Netherlands.]

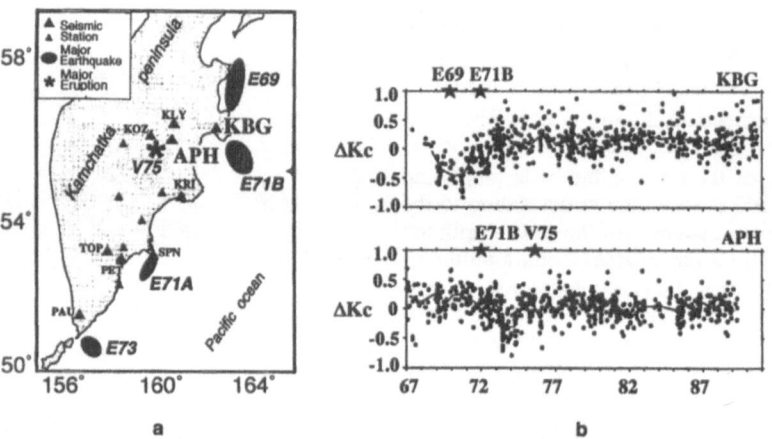

a   b

FIGURE 3.21. (a) Locations of Kamchatka seismic stations, major earthquakes (E69, E71A, E71B, and E73), and a volcanic eruption (V75). (b) Coda magnitude residual ΔK_c vs. time at two stations, KBG and APH. [From Gusev, 1995b, copyright by the American Geophysical Union.]

main shock, coda duration (open circles) was longer for a given local magnitude than the average for the whole period (solid line). The lengthening of coda duration can be interpreted as an increase of scattering strength during the earthquake preparation stage.

Malamud [1974] monitored the temporal change in the ratio of coda duration on a horizontal component of motion to that on the vertical component. He found the ratio took lower values for periods lasting up to a few months preceding moderate earthquakes in Central Asia. Yan and Mo [1984] reported a decrease of the ratio of coda duration on the horizontal component to that on the vertical component during the few days before the Jianchuan, China earthquake

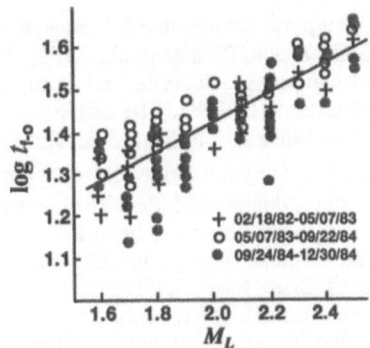

FIGURE 3.22. Logarithm of coda duration [s] vs. local magnitude before and after the western Nagano earthquake, Japan of $Ms=6.8$, where the solid line is the regression line for all the data. [From Sato, 1987, copyright by the American Geophysical Union.]

($M_L=5.3$, July 3, 1982). These reports suggest the possibility of changes in scattering and attenuation due to aligned cracks.

Including the reports described here, Jin and Aki [1991] cited 12 cases where precursor-like temporal changes in Q_c^{-1} had been reported to be associated with moderate to large earthquakes [e.g. Novelo–Casanova et al., 1985; Tsukuda, 1988; Sato, 1986].

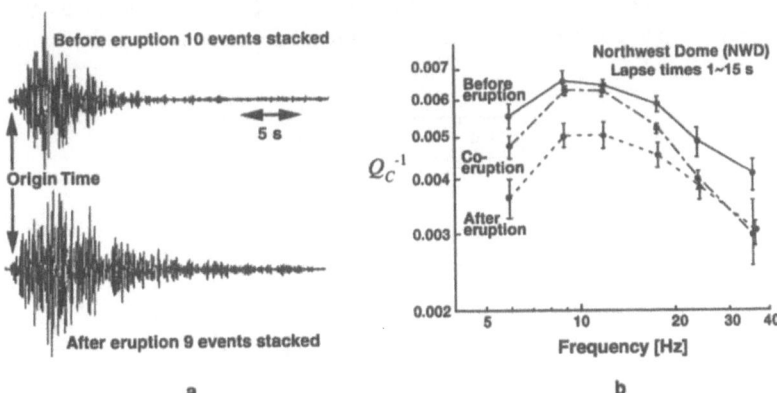

a

b

FIGURE 3.23. (a) Stacked seismograms before and after an eruption of Mt. St. Helens showing the different coda characteristics for the two periods. (b) Coda attenuation before, during, and after the eruption. [From Fehler et al., 1988, copyright by the American Geophysical Union.]

Temporal change in Q_c^{-1} was reported before and after an eruption of Mt. St. Helens volcano [Fehler et al., 1988]. Figure 3.23a shows stacked seismograms of local earthquakes recorded close to the summit before and after the eruption on September 2–6, 1981. The difference between their envelopes is characterized by Q_c^{-1} as illustrated in Figure 3.23b: Q_c^{-1} was 20–30% larger before the eruption than after. They interpreted the changes in Q_c^{-1} as having been caused by volcanic inflation-induced crack density changes.

Although there have been many studies that indicate a positive correlation between temporal change in coda characteristics and the occurrence of large earthquakes, there were significant criticisms of the studies in the mid-1980s [Sato, 1988c; Frankel, 1991; Ellsworth, 1991]. Among the most serious criticisms are the possible influences of using different lapse times, different earthquake focal regions, and earthquakes having differing focal mechanisms to establish the temporal change in coda characteristics. The variation of Q_c^{-1} with lapse time was discussed in Section 3.3.2. When studying the temporal variation of Q_c^{-1} as a possible precursor to large earthquakes, changes in the lapse times of data used could lead to an erroneous conclusion about the temporal variation of Q_c^{-1}. Since many of the investigators who have studied temporal variation have not reported the lapse times used in their studies, there is concern that their results may be influenced by variations in the lapse times of the data analyzed. In addition, if data from the early coda are used, the results can be significantly influenced by the focal mechanism of the earthquakes studied and the angle between the normal to the focal plane and observation station, which will be discussed more in Chapter 6. Thus, studies that used earthquakes from a wide range of locations and with a variety of focal mechanisms to establish temporal variations in Q_c^{-1} may be suspect. The signs of Q_c^{-1} changes are not systematic among the studies reported; in some cases increases were reported and in other cases decreases before large earthquakes. As a result of these criticisms, recent analysis of the temporal change in Q_c^{-1} has been conducted with more attention to the lapse-time window and focal depths of earthquakes studied.

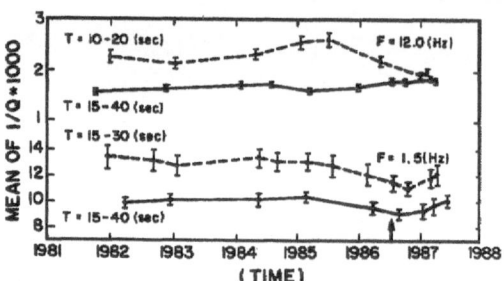

FIGURE 3.24. Temporal variation in coda attenuation Q_c^{-1} for two lapse-time windows for each of two frequency bands measured before and after the 1986 North Palm Springs earthquake, California (indicated by an arrow). [From Su and Aki, 1990, with permission from Birkhäuser Verlag AG, Switzerland.]

Su and Aki [1990] studied data collected from 1981 to 1987 in the epicentral area of the 1986 North Palm Springs earthquake. They measured Q_c^{-1} for various lapse times corresponding to the sizes of regions the waves had sampled. For lapse times less than 20 s, they argued that waves sampled primarily the epicentral region whereas lapse times close to 40 s sampled a region much larger than the epicentral area. They found that, beginning in 1983, there was a gradual increase in Q_c^{-1} measured for 10–20 s lapse times that was followed by a rapid drop during the one year period before the main shock in the 12 Hz band (see Figure 3.24). Q_c^{-1} of the surrounding area, corresponding to lapse times up to 40 s, did not change during the same period.

Analyzing short-period seismograms recorded at Riverside, California for the 55-year period between 1933 to 1987, Jin and Aki [1989] found a temporal variability in Q_c^{-1} at frequency of about 1.6 Hz in southern California as shown in Figure 3.25. This is the longest baseline study published. They found a positive correlation between Q_c^{-1} and the seismic b-value calculated for $M_L > 3$ earthquakes within a 180 km radius. Seismic b-value is a measure of the ratio between the number of small to large earthquakes; smaller b-values mean there are relatively fewer small earthquakes compared to the number of larger ones [Lee and Stewart, 1981]. The positive correlation between Q_c^{-1} and the b-value implies that Q_c^{-1} increased when the number of small earthquakes increased relative to the number of large earthquakes. They found that the cross-correlation of Q_c^{-1} and the b-value is highest at 0.79 for zero time lag and decays with increasing time shift. They also observed a small negative correlation between Q_c^{-1} and the b-value with variation of the b-value preceding that of coda attenuation by 5–12 years. These two quantities were determined by independent observation. Therefore, the temporal changes seem to be reliable. A similar correlation between two values was found in central

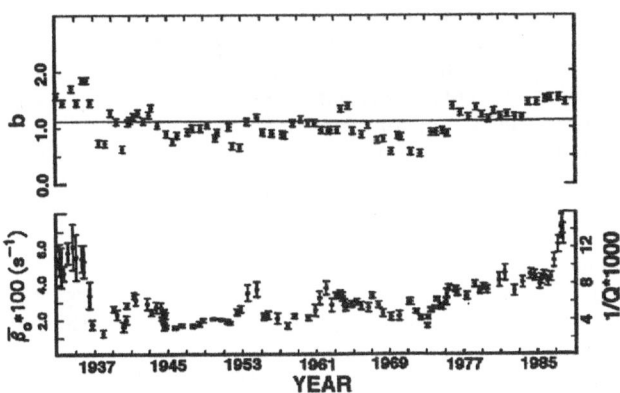

FIGURE 3.25. Measurements of coda attenuation Q_c^{-1} (lower) and b-value (upper) in southern California as a function of time. [From Jin and Aki, 1989, copyright by the American Geophysical Union.]

California [Jin and Aki, 1993; Aki, 1995]. However, other studies report an anti-correlation between Q_C^{-1} and the b-value [e.g., Jin and Aki, 1986, for the Tangshan earthquake in China; Novelo–Casanova et al., 1985, for the 1979 Petatlan earthquake, Mexico]. Hiramatsu et al. [1992] examined the temporal change in Q_C^{-1} of crustal earthquakes in the Hida region, central Japan for the period from 1985 to 1991. They found a small increase in Q_C^{-1} in the volcanic area, especially for frequencies of 12–24 Hz, that was related to earthquake swarms that occurred in 1990.

Roughly speaking, coda shape is stable and independent of focal mechanism and source location in a stochastic sense. However, when we are looking for very small changes, source characteristics may be important and should not be neglected. The strong criticism about the effects of differing focal mechanisms and earthquake locations on the conventional Q_C^{-1} measurements was investigated by Got et al. [1990]. They showed that small changes in rupture kinematics can create variations in coda excitation that mimic significant changes in Q_C^{-1}. They proposed using doublets, two earthquakes having the same location and focal-plane solution but occurring at different times. The temporal change in the ratio of squared spectral amplitude of coda particle velocity $\ddot{u}_{a,b}^{S\,Coda}$ at frequency f and lapse time t for earthquakes a and b is written as

$$\ln \frac{\left\langle \left| \ddot{u}_a^{S\,Coda}(t;f) \right|^2 \right\rangle_T}{\left\langle \left| \ddot{u}_b^{S\,Coda}(t;f) \right|^2 \right\rangle_T} = -2\pi f \left(Q_{Ca}^{-1} - Q_{Cb}^{-1} \right) t + \text{Const.} \left(W_a^S g_{0a} / W_a^S g_{0b} \right) \quad (3.42)$$

from (3.35) and (3.36). Assuming no change in total scattering coefficient with respect to the difference in origin times, they proposed to measure the difference in coda attenuation $\Delta Q_C^{-1} \equiv Q_{Ca}^{-1} - Q_{Cb}^{-1}$ from the plot of the spectral ratio of the left hand of (3.42) against lapse time for each frequency. Analyzing doublets in the vicinity of and close in time to the Coyote Lake earthquake (August 1979, M=5.9), they reported no major change in the coda attenuation in the crust preceding this shock. Taking the same procedure for 21 doublets recorded between 1978 and 1991 in three frequency bands from 2 to 15 Hz, Beroza et al. [1995] examined the temporal change in Q_C^{-1} for the Loma Prieta earthquake sequence in California whose main shock had magnitude $M_w=6.9$ and occurred on October 1989. They found that the upper bound of the change in Q_C^{-1} was about 5%. There might have been a precursory decrease in Q_C^{-1} during the two years before the main shock, as shown in Figure

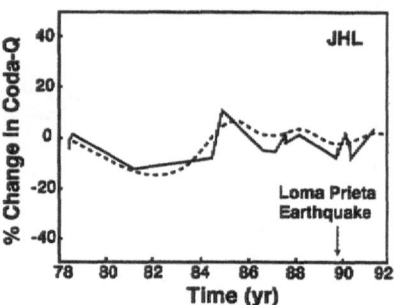

FIGURE 3.26. Temporal change in coda attenuation (5–10 Hz) for the Loma Prieta earthquake sequence in California. [From Beroza et al., 1995, copyright by the American Geophysical Union.]

3.26. Aster et al. [1996] and Antolik et al. [1996] have used a method similar to the one introduced by Got et al. [1990] to investigate variations in Q_C^{-1} in two regions, the Anza seismic gap and Parkfield, California. By using a large number of doublets that all have similar waveforms and whose locations are estimated to be within 20 m, Antolik et al. [1996] estimate that Q_C^{-1} in the Parkfield region was constant to within 5% over a frequency range 3–30 Hz during the study period from 1987–1994. No major earthquakes occurred at Parkfield during the study period.

3.4 CODA-NORMALIZATION METHOD

The coda-normalization method provides a reliable way to estimate the frequency dependence of important parameters quantifying the seismic source radiation and receiver site amplification, both of which are used in seismic risk assessment. It also allows the investigation of propagation effects. Most of seismology is focused on characterizing one of these three influences, source radiation, propagation, and site amplification, on seismograms. Usually, this is accomplished by eliminating the influence of two of them so that the one of interest can be isolated and studied in more detail. Source radiation quantifies the frequency dependence of the wavefield leaving the vicinity of the seismic source. The propagation effect combines all the influences on the seismic wavefield between the seismic source and the recording site. This includes the effect of deterministic structure and other influences along the source–receiver path. The site amplification effect includes influences of near-surface geology that modifies the character of the recorded waveform only near the recording site. Near-surface geology may cause reverberations, local amplification of signal, or introduction of additional complexity in the waveform that cannot be modeled deterministically from available information.

Estimates of the source radiation are most important for quantifying the size of earthquakes and explosions. As a first-order description of the size of a seismic source, seismologists wish to characterize the radiation as a function of frequency. The site effect is important in seismic hazards analysis. Obtaining reliable estimates of the relative ground motion as a function of spatial location in seismically active zones is essential for establishing building codes and in estimating which areas will be most prone to seismic hazards. Such estimates are most useful if they can be given as a function of frequency since the response of buildings to ground motion varies with frequency.

Conventional methods for estimating source and site amplification factors at regional distances, less than about 100 km, rely on use of the direct P or S-arrival on the seismic trace. Since regional networks are usually designed to reliably record the time of the first arrival, the first arrival waveform is often clipped. The later portions of the waveform are usually not clipped. This means that coda methods may be the only means by which site and source information can be obtained from high-gain regional seismic networks.

The coda-normalization method is based on the idea that at some lapse time, the seismic energy is uniformly distributed in some volume surrounding the source. The limits of this assumption have recently been investigated theoretically in the framework of the multiple scattering process based on the radiative transfer theory as discussed in Chapter 7, but the reliability of the results obtained by the coda-normalization method helps confirm the validity of the assumption. The assumption is consistent with that used to develop the energy-flux model in Section 3.2.2. The idea of the coda normalization method grew out of the empirical observation discussed in Sections 2.4.1 and 3.3.2 that the length of a seismogram recorded by a regional seismic network is proportional to the magnitude of the event. Another key observation in support of the coda-normalization method is that, for local earthquakes recorded at times greater than roughly twice the travel time of an S-wave from a source to a receiver, the envelope of a bandpass-filtered seismogram has a common shape that is independent of the source-receiver distance as shown in Figure 2.23a. The amplitude of the envelope varies with source size and recording site amplification. Figure 2.23b shows the similarity in the shape of the coda for earthquakes of different sizes recorded by a single station. It is important to note that, contrary to a popular misconception, the coda-normalization method is not founded on any theoretical model of wave propagation in the earth. In particular, it does not rely on the validity of the single scattering model. The original foundation of the approach is empirical.

Interpreting S-coda as an incoherent superposition of scattered S-waves, we may explicitly write the S-coda power as a convolution of the source, propagation, and site effects using (3.36) with either (3.25), (3.26), or (3.30) as

$$\left\langle \left| \dot{u}_{ij}^{S\,\text{Coda}}(t;f) \right|^2 \right\rangle_T \propto W_i^S(f) \left| N_j^S(f) \right|^2 \frac{e^{-Q_c^{-1} 2\pi f t}}{t^n} \tag{3.43}$$

where $\dot{u}_{ij}^{S\,\text{Coda}}(t;f)$ is the S-coda velocity wavefield at receiver j filtered in a frequency band having center frequency f, $W_i^S(f)$ is the energy radiation from source i in the same frequency band, $N_j^S(f)$ is the S-wave site amplification factor for site j, and power n is 1–2 depending on the dominance of surface, diffusive, or body waves. Here in Section 3.4, we suppose that the proportionality factor in (3.43), which characterizes the coda-wave excitation as a function of total scattering coefficient of S-waves $g_0(f)$, is constant irrespective of source and site locations.

3.4.1 Site Amplification Measurements

At lapse times t_c large enough that energy is uniformly distributed in some volume surrounding the seismic source that contains two recording sites, the relative amplitude of the seismograms recorded at the two sites should be the same except for the influence of the near-recording site amplification. The relative amplitude of the two recording sites can thus be obtained from (3.43) by dividing the amplitude of the seismogram at one site by the amplitude at another site taken at the same absolute time for the same source i:

$$\frac{N_j^S(f)}{N_k^S(f)} = \sqrt{\frac{\left\langle \left| \ddot{u}_{ij}^{S\,Coda}(t_c;f) \right|^2 \right\rangle_T}{\left\langle \left| \ddot{u}_{ik}^{S\,Coda}(t_c;f) \right|^2 \right\rangle_T}} \tag{3.44}$$

Here, station k is chosen as the reference site. To obtain a more robust measure of the relative amplitude, we usually take the amplitudes of the seismogram envelopes by using the Hilbert transform or the RMS of the bandpass-filtered trace. The envelope can be calculated for selected frequency bands so that the frequency dependence of the site amplification can be measured. The estimate of the relative amplitude can be stabilized by taking the average of the ratio determined for many time windows. The relative amplification factors for an array of recording sites can be determined by computing the ratios relative to one reference site. The reference site is usually chosen as one on solid rock in simple terrain or in a deep borehole where one can consider that the near-recording site amplification is minimal.

Tsujiura [1978] first demonstrated the reliability of the coda-normalization method for finding relative site amplifications using the coda approach and compared with measurements on direct waves. He analyzed many bandpass-filtered seismograms of local earthquakes recorded at four sites at Mt. Dodaira, Japan having differing lithology. He plotted the spectral ratios of S-coda waves measured on horizontal component seismograms recorded at pairs of sites taking hard rock site H5 as the reference (see Figure 3.27). He also took the spectral ratio of direct S-

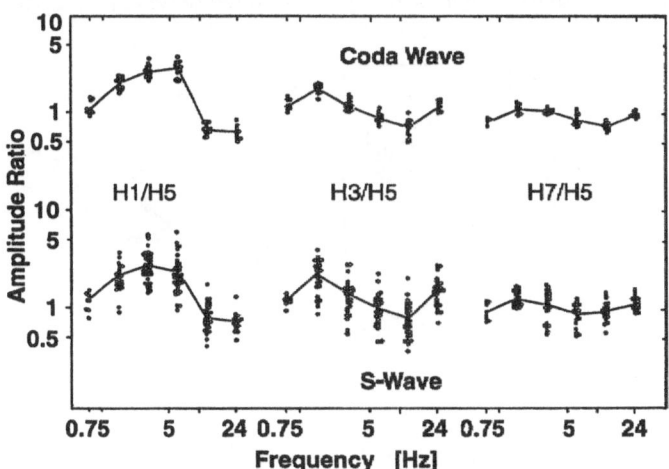

FIGURE 3.27. Amplification factors of direct S-wave and S-coda waves at seismic stations at Mt. Dodaira, Japan having differing lithology relative to site H5; H1 and H3 are on low-grade metamorphic rocks like slate and chert, and H5 and H7 are on crystalline rocks. The aperture of the Dodaira array is about 2 km. [From Tsujiura, 1978, copyright by Earthquake Research Institute, University of Tokyo.]

waves. The resultant spectral ratio of S-coda is almost the same as that of direct S-waves for frequencies from 0.75 to 24 Hz. The S-coda thus has the same site amplification factors as that of the S-wave. The similarity of site amplifications measured using S-coda waves and direct S-waves provides direct observational support of the basic hypothesis that S-coda is composed mostly of scattered S-waves. Tsujiura also reported that the S-coda wave amplitude ratios are different from the amplitude ratios obtained using direct P-waves. These results established the viability of the coda-normalization method for determining S-wave site amplification factors. As seen in Figure 3.27, S-coda measurements have less scatter than direct S-wave measurements. Thus, when few data are present, the coda method is likely to give more reliable results than the spectral ratio method applied to direct waves.

In a study similar to that of Tsujiura [1978], Tucker and King [1984] compared the site amplification factors obtained by the coda-normalization method with those obtained using measurements of direct S-arrivals. They reported good agreement between the measurements on direct and coda waves but also concluded that the results obtained using the coda method had less scatter than the results obtained using direct arrivals.

Phillips and Aki [1986] presented a method for inverting relative amplitudes determined at a suite of sites using a series of time windows to determine a site amplification factor for each site relative to the array of sites. They assume that the shape of the coda decay curves is the same at all sites for all sources and that only the amplitude of the curves differs depending on the site and source factors. By assuming that the source factors for all seismograms from a given earthquake are the same and that the site amplification is the same for all events recorded at a given site, they arrive at an expression relating the source factors, site amplification factors, and the shape of the common decay curve to the observed data. They applied

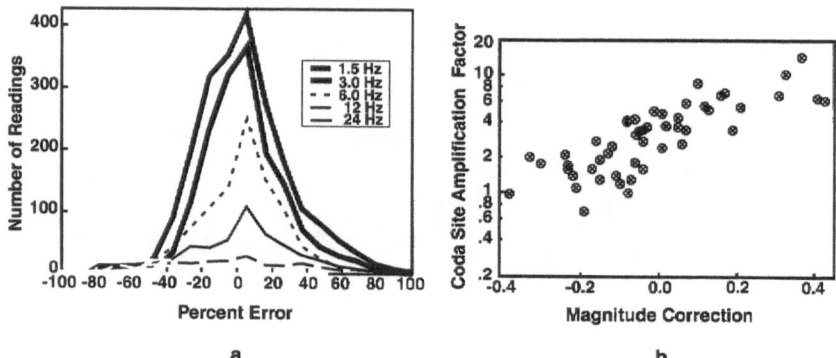

Figure 3.28. (a) Distribution of site amplification measurements using many 10 s windows of coda from many events in the Kanto–Tokai region, Japan. Percent differences between individual measurements using one time window for a station-event pair and mean value for all data at the station are shown. (b) Site amplification factor for the 6 Hz band estimated from the coda-normalization method vs. local magnitude residual. [From Fehler et al., 1992, with permission from Blackwell Science, United Kingdom.]

their method to determine site amplifications for 150 stations in northern California. They found that the site amplifications determined were independent of the earthquakes or lapse times used in their analysis for most stations. For a few stations, the results were lapse-time dependent. They attributed this lapse-time dependence to the existence of energy trapped near the recording sites caused by local geological structures. This trapped energy violates the basic assumptions of the coda-normalization method since the trapped energy dissipates more slowly than the energy in the nearby earth medium.

Mayeda et al. [1991] measured site amplification factors in and around the Long Valley Caldera area of California. They found that relative site amplifications computed for octave frequency bands centered at 1.5 and 3.0 Hz agreed well with magnitude residuals for each site found using both amplitude and coda-duration measures of magnitude. The conventional amplitude and coda-duration measures of magnitude are made on raw, unfiltered seismograms. At higher frequencies, the agreement between coda results and the other methods was not so good. They concluded that the magnitude residuals measured using the conventional amplitude and coda duration measures were dominated by the 1.5-3 Hz frequency range in the Long Valley region.

Fehler et al. [1992] developed a map of site amplification factors for the Kanto–Tokai region of Japan using coda-wave data. They analyzed the spread of the measurements from individual time windows from many events and showed that the spread was small for frequencies below about 8 Hz where signal-to-noise was high. Figure 3.28a shows the spread of their site amplification measurements obtained using many earthquakes and many lapse times. This figure shows that the coda-normalization method can provide consistent measures of site amplification independent of the earthquakes or lapse times used. Figure 3.28b shows that the site amplification factors obtained from the coda-normalization method for the octave frequency band centered at 6 Hz agreed well with local magnitude residuals, which were obtained by Noguchi [1990] from maximum amplitude measurements for the same stations.

Kato et al. [1995] compared amplification factors obtained using direct S-waves and S-coda waves for three stations in southern California. They used high-quality digital recordings of aftershocks of the 1992 Landers earthquake. They found that the site amplifications obtained using the two methods over the six frequency bands studied agreed to within a factor of 1.5 as shown in Figure 3.29.

Air-gun sources are considered to generate P-waves efficiently. Using array measurements of air-gun generated

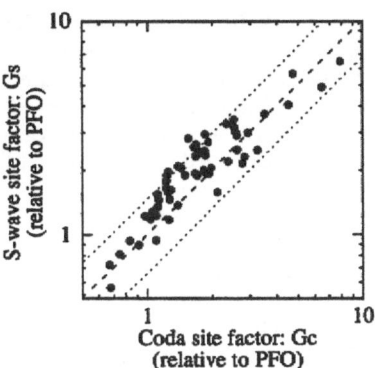

FIGURE 3.29. Comparison of direct S-wave site factor with S-coda site factor in southern California for 0.75–6 Hz. [From Kato et al., 1995, copyright by the Seismological Society of America.]

signals in the Bandai volcanic area, Japan, Matsumoto and Hasegawa [1991] reported that there is a strong correlation between S-coda amplitude on horizontal components at a fixed lapse time and direct P-wave site amplification factor measured on the vertical component in the 10 Hz band. This result suggests that there is a large contribution of conversion scattering from P to S for the excitation of S-coda in this volcanic area even when there is no source radiation of S-waves.

The coda-normalization method has often been applied to data recorded on rock sites. There are reports of nonlinear amplitude dependence of site amplification factors for soil sites. This nonlinear effect means that it may not be possible to estimate shaking during strong earthquakes by using simple scaling of amplitudes measured for small earthquakes. From the analysis of strong motion records of the Loma Prieta earthquake in California, Chin and Aki [1991] reported that peak accelerations predicted from site factors, which were determined using the coda-normalization method and small amplitude seismograms of local microearthquakes, overestimated the accelerations observed for sediment sites and underestimated it for the Franciscan formation. The systematic disagreement occurs for stations within 50 km from the mainshock epicenter, where the acceleration level was above 0.1 to 0.3 g.

3.4.2 Source Radiation Measurements

Aki [1967] and Brune [1970] first showed how to quantify the size of an earthquake as a function of frequency. They argued that the source-radiation displacement spectrum of earthquakes is flat at low frequencies. Above a corner frequency, the spectrum rolls off like a negative power of frequency. The corner frequency is related to a physical dimension of the fault. The amplitude of the low-frequency portion of the displacement spectrum is related to a parameter called the seismic moment, which is a function of the amount of slip along the fault [Aki and Richards, 1980]. Seismic moment is known to be a better measure of earthquake size than magnitude [Kanamori, 1977]. Spectral analysis of direct waves can be used to determine the spectral shapes of source radiation and the relative source radiation vs. frequency for different earthquakes [Tucker and Brune, 1977]. To determine relative or absolute sizes of earthquakes, spectral measurements on direct-arriving phases need to be corrected for the nonisotropic radiation from the source and propagation effects. These corrections are difficult to make. Often, there are too few stations to allow an estimation of the directional radiation pattern for regional earthquakes. The coda-normalization method provides an easy method for characterizing the spectral differences in source radiation among nearby seismic sources without requiring knowledge of source radiation pattern or propagation effects. Thus, the relative seismic moment, or magnitude, can be reliably determined using the coda-normalization method.

From (3.43), we can find relative source radiation as a function of frequency by dividing the amplitude of the seismogram recorded at one site for a given earthquake by the amplitude at the same site for a different earthquake taken at the same absolute lapse time t_c at site j:

$$\frac{W_i^S(f)}{W_k^S(f)} = \frac{\left\langle \left| \dot{u}_{ij}^{S\,coda}(t_c; f) \right|^2 \right\rangle_T}{\left\langle \left| \dot{u}_{kj}^{S\,coda}(t_c; f) \right|^2 \right\rangle_T} \tag{3.45}$$

where event k has been chosen as the reference event. Since coda waves are scattered, waves leaving the source region in all directions arrive at a given receiver. Thus, to first order, coda waves average over the angular radiation from the source and no radiation-pattern correction is necessary for estimating source excitation when using the coda-normalization method.

Reliable relative source radiation measurements can be made using data from only a single station. Figure 3.30 shows Lg-wave amplitude measures at two stations in Nevada, U. S. A. made on (a) coda waves and (b) direct waves by Mayeda and Walter [1996]. Lg-waves are common in continental regions and are composed of multiply reflected S-waves in the crust. Although considerable effort has been made to correct the direct-wave measurements for propagation and site amplification effects, the figure shows clearly that there is a more consistent relationship between the measurements made on the coda waves than on the direct arrivals even though there is a large station separation of 500km.

Biswas and Aki [1984] used seismic moments from two well-analyzed earthquakes in Alaska to develop a coda-amplitude vs. seismic-moment scale for earthquakes in Alaska. They found a relationship between coda amplitude measured at

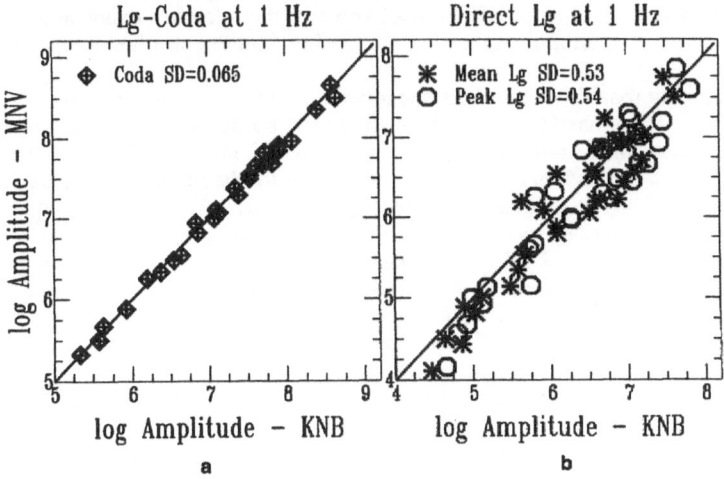

FIGURE 3.30. (a) Amplitude measurements made using Lg-coda waves at two stations, KNB and MNV in Nevada, U. S. A. for regional earthquakes, where the station separation is about 500 km. (b) Measurements made from direct Lg-waves with path corrections at the same two stations for the same set of earthquakes. [From Mayeda and Walter, 1996, copyright by the American Geophysical Union.]

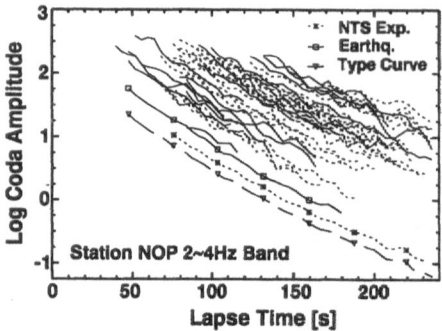

FIGURE 3.31. Coda decay curves for explosions at Nevada Test Site and earthquakes in the Southern Great Basin, U. S. A. [From Hartse et al., 1995, copyright by the Seismological Society of America.]

some lapse time after the earthquake origin-time and the seismic moment for the earthquake. This relationship can be used to make a fast and reliable determination of the size of an earthquake using data from only one station. The advantage of the coda amplitude to moment relationship developed for Alaska is that it uses coda data occurring well after the arrival time of the direct waves. Since direct waves are often clipped, conventional measures of seismic moment cannot be applied. Mayeda [1991] used spectral ratios of coda waves from earthquakes in California to examine the departure of source spectra from the well-known ω^2 model. Dewberry and Crosson [1995] performed a detailed analysis of the seismic source spectrum of earthquakes in the U. S. Pacific Northwest using data from coda waves. They employed a formal inversion technique and data from many stations to find the differences in source radiation among events as a function of frequency for 78 events. They scaled their results to an absolute scale by choosing one of the events as a reference event and estimated the seismic moment and corner frequency for that event using other information. Then the differences between the reference event and the other events determined from the inversion are used to find the absolute spectra for each event.

Obtaining reliable relative source radiation estimates for many events as a function of frequency has been a method employed by seismologists to discriminate natural from man-made seismic events. Su et al. [1991] used the coda-normalization method as one part of a technique to distinguish quarry blasts from earthquakes. They found that quarry blasts excite relatively less high-frequency energy than earthquakes having the same low-frequency excitation. An observed rapid temporal decay of 1.5–3 Hz coda for quarry blasts for lapse times less than 30 s is attributed to the larger portion of surface waves in the early coda of seismograms from quarry blasts. The difference disappears as lapse time increases beyond 30s due to the contribution of scattering from deeper regions. Hartse et al. [1995] used the coda-normalization method to determine source radiation as a function of frequency for earthquakes and nuclear explosions in the Basin and Range province of the western U. S. A. Instead of using the spectral ratio of a single time window, they developed type curves for the shape of the coda envelope vs. lapse time. This allowed them to effectively use data from a large portion of the seismic coda to obtain a more robust measure of relative source size using data from only a single station. Figure 3.31 shows examples of the type curves developed for a single station located in Nevada, U. S. A. Measures of the envelope shape vs. lapse time are plotted for both earthquakes and explosions. Also shown are the average shapes for earthquakes, explosions, and earthquakes and explosions combined. These average-shape curves are determined by inversion using the method of Phillips and Aki [1986]. The curves show a remarkable similarity in shape that is independent of source size and source–receiver distance. Hartse et al. [1995] found that the frequency dependence of the source radiation obtained at a single station provided a reliable means of distinguishing between the explosions and the earthquakes. Their method used frequency characteristics that had previously been developed using measurements on direct arrivals. The coda method provided reliable results using data from only a single station; the direct-arrival method required averaging over many stations to average the angular dependence of the source radiation.

3.4.3 Attenuation Measurements

Aki [1980a] proposed a coda-normalization approach for measuring the amplitude attenuation of direct S-waves with travel distance in the lithosphere. The square of the direct S-wave particle velocity amplitude at station j in a frequency band having central frequency f for local earthquake source i is written as

$$\left|\dot{u}_{ij}^{\,S\,\text{Direct}}(f)\right|^2 \propto \frac{W_i^S(f)}{r_{ij}^{\,2}}\left|N_j^S(f)\right|^2 e^{-Q_S^{-1}\,2\pi f\,r_{ij}/\beta_0} \tag{3.46}$$

where Q_S^{-1} is the S-wave attenuation. From (3.43), the time average squared S-coda amplitude around a fixed lapse time t_c at station j is

$$\left\langle \left|\dot{u}_{ij}^{\,S\,\text{Coda}}(t_c;f)\right|^2 \right\rangle_{\text{T}} \propto \frac{W_i^S(f)}{t_c^{\,n}}\left|N_j^S(f)\right|^2 e^{-Q_C^{-1}\,2\pi f t_c} \tag{3.47}$$

where the site amplification of S-coda waves is the same as that for direct S-waves as discussed in Section 3.4.1. Taking the natural logarithm of the ratio of the product of hypocentral distance and the direct S-wave amplitude to the averaged coda amplitude, the site amplification and source terms cancel, and we get

$$\ln \frac{r_{ij}\left|\dot{u}_{ij}^{\,S\,\text{Direct}}(f)\right|}{\sqrt{\left\langle \left|\dot{u}_{ij}^{\,S\,\text{Coda}}(t_c;f)\right|^2 \right\rangle_{\text{T}}}} = -\left(Q_S^{-1}(f)\pi f/\beta_0\right)r_{ij} + \text{Const.} \tag{3.48}$$

where we suppose that focal mechanisms are random. We may smooth out the radiation pattern differences when the measurements are made over a large enough number of earthquakes. Plotting the left-hand side against hypocentral distances for many earthquakes, the gradient gives the attenuation of direct S-waves per travel distance. Aki [1980a] applied this method to high-frequency seismograms of 900 earthquakes recorded in Kanto, Japan. Figure 3.32a is a plot of the left-hand side of (3.48) against travel distance for several frequency bands. Regression analysis gives an estimate of Q_S^{-1} for frequencies from 1.5 to 24 Hz, which are plotted in

Figure 3.32b. This plot shows that Q_S^{-1} decreases according to the negative power of frequency for $f > 1$ Hz. Aki's [1980a] method has become known as the single station method or the coda-normalization method. Aki's work stimulated measurements of attenuation in many areas of the world and the development of the theoretical study of scattering attenuation caused by heterogeneous structure. We will compile results from some of these studies in Chapter 5. Combining the fact that Q_S^{-1} for very low frequencies estimated from surface wave analysis is less than about 2×10^{-3}, Aki [1980a, b] first conjectured that Q_S^{-1} has a peak around 0.5 Hz as indicated in Figure 3.32b.

Yoshimoto et al. [1993] extended the conventional coda-normalization method to measure the attenuation of the direct P-wave with travel distance. They assumed that the ratio of P- to S-wave radiated energy $W_i^P(f)/W_i^S(f)$ is independent of magnitude for earthquakes within a small magnitude range. Similar to the S-wave case, the square of the direct P-wave amplitude at station j is written as

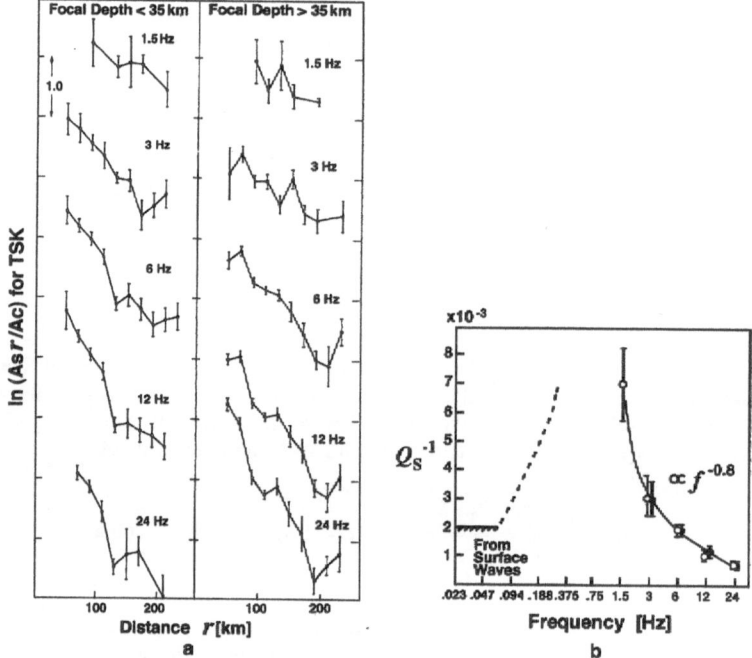

FIGURE 3.32. (a) The average of the natural logarithm of S- to coda-amplitude ratio multiplied by the source–receiver distance (left-hand side of (3.48)) plotted against the source-receiver distance for station TSK in Kanto, Japan. (b) Frequency dependence of Q_S^{-1} for earthquakes in the southeast part of Kanto, Japan. Open circles are for data from earthquakes having focal depth less than 35 km and solid circles for deeper than 35 km. [From Aki, 1980a, with permission from Elsevier Science - NL, Sara Burgerhartsraat 25, 1055 KV, Amsterdam, The Netherlands.]

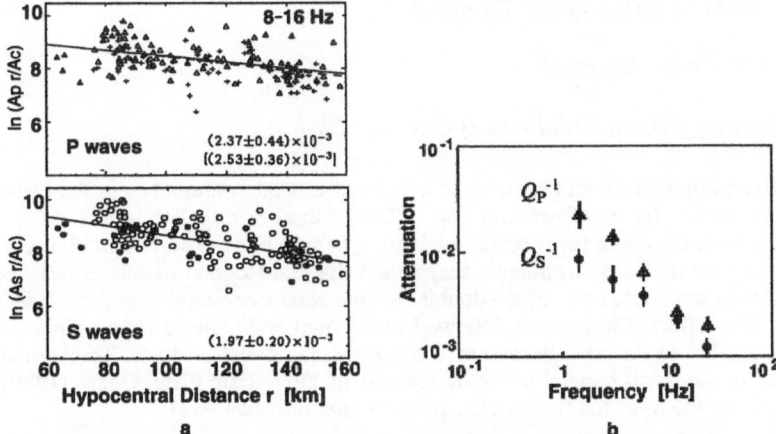

FIGURE 3.33. (a) Plot of the left-hand side of (3.50) (upper) and (3.48) (lower) against the source–receiver distance for data from Kanto, Japan, where solid lines are regression lines. (b) Frequency dependence of Q_P^{-1} (triangle) and Q_S^{-1} (circle) for Kanto, Japan, where the vertical bar denotes one standard deviation. [From Yoshimoto et al., 1993, with permission from Blackwell Science, United Kingdom.]

$$\left| \dot{u}_{ij}^{\,P\,\text{Direct}}(f) \right|^2 \propto \frac{W_i^P(f)}{r_{ij}^2} \left| N_j^P(f) \right|^2 e^{-Q_P^{-1} 2\pi f\, r_{ij}/\alpha_0} \tag{3.49}$$

Proceeding as above for S–waves, we get

$$\ln \frac{r_{ij} \left| \dot{u}_{ij}^{\,P\,\text{Direct}}(f) \right|}{\sqrt{\left\langle \left| \dot{u}_{ij}^{\,S\,\text{Coda}}(t_c, f) \right|^2 \right\rangle_{\text{T}}}} = -\left(Q_P^{-1}(f)\pi f / \alpha_0 \right) r_{ij} + \text{Const.} \tag{3.50}$$

since the ratio of P- to S-wave site amplification factors $\left| N_j^P \right| / \left| N_j^S \right|$ is constant at the jth station. From regression analysis, Q_P^{-1} can be evaluated in each frequency band. Yoshimoto et al. [1993] applied this method for measuring both Q_P^{-1} and Q_S^{-1} to seismograms from 174 small earthquakes that took place around Kanto, Japan. The top portion of Figure 3.33a shows a plot of the left-hand side of (3.50) for P-waves and the bottom part shows the left-hand side of (3.48) for S waves for data in the 8–16 Hz frequency band. The magnitude range of earthquakes used for the S-wave analysis was from 2 to 5.5. However, the P-wave analysis was restricted to earthquakes having magnitudes between 2.5 and 3.5 to satisfy the assumption about the constant ratio of P- to S-wave radiated energy. Figure 3.33b shows the resultant plots of attenuation against frequency for both P- and S-waves showing that Q_P^{-1} is higher than Q_S^{-1} for frequencies from 1 to 32 Hz.

3.5 RELATED CODA STUDIES

3.5.1 S-Coda Anomalies

Reflection from a Subducting Oceanic Slab

Examining the envelope decay of S-coda of crustal earthquakes in the southern Kanto district, Japan, Obara and Sato [1988] found a very late reflected S-phase arrival from the upper layer of the subducting Pacific plate as shown in Figure 3.34. By using an inversion technique, they found that the location of the reflector coincides with the upper layer of the double seismic zone associated with the subducting slab. The reflected S-phase is followed by its own coda whose decay curve is almost parallel to the coda decay curve following the direct S-phase. Tracing the reflector to the north parallel to the trench axis of the Pacific plate, Obara [1989] reported that the reflected S-phase disappears under the coda level.

Surface Waves in S-Coda

Phillips et al. [1993] observed large amplitude phases on horizontal components of motion in the S-coda of earthquakes recorded by the Fuchu array in the Kanto region, Japan. These phases were strongest in the 1 Hz band and were observed most strongly on recordings at the surface. They were much weaker on borehole recordings at depths as shallow as 150 m. Since the phases were recorded only on horizontal components of motion and were small or nonexistent in borehole recordings, they concluded that the phases were Love waves. Polarization and slant

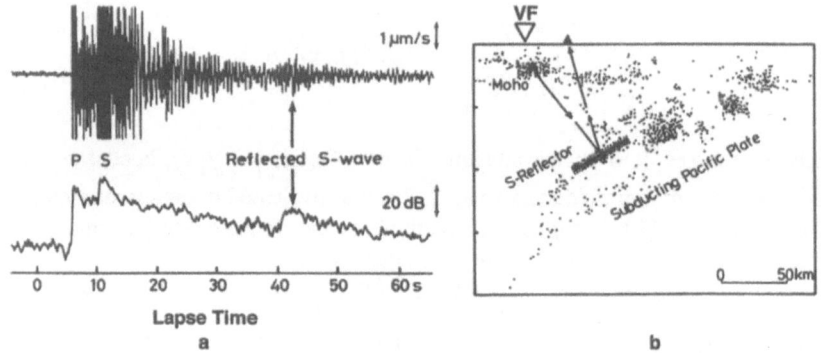

FIGURE 3.34. (a) Upper, horizontal component seismogram of a crustal earthquake recorded at a nearby station showing reflected S-wave from the subducting Pacific plate; lower, RMS-log trace of seismogram. (b) Vertical cross section showing geometry of source (solid circle), S reflector (bold line), and receiver (solid triangle) in Kanto, Japan, where small dots are microearthquakes and VF indicates the trace of the volcanic front. [From Obara and Sato, 1988, copyright by the American Geophysical Union.]

stack analysis led them to conclude that they were waves scattered from the portion
of the Kanto basin boundary located closest to the array. In S-coda recorded on soft
deposits, we may expect a considerable contribution of surface waves which are
converted at surface topographic variations and the irregular boundary between hard
rock and deposits.

Inversion for the Spatial Variation in the Total Scattering Coefficient

The models for S-coda envelope synthesis developed in Sections 3.1 and 3.2
are based on the assumption of a homogenous distribution of isotropic scatterers
and predict results consistent with the observed characteristics of coda as outlined at
the beginning of this chapter. However, detailed observations show that there may
be departures from the observed characteristics of S-coda waves. If the distribution
of scatterers is inhomogeneous, we may introduce a coordinate dependence of total
scattering coefficient $g_0(\mathbf{z})$ in (3.12). Then, we get the single scattering energy density at frequency f as

$$E^1_{\text{Inhomo.}}(\mathbf{x},t) = \frac{W}{(4\pi)^2} e^{-Q_c^{-1} 2\pi f t} \int\limits_{-\infty}^{\infty}\int\limits_{-\infty}^{\infty}\int\limits_{-\infty}^{\infty} g_0(\mathbf{z}) \frac{\delta(r_a + r_b - V_0 t)}{r_a^2 r_b^2} d\mathbf{z} \qquad (3.51)$$

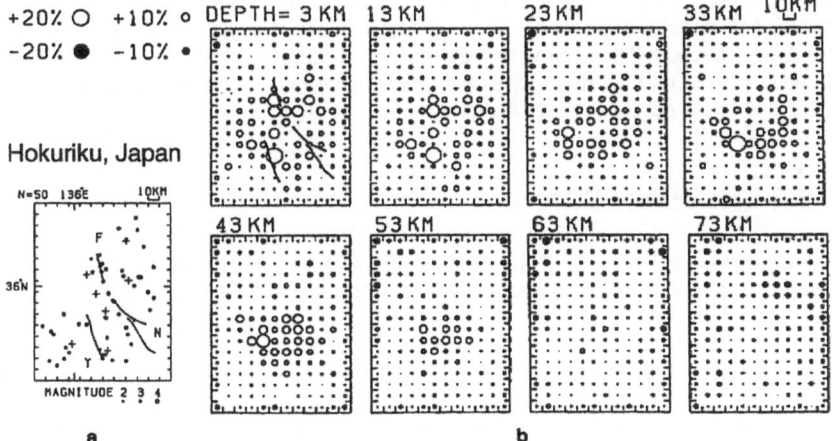

FIGURE 3.35. (a) Epicenters of earthquakes (circles) and stations (crosses) in the
Hokuriku district, Japan used in the inversion for the S-wave total scattering coefficient. Traces of major active faults are shown by solid lines. The region analyzed has
dimensions 120 km (EW) x160 km (NS) x75 km (depth). (b) Plan view cross sections
at various depths showing the spatial distribution of the fractional fluctuation of the
total scattering coefficient from S-coda analysis around 10 Hz. Symbol size represents the perturbation of the total scattering coefficient. [From Nishigami, 1991, copyright by the American Geophysical Union.]

where coda attenuation has been included. The difference between (3.12) and (3.51) is characterized by the spatial inhomogeneity in the total scattering coefficient, $g_0(\mathbf{z}) - g_0$. Then, small variations in the temporal dependence of observed coda envelopes at many stations may be used to estimate the spatial change in the total scattering coefficient [Kostrov, 1980]. Nishigami [1991] developed an inversion scheme to estimate the spatial inhomogeneity of the total scattering coefficient from readings at many stations of the temporal difference between the S-coda MS amplitude and the average long-term decay curve. Figure 3.35 shows the estimated spatial distribution of the fractional fluctuation of the total scattering coefficient $[g_0(\mathbf{z}) - g_0]/g_0$ in the Hokuriku district, Japan determined from analysis of records of 50 earthquakes at 7 stations for the 10 Hz band. Nishigami [1991] found that the locations of some of the strong scatterers found in the upper crust are close to the locations of major active faults and that the horizontal variation of total scattering coefficient is smaller in the uppermost mantle than in the crust.

Nishimura et al. [1997] used a single scattering approach to model three-component seismic data from an active experiment using explosion sources that was conducted across the Jemez volcanic field in New Mexico, U. S. A. (see Figures 2.9 and 2.11). They divided the subsurface into layers and considered that each layer has different total scattering coefficients, which are chosen to be those for isotropic scattering for P-to-P waves and P- to S-waves. They numerically calculated a Green function for scattering from each layer, where the geometrical spreading factor was calculated appropriately for the velocity structure assumed. Using the Green functions, they performed an inversion to find the ratio of total scattering coefficients for each layer relative to a reference layer that gave the best fit to the observed MS envelope shapes on all three components of motion from P-wave onset until early S-coda. They found that the midcrust under most of the region was fairly transparent but that the lower crust was heterogeneous. The strongest scattering occurs at shallow depths beneath the center of the caldera.

3.5.2 Teleseismic P-Coda

Cessaro and Butler [1987] examined the partition of teleseismic P-coda wave energy into radial and transverse components. 3-D heterogeneity is necessary to explain the excitation of P-waves on the transverse component. They speculated that the low level of transverse energy observed for many teleseisms may be caused by near-receiver scattering, but that higher levels observed at all frequencies for regional earthquakes may be attributed to the contribution of near-source scattering. Analyzing teleseismic P waves, Langston [1989] found high coda levels and slower decay rates in California compared to Pennsylvania where lower coda levels and more rapid decay were observed. He proposed an extension of the energy-flux model [Frankel and Wennerberg, 1987] for 1-D propagation in a medium consisting of an inhomogeneous layer over a homogeneous half-space. In addition to the uniform distribution of multiple scattered energy in the inhomogeneous layer, he considered the energy transferred into the homogenous lower space to arrive at a model for the teleseismic P-coda at frequency f:

$$\sqrt{\left\langle \left| \dot{u}^{P\,Coda}(t;f) \right|^2 \right\rangle_T} \propto \frac{1}{\sqrt{t}} \sqrt{1 - e^{-{}^{Sc}Q_P^{-1} 2\pi f t}}\, e^{-{}^I Q_P^{-1} \pi f t} \tag{3.52}$$

where the $\sqrt{t}$ dependence is due to the linearly increasing volume with time of the region of the layer over which energy is distributed. The last exponential term is for intrinsic attenuation and includes the radiation loss into the lower half-space. Figure 3.36 shows stacked P-wave envelopes of teleseisms recorded in California and Pennsylvania. The difference in the character of the codas indicates stronger inhomogeneities in California than in Pennsylvania. Considering the diffusion of energy from the surface scattering layer to the adjacent semi-infinite homogeneous medium, Korn [1990] proposed a similar model for the coda of vertically incident teleseismic P-waves. Applying the extended energy-flux model to explain scattering of high-frequency teleseismic P-wave data collected by stations of the Global Digital Seismic Network (GDSN) in the circum-Pacific area, Korn [1993] found strong scattering ($^{Sc}Q_P^{-1} = 0.005 - 0.01$) at island arcs and smaller scattering ($^{Sc}Q_P^{-1} < 0.002$) on stable continental areas like Australia.

Key [1967] investigated the contribution of topographic features to P-coda in the 1–2 Hz band using array data at Eskdalemuir, Scotland. He concluded that an observed elliptic particle motion of the P-coda was caused by Rayleigh waves that were generated from a deep river valley located 13 km from the array. Wagner and Langston [1992a] used a finite difference method to simulate scattering of vertically incident P-waves by an inhomogeneous layer having a nonisotropic ACF. Analyzing teleseismic P-wave records obtained at the NORESS array in Norway, they found a large-amplitude phase in the 0.5 Hz band that arrived 15 s after the initial P

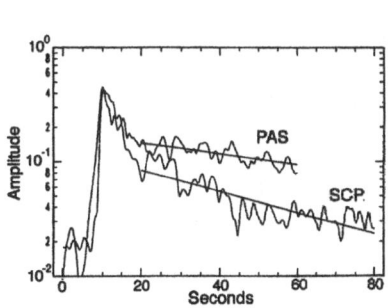

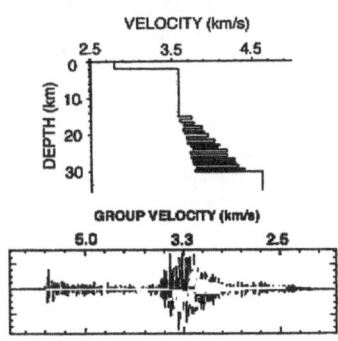

FIGURE 3.36. Stacked teleseismic P-coda envelopes obtained in California (PAS) and Pennsylvania (SCP). [From Langston, 1989, copyright by the American Geophysical Union.]

FIGURE 3.37. Synthesis of Lg-waves at a distance of 370 km, where the lower crust contains layers having random thicknesses between 100 m and 1000 m. [Reprinted from Campillo and Paul, 1992, copyright by the American Geophysical Union.]

phase. This late phase had a slow apparent velocity of 3 km/s [Wagner and Langston, 1992b]. By using *f-k*, particle motion analyses, and comparison with numerical simulations, they identified the Mjosa lake and nearby hills southwest of the array as the scattering point for P-to-Rayleigh scattering. They also suggested that the correlation distance is much longer in the horizontal than in the vertical direction. This result is consistent with other studies in this area [Gupta et al., 1990; Bannister et al., 1990]. Using ideas similar to those used to investigate near-receiver scattering, there was an attempt to resolve near-source scattering especially for underground nuclear explosions [Lay, 1987].

3.5.3 Lg and Lg-Coda

The regional phase Lg is a characteristic feature of high-frequency seismograms recorded in continental regions. It is observed at epicentral distances ranging from as close as 150 km up to several thousand kilometers. Campillo [1990] numerically simulated Lg-waves as multiply reflected waves within the crust. Assuming strong inhomogeneity of the lower crust, Campillo and Paul [1992] numerically simulated well-developed early Lg-coda as illustrated in Figure 3.37. Their model predicts that layering in the lower crust increases the duration of the Lg-phase on both the vertical and radial components and that the amplitude of the early Lg-coda depends on the distribution of layer thicknesses in the lower crust.

Toksöz et al. [1991] examined how Lg-arrivals lose coherence over propagation distances on the order of 10 km. Even for the direct Lg-window, Toksöz et al. [1991] reported that coherency declines with increasing separation and declines faster for higher frequencies as illustrated in Figure 3.38. Examining the coherence and the amplitude level of Lg-wave trains in different time windows, Der et al. [1984] and Dainty and Toksöz [1990] found that the early part consists of forward scattered waves, but the later portion consists of omnidirectionally scattered waves,

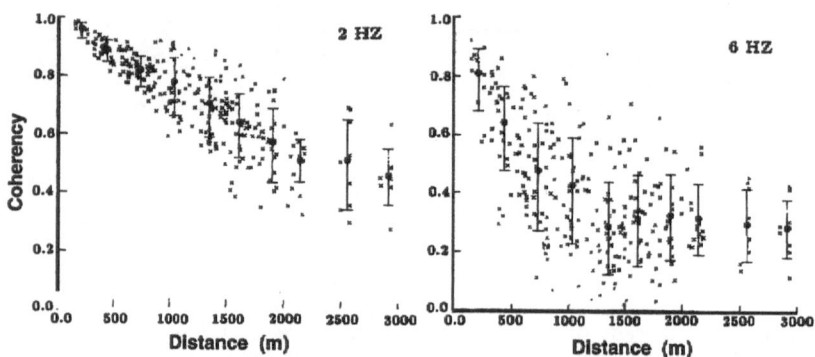

FIGURE 3.38. Coherency of Lg-waves in two frequency bands for a quarry blast. [From Toksöz et al., 1991, with permission from Elsevier Science - NL, Sara Burgerhartsraat 25, 1055 KV, Amsterdam, The Netherlands.]

as shown in Figure 2.26 for S-coda. Layered structure models cannot explain the omnidirectional propagation characteristics of late Lg-coda waves and 3-D heterogeneities are necessary.

CHAPTER 4

Born Approximation for Wave Scattering in Inhomogeneous Media

As shown by well-log data introduced in Chapter 2, the earth's lithosphere can be characterized as a randomly inhomogeneous elastic medium using an autocorrelation function (ACF). Characteristics of seismograms discussed in Chapter 2 support the view of treating wave propagation in the earth with a scattering approach. The phenomenological approaches discussed in Chapter 3 provide reasonable models of some features of the observed seismograms. We will now present the mathematical basis for the study of wave propagation and/or scattering in inhomogeneous media based on the first-order perturbation method known as the Born approximation. We will begin with the scalar wave equation and investigate the Born approximation for a plane wave incident on a localized inhomogeneity. Then we will present the concepts of an ensemble of random media and the stochastic average. Finally, we will study scattering of elastic vector waves in inhomogeneous elastic media.

4.1 SCALAR WAVES

4.1.1 Born Approximation for a Localized Velocity Inhomogeneity

We begin our study using the scalar wave equation in inhomogeneous media to simplify the development and allow us to gain better insight into the scattering characteristics. The wave propagation velocity is written as a sum of an average background velocity V_0 and a spatially varying fluctuation of coordinate $\mathbf{x}$:

$$V(\mathbf{x}) \equiv V_0 + \delta V(\mathbf{x}) = V_0 \left[1 + \xi(\mathbf{x}) \right] \tag{4.1}$$

where we assume that the fractional fluctuation is small, $|\xi| \ll 1$. Initially, we suppose a localized inhomogeneity around the origin having dimension L as illustrated in Figure 4.1:

$$\xi(\mathbf{x}) \neq 0 \quad \text{only for} \quad -L/2 < x_i < L/2 \quad (i = 1 \sim 3) \tag{4.2}$$

The scalar wave equation is

$$\left[\Delta - \frac{1}{V(\mathbf{x})^2} \partial_t^2 \right] u(\mathbf{x}, t) = 0 \tag{4.3}$$

which, for small fluctuation in velocity, may be written as

$$\left(\Delta - \frac{1}{V_0^2} \partial_t^2 \right) u + \frac{2}{V_0^2} \xi(\mathbf{x}) \partial_t^2 u = 0 \tag{4.4}$$

We solve wave equation (4.4) for the localized inhomogeneity given by (4.2) for a given frequency using the first-order perturbation method. The total wavefield is written as a sum of an incident plane wave u^0 and a scattered wave u^1:

$$u = u^0 + u^1 \tag{4.5}$$

where we assume $|u^1| << |u^0|$. The incident wave obeys the homogenous equation

$$\left(\Delta - \frac{1}{V_0^2} \partial_t^2 \right) u^0 = 0 \tag{4.6}$$

Substituting (4.5) in (4.4) and neglecting the cross term $\xi \partial_t^2 u^1$, we arrive at a wave equation for u^1:

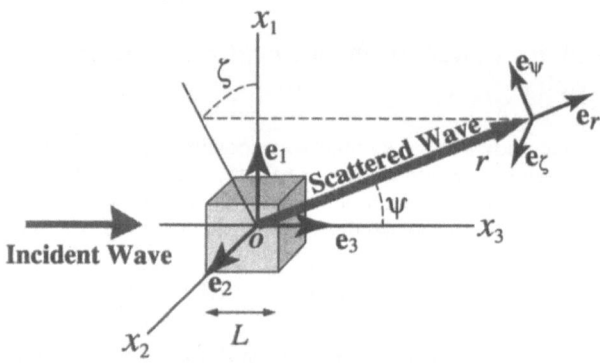

FIGURE 4.1. Geometry of scattering by a localized inhomogeneity of extent L, where (r, ψ, ζ) are spherical coordinates and (e_r, e_ψ, e_ζ) are unit base vectors.

$$\left(\Delta - \frac{1}{V_0^2}\partial_t^2\right)u^1 = -\frac{2}{V_0^2}\xi\partial_t^2 u^0 \tag{4.7}$$

where the incident wave appears on the right-hand side as a source term: the interaction between the incident wave and the inhomogeneity generates the scattered wave. Taking the unit base vectors in Cartesian coordinates as (e_1, e_2, e_3) and the incident wave of unit amplitude ($A_0 = 1$) at angular frequency ω propagating along the third direction as $u^0(\mathbf{x}, t) = A_0 e^{i(ke_3 \mathbf{x} - \omega t)}$,

$$\left(\Delta - \frac{1}{V_0^2}\partial_t^2\right)u^1 = 2k^2\xi A_0 e^{i(ke_3 \mathbf{x} - \omega t)} \tag{4.8}$$

where the wavenumber $k = \omega/V_0$. Figure 4.1 schematically illustrates the generation of scattered waves by the inhomogeneity localized around the origin, where (r, ψ, ζ) are spherical coordinates and (e_r, e_ψ, e_ζ) are unit base vectors.

To solve (4.8), we use the scalar Green function G in 3-D homogeneous space with background velocity V_0, which satisfies

$$\left(\Delta - \frac{1}{V_0^2}\partial_t^2\right)G(\mathbf{x}, t) = \delta(\mathbf{x})\delta(t) \tag{4.9}$$

where δ is a delta function. The causal retarded solution is given by [Morse and Feshbach, 1953, p. 838]

$$G(\mathbf{x}, t) = \frac{-1}{4\pi|\mathbf{x}|}\delta\left(t - \frac{|\mathbf{x}|}{V_0}\right)H(t) \tag{4.10}$$

By using the Green function, we may explicitly write the scattered wave as a convolution integral:

$$\begin{aligned}
u^1(\mathbf{x}, t) &= \int_{-\infty}^{\infty} dt' \int\!\!\int\!\!\int_{-\infty-\infty-\infty}^{\infty\,\infty\,\infty} d\mathbf{x}' \, G(\mathbf{x} - \mathbf{x}', t - t') \, 2k^2 \, \xi(\mathbf{x}') A_0 e^{i(ke_3\mathbf{x}' - \omega t')} \\
&= \frac{-k^2}{2\pi}A_0 e^{-i\omega t} \int\!\!\int\!\!\int_{-\infty-\infty-\infty}^{\infty\,\infty\,\infty} d\mathbf{x}' \, \frac{\xi(\mathbf{x}')}{|\mathbf{x} - \mathbf{x}'|} \, e^{ik(e_3\mathbf{x}' + |\mathbf{x} - \mathbf{x}'|)}
\end{aligned} \tag{4.11}$$

The 3-D integral is nonzero only over the volume of extent L^3. When the observation distance $r \equiv |\mathbf{x}|$ is in the far field $r \gg L$, we may approximate $|\mathbf{x} - \mathbf{x}'| \approx r$ in the denominator of (4.11). Furthermore, when

$$r >> \frac{1}{\pi} L^2 k \qquad (4.12)$$

we can approximate $|\mathbf{x} - \mathbf{x'}| \approx r - \mathbf{e}_r \mathbf{x'}$ in the exponent, where $\mathbf{e}_r = \mathbf{x}/r$. This is the condition for the Fraunhofer zone [Chernov, 1960, p. 45]. Then, (4.11) can be written as an outgoing spherical wave having wavenumber vector $k\mathbf{e}_r$:

$$u^1(\mathbf{x},t) = \frac{e^{i(kr-\omega t)}}{r} \left(\frac{-k^2}{2\pi} \right) \tilde{\xi}(k\mathbf{e}_r - k\mathbf{e}_3) A_0 = \frac{e^{i(kr-\omega t)}}{r} F A_0 \qquad (4.13)$$

where the tilde means the Fourier transform in 3-D space:

$$\tilde{\xi}(\mathbf{m}) \equiv \int_{-\infty}^{\infty} \int_{-\infty}^{\infty} \int_{-\infty}^{\infty} \xi(\mathbf{x}) \, e^{-i\mathbf{m}\mathbf{x}} d\mathbf{x} \qquad (4.14)$$

The argument of $\tilde{\xi}$ in (4.13) is the exchange wavenumber between the scattered wave and the incident plane wave. The scattering pattern is characterized by the factor

$$F = \frac{-k^2}{2\pi} \tilde{\xi}(k\mathbf{e}_r - k\mathbf{e}_3) \qquad (4.15)$$

which is called the scattering amplitude for a localized inhomogeneity of extent L^3. The scattering amplitude depends on frequency and is generally nonisotropic.

For incident plane waves of angular frequency ω, the energy-flux density in the third direction can be written as a product of wave velocity V_0 and energy density E^0 [see Howe, 1973]:

$$J^0 \equiv -\frac{\rho_0 V_0^2}{2} \left(\partial_t u^{0*} \nabla u^0 + \partial_t u^0 \nabla u^{0*} \right)\Big|_3 = V_0 \rho_0 \omega^2 |A_0|^2 = V_0 E^0 \qquad (4.16a)$$

where the asterisk means a complex conjugate and

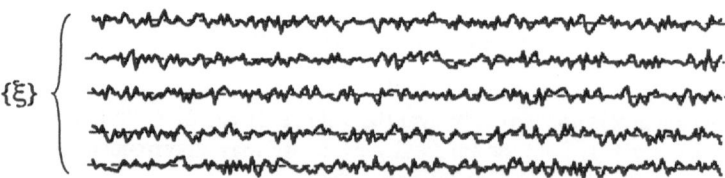

FIGURE 4.2. Concept of an ensemble of random fluctuations.

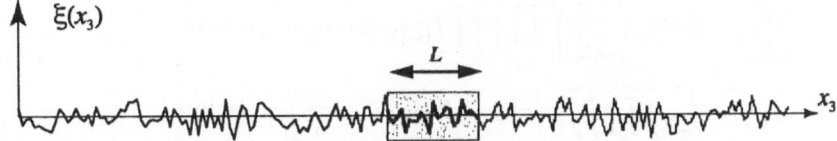

FIGURE 4.3. One-dimensional schematic illustration of velocity fractional fluctuation having continuous spatial extent. One block of extent L from the random medium is sampled.

$$E^0 \equiv \frac{\rho_0}{2}\left(\left|\partial_t u^0\right|^2 + V_0^2 \left|\nabla u^0\right|^2\right) = \rho_0 \omega^2 \left|A_0\right|^2 \qquad (4.16b)$$

We have introduced mass density ρ_0 of the background medium for comparison with the elastic vector theory in the following section. The energy-flux density of the outgoing spherically scattered waves in the far field is given by

$$J^1 = \rho_0 \omega^2 V_0 \left|u^1\right|^2 = \frac{\rho_0 \omega^2 V_0 \left|A_0\right|^2 \left|F\right|^2}{r^2} = \frac{\left|F\right|^2}{r^2} J^0 \qquad (4.17)$$

The amount of scattered energy per time within solid angle $d\Omega$ is $J^1 r^2 d\Omega$. As given by (3.1), the differential scattering cross section of this localized inhomogeneity is given by

$$\frac{d\sigma}{d\Omega} = \left|F\right|^2 \qquad (4.18)$$

This is just a square of the scattering amplitude.

4.1.2 Scattering by Distributed Velocity Inhomogeneities

Now we introduce the concept of an ensemble, and we make a statistical study of the scattering power per unit volume for media having distributed velocity inhomogeneities. We imagine an ensemble of random functions $\{\xi(\mathbf{x})\}$, where $\langle\xi\rangle = 0$ as illustrated in Figure 4.2. We suppose that the inhomogeneous media are homogeneous and isotropic random media having ACF and PSDF that are characterized by MS fractional fluctuation ε^2 and correlation distance a. As illustrated in Figure 4.3, we divide the inhomogeneous medium into blocks of dimension L with $L > a$. Replacing L with a in condition (4.12), we have the least restrictive condition for the mutual relationship between distance, wavenumber, and correlation distance as $r \gg a^2 k / \pi$, since the minimum dimensions of the scattering volume are of the order of a [see Chernov, 1960, p. 45]. Then, the ensemble average of the scattering cross section due to one block is given by

$$\left\langle \frac{d\sigma}{d\Omega} \right\rangle \equiv \left\langle |F|^2 \right\rangle = \left(\frac{-k^2}{2\pi} \right)^2 \int\int\int\int\int\int \left\langle \xi(\mathbf{x}')\xi(\mathbf{x}'') \right\rangle e^{-i(k\mathbf{e}_r - k\mathbf{e}_3)(\mathbf{x}' - \mathbf{x}'')} d\mathbf{x}' \, d\mathbf{x}''$$

$$= \frac{k^4}{4\pi^2} \int\int\int\int\int\int \left\langle \xi\left(\mathbf{x}_c + \frac{\mathbf{x}_d}{2}\right)\xi\left(\mathbf{x}_c - \frac{\mathbf{x}_d}{2}\right) \right\rangle e^{-i(k\mathbf{e}_r - k\mathbf{e}_3)\mathbf{x}_d} d\mathbf{x}_c \, d\mathbf{x}_d \quad (4.19)$$

$$= \frac{k^4}{4\pi^2} \int\int\int R(\mathbf{x}_d) e^{-i(k\mathbf{e}_r - k\mathbf{e}_3)\mathbf{x}_d} d\mathbf{x}_d \times L^3 = \frac{k^4}{4\pi^2} P(k\mathbf{e}_r - k\mathbf{e}_3) \times L^3$$

where $\mathbf{x}' = \mathbf{x}_c + \mathbf{x}_d/2$, and $\mathbf{x}'' = \mathbf{x}_c - \mathbf{x}_d/2$, and we have replaced the integral over the center-of-mass coordinate $\mathbf{x}_c$ with L^3. The argument of the PSDF in (4.19) is the exchange wavenumber vector, which is the difference between the wavenumbers of the incident plane wave and the outgoing scattered wave. The scattering coefficient defined by (3.2) as the scattering power per unit volume is given by

$$g(\psi, \zeta; \omega) \equiv 4\pi \frac{1}{L^3} \left\langle \frac{d\sigma}{d\Omega} \right\rangle = 4\pi \frac{1}{L^3} \left\langle |F|^2 \right\rangle$$

$$= \frac{k^4}{\pi} P(k\mathbf{e}_r - k\mathbf{e}_3) = \frac{k^4}{\pi} P\left(2k \sin\frac{\psi}{2}\right) \quad (4.20)$$

where the exchange wavenumber appearing as the argument of the PSDF is written using the scattering angle ψ. Thus, the scattering coefficient is directly related to the PSDF of the fractional fluctuation. This makes a bridge between the distribution of point-like scatterers and the random inhomogeneity that are schematically illustrated in Figure 3.1. When the random media are statistically homogeneous and isotropic, the scattering coefficient is axially symmetric with respect to the incident direction since the PSDF depends only on the absolute value of the exchange wavenumber. The scattering pattern is not necessarily isotropic even though the statistical characterization of the random media is isotropic. Combining (4.20) with (4.16), (4.17), and (4,18), we get the ratio of the square of the scattered wavefield from the volume L^3 to the square of the incident wavefield as

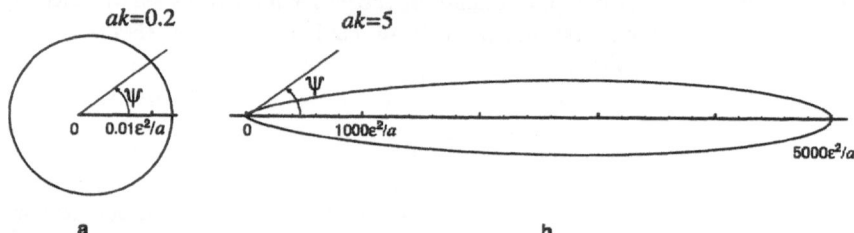

FIGURE 4.4. Angular dependence of scattering coefficient (4.22) for scalar waves in 3-D random media characterized by an exponential ACF: (a) low wavenumber $ak = 0.2$; (b) high wavenumber $ak = 5$.

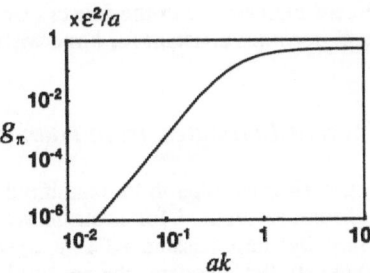

FIGURE 4.5. Wavenumber dependence of the backscattering coefficient for scalar waves in 3-D random media characterized by an exponential ACF, where k is the wavenumber.

$$\left\langle \left|u^1\right|^2\right\rangle \Big/ \left\langle \left|u^0\right|^2\right\rangle \approx g(\psi,\zeta;\omega)\frac{L^3}{4\pi r^2} \qquad (4.21)$$

Exponential ACF

When ACF is an exponential type, we can substitute the PSDF given by (2.10) in (4.20) to get

$$g(\psi,\zeta;\omega) = \frac{8\varepsilon^2 a^3 k^4}{\left(1 + 4a^2 k^2 \sin^2 \dfrac{\psi}{2}\right)^2} \qquad (4.22)$$

Scattering is isotropic for low frequencies $ak \ll 1$, as illustrated in Figure 4.4a. For high frequencies $ak \gg 1$, scattering is large in a narrow cone around the forward direction defined by $\psi < 1/ak$, as shown in Figure 4.4b. That is, scattering is strong in the forward direction for high frequencies. Even for the case of Gaussian ACF with correlation distance a, there is a large scattering within the cone $\psi < 1/ak$, as is the case for the exponential ACF. This means that the Born approximation is violated for large wavenumbers.

We illustrate the change in the scattering coefficient with a wavenumber by plotting the backscattering coefficient for the exponential ACF in Figure 4.5:

$$g_\pi(\omega) \equiv g(\psi = \pi, \zeta; \omega) = \frac{k^4}{\pi} P(2k) = \frac{8\varepsilon^2 a^3 k^4}{\left(1 + 4a^2 k^2\right)^2}$$

$$\approx \begin{cases} 8\varepsilon^2 a^3 k^4 & \text{for } ak \ll 1 \\ \varepsilon^2/2a & \text{for } ak \gg 1 \end{cases} \qquad (4.23)$$

The backscattering coefficient increases with the fourth power of the wavenumber for small wavenumbers and becomes constant for large wavenumbers.

Backscattering Coefficient Estimated from Numerical Simulation

It is possible to use numerical simulation to examine the validity of the Born approximation. A finite difference method for acoustic wave propagation in 2-D random media characterized by the Gaussian ACF was used by Jannaud et al. [1991a] to model coda waves. In the 2-D case, the ensemble average of the square of the scattered wavefield due to localized inhomogeneities of extent L for incident plane waves is written using the scattering coefficient as

$$\left\langle \left| u' \right|^2 \right\rangle \approx g(\psi; \omega) \frac{L^2}{2\pi r} \left\langle \left| u^0 \right|^2 \right\rangle \tag{4.24}$$

where the backward scattering coefficient for 2-D random media having Gaussian ACF is given by

$$g_\pi(\omega) \equiv g(\psi = \pi; \omega) = k^3 P(2k) = \pi \varepsilon^2 a^2 k^3 e^{-a^2 k^2} \tag{4.25}$$

The backscattering coefficient has a peak at $ak = \sqrt{3/2}$. Jannaud et al. [1991a] compared the ratio of the spectral of coda wave to that of the direct wave with the backscattering coefficient given by (4.25), as shown in Figure 4.6. The smooth curves, showing the theoretical backscattering coefficient from (4.25), decay with increasing frequency because the Gaussian ACF is poor in short-wavelength components. They found that the backscattering coefficients inverted from the coda power spectrum, shown by the irregular curves, are reasonably coincident with the theoretical curves for small velocity fluctuations $\varepsilon \leq 5\%$. However, the discrepancy between model and theory becomes clear for higher frequencies for $\varepsilon = 20\%$,

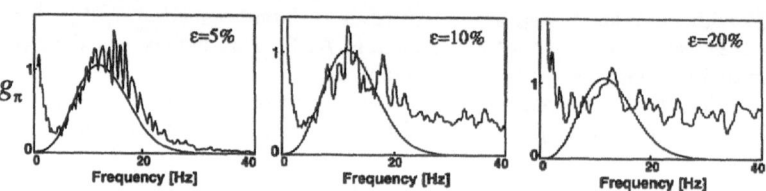

FIGURE 4.6. Backscattering coefficient estimated from the power spectral ratio of coda to that of the direct wave for 2-D random acoustic media characterized by a Gaussian ACF with $a=25$ m, $V_0=1.5$ km/s for different values of ε. Irregular curves show results for numerical experiments, and regular curves show theoretical prediction using (4.25). Plots are normalized by the peak value of the theoretical curve. [From Jannaud, et al., 1991a, copyright by the American Geophysical Union.]

where scattering is strong, because of the breakdown of the Born approximation. Coda excitation in 2-D nonisotropic random media was numerically studied by Ikelle et al. [1993] and Jannaud et al. [1992].

The Born approximation does not account for the feedback of scattered waves into incident waves. This means that total energy is not conserved when using the approximation. The basic assumption of the Born approximation is that scattered wave amplitude is small. A small velocity fluctuation for the random inhomogeneity is a necessary, but not a sufficient condition: scattering amplitude becomes large enough to violate the smallness condition near the forward direction for high frequencies even though the velocity fluctuation is small. In Chapter 5, we will further discuss the mechanism of large forward scattering for high frequencies and seek a way to use the Born approximation for high frequencies.

4.2 ELASTIC VECTOR WAVES

4.2.1 Born Approximation for a Localized Elastic Inhomogeneity

We apply the same procedure that was introduced for scalar waves in the preceding section to vector wave scattering by a localized inhomogeneity [Miles, 1960; Knopoff and Hudson, 1964; Sato, 1984a, b, 1990; Wu and Aki, 1985b; Wu, 1989]. There have been studies of wave scattering in anisotropic elastic media [Gibson and Ben-Menahem, 1991]; however, we will restrict ourselves to scattering in locally isotropic media. Following the procedure used for scalar waves, we write the spatial variation in Lamé coefficients and mass density as

$$\lambda(\mathbf{x}) = \lambda_0 + \delta\lambda(\mathbf{x}), \ \mu(\mathbf{x}) = \mu_0 + \delta\mu(\mathbf{x}), \ \text{and} \ \rho(\mathbf{x}) = \rho_0 + \delta\rho(\mathbf{x}) \quad (4.26)$$

Fractional fluctuations are assumed to be small: $|\delta\lambda/\lambda_0|, |\delta\mu/\mu_0|, |\delta\rho/\rho_0| \ll 1$. The wave equation for the displacement vector wavefield $\mathbf{u}(\mathbf{x}, t)$ is

$$\rho(\mathbf{x})\ddot{u}_i(\mathbf{x}, t) - \partial_j T_{ij}(\lambda, \mu; u_k) = 0 \quad (4.27)$$

where an overdot means time derivative. The stress tensor is given by

$$T_{ij}(\lambda, \mu; u_k) = \lambda(\mathbf{x})\delta_{ij}\partial_l u_l(\mathbf{x}, t) + \mu(\mathbf{x})\left[\partial_i u_j(\mathbf{x}, t) + \partial_j u_i(\mathbf{x}, t)\right] \quad (4.28)$$

Initially, we focus on scattering from a block of dimension L located around the origin:

$$\delta\lambda(\mathbf{x}), \ \delta\mu(\mathbf{x}), \ \delta\rho(\mathbf{x}) \neq 0 \quad \text{only for} \ -L/2 < x_i < L/2 \quad (i = 1 - 3) \quad (4.29)$$

As an extension of (2.1), P- and S-wave velocities are given by

$$\alpha(\mathbf{x}) \equiv \sqrt{\frac{\lambda(\mathbf{x}) + 2\mu(\mathbf{x})}{\rho(\mathbf{x})}} \quad \text{and} \quad \beta(\mathbf{x}) \equiv \sqrt{\frac{\mu(\mathbf{x})}{\rho(\mathbf{x})}} \tag{4.30}$$

We note the following relationship:

$$\frac{\delta\lambda + 2\delta\mu}{\lambda_0 + 2\mu_0} = 2\frac{\delta\alpha}{\alpha_0} + \frac{\delta\rho}{\rho_0} \quad \text{and} \quad \frac{\delta\mu}{\mu_0} = 2\frac{\delta\beta}{\beta_0} + \frac{\delta\rho}{\rho_0} \tag{4.31}$$

We can solve (4.27) by using a first-order perturbation, the Born approximation. The total vector wavefield is written as a sum of the incident plane wave and the scattered wave:

$$\mathbf{u} = \mathbf{u}^0 + \mathbf{u}^1 \tag{4.32}$$

where we expect the scattered wave amplitude to be much smaller than the incident wave amplitude: $|\mathbf{u}^1| \ll |\mathbf{u}^0|$. The incident plane wave satisfies the homogeneous wave equation

$$\rho_0 \ddot{u}_i^0 - \partial_j T_{ij}(\lambda_0, \mu_0; u_k^0) = 0 \tag{4.33}$$

Substituting (4.32) in (4.27) and using (4.33), we get the wave equation for the scattered wave:

$$\rho_0 \ddot{u}_i^1 - \partial_j T_{ij}(\lambda_0, \mu_0; u_k^1) = \delta f_i(\mathbf{x}, t) \tag{4.34}$$

where we have neglected cross terms of $(\delta\lambda, \delta\mu, \delta\rho) \times u_i^1$ because they are assumed to be small. The right-hand side is called the equivalent body force and represents the interaction of the incident wave with the inhomogeneity

$$\delta f_i(\mathbf{x}, t) = -\delta\rho\, \ddot{u}_i^0 + \partial_i \delta\lambda\, \partial_j u_j^0 + \partial_j \delta\mu \left(\partial_i u_j^0 + \partial_j u_i^0 \right)$$
$$+ \delta\lambda\, \partial_i \partial_j u_j^0 + \delta\mu\, \partial_j \left(\partial_i u_j^0 + \partial_j u_i^0 \right) \tag{4.35}$$

If we consider an incident plane P-wave of unit amplitude propagating in the third direction given by

$$\mathbf{u}^{0P} = \mathbf{e}_3\, e^{i(k e_3 \mathbf{x} - \omega t)} \quad \text{where} \quad k = \frac{\omega}{\alpha_0} \tag{4.36}$$

we have the corresponding equivalent body force

$$\delta f_k^P(\mathbf{x}, t) = \left\{ \left[\omega^2 \delta\rho - k^2 (\delta\lambda + 2\delta\mu) \right] \delta_{k3} + 2ik\partial_3 \delta\mu\, \delta_{k3} + ik\partial_k \delta\lambda \right\} e^{i(k e_3 \mathbf{x} - \omega t)} \tag{4.37}$$

For an incident plane S-wave of unit amplitude propagating in the third direction having polarization in the first direction,

$$\mathbf{u}^{0S} = \mathbf{e}_1 \, e^{i(l e_3 \mathbf{x} - \omega t)} \quad \text{where} \quad l = \frac{\omega}{\beta_0} \tag{4.38}$$

with equivalent body force

$$\delta f_k^S(\mathbf{x}, t) = \left[\left(\omega^2 \delta \rho - l^2 \delta \mu + i l \partial_3 \delta \mu \right) \delta_{k1} + i l \delta_{k3} \partial_1 \delta \mu \right] e^{i(l e_3 \mathbf{x} - \omega t)} \tag{4.39}$$

To solve (4.34) with (4.37) or (4.39), we need the explicit form of the Green function of elastic vector waves in 3-D space, which satisfies

$$\rho_0 \ddot{G}_{ik}(\mathbf{x}, t) - \partial_j T_{ij}(\lambda_0, \mu_0; G_{lk}) = \delta_{ik} \delta(t) \delta(\mathbf{x}) \tag{4.40}$$

The causal retarded solution [Aki and Richards, 1980, p. 73] is given by

$$G_{ik}(\mathbf{x}, t) = \left[\frac{e_{ri} e_{rk}}{4\pi\rho_0 \alpha_0^2 r} \delta\left(t - \frac{r}{\alpha_0} \right) + \frac{(\delta_{ik} - e_{ri} e_{rk})}{4\pi\rho_0 \beta_0^2 r} \delta\left(t - \frac{r}{\beta_0} \right) \right.$$
$$\left. + \frac{(3 e_{ri} e_{rk} - \delta_{ik})}{4\pi\rho_0 r^3} \int_{r/\alpha_0}^{r/\beta_0} t' \delta(t - t') dt' \right] H(t) \tag{4.41}$$

where $r = |\mathbf{x}|$ and e_{ri} means the ith component of the unit radial vector $\mathbf{e}_r$. The first term in brackets represents the far-field P-wave, the second term the far-field S-wave, and the last term shows the near-field term that becomes extremely small in the far field. In the far field $r \gg L$, we use only the first two terms.

Scattered P-waves in the far field for the incident plane P-wave of unit amplitude are given by the convolution of the far-field component of the Green function with the equivalent body force to obtain

$$u_i^{1PP}(\mathbf{x}, t) = \frac{e_{ri} e_{rk}}{4\pi\rho_0 \alpha_0^2 r} \int_{-\infty}^{\infty} \int_{-\infty}^{\infty} \int_{-\infty}^{\infty} \delta f_k^P\left(\mathbf{x}', t - \frac{|\mathbf{x} - \mathbf{x}'|}{\alpha_0} \right) d\mathbf{x}' \tag{4.42}$$

where superscript PP represents the P-to-P scattering mode. When the observation distance is large enough so that $r \gg L$ and $r \gg L^2 k / \pi$, we may use the approximation $|\mathbf{x} - \mathbf{x}'| \approx r - \mathbf{e}_r \mathbf{x}'$ in the argument of equivalent body force. Then, we have

$$u_i^{1PP}(\mathbf{x},t) = \frac{e_{ri}e_{rk}}{4\pi\rho_0\alpha_0^2 r} \int_{-\infty}^{\infty}\int_{-\infty}^{\infty}\int_{-\infty}^{\infty} \{[\omega^2\delta\rho - k^2(\delta\lambda + 2\delta\mu) + 2ik\partial_3\delta\mu]\delta_{k3} + ik\partial_k\delta\lambda\}$$

$$\times e^{ike_3\mathbf{x}' - i\omega\left(t - \frac{r}{\alpha_0} + \frac{e_r\mathbf{x}'}{\alpha_0}\right)} d\mathbf{x}'$$

$$= \frac{e^{i(kr - \omega t)}}{r} \frac{k^2 e_{ri}}{4\pi\rho_0\alpha_0^2} \int_{-\infty}^{\infty}\int_{-\infty}^{\infty}\int_{-\infty}^{\infty} [e_{r3}\alpha_0^2\delta\rho - \delta\lambda - 2e_{r3}^2\delta\mu] e^{-i(ke_r - ke_3)\mathbf{x}'} d\mathbf{x}'$$

$$\quad\quad\quad\quad\quad\quad\quad\quad\quad\quad\quad\quad\quad\quad\quad\quad\quad\quad\quad\quad (4.43a)$$

$$= \frac{e^{i(kr - \omega t)}}{r} \frac{l^2 e_{ri}}{4\pi\gamma_0^2} \int_{-\infty}^{\infty}\int_{-\infty}^{\infty}\int_{-\infty}^{\infty} \left\{\left[-1 + e_{r3} + \frac{2}{\gamma_0^2}(1 - e_{r3}^2)\right]\frac{\delta\rho(\mathbf{x}')}{\rho_0}\right.$$

$$\left. - 2\frac{\delta\alpha(\mathbf{x}')}{\alpha_0} + \frac{4}{\gamma_0^2}(1 - e_{r3}^2)\frac{\delta\beta(\mathbf{x}')}{\beta_0}\right\} e^{-i(ke_r - ke_3)\mathbf{x}'} d\mathbf{x}'$$

$$= \frac{e^{i(kr - \omega t)}}{r} F_i^{PP}$$

where we used partial integration for a finite distribution of the inhomogeneity and defined the velocity ratio $\gamma_0 \equiv \alpha_0/\beta_0$. The P-to-P scattering amplitude vector can be explicitly written as

$$F_i^{PP} = \frac{l^2 e_{ri}}{4\pi\gamma_0^2} \left\{\left[-1 + e_{r3} + \frac{2}{\gamma_0^2}(1 - e_{r3}^2)\right]\frac{\delta\tilde{\rho}(ke_r - ke_3)}{\rho_0}\right.$$

$$\left. - 2\frac{\delta\tilde{\alpha}(ke_r - ke_3)}{\alpha_0} + \frac{4}{\gamma_0^2}(1 - e_{r3}^2)\frac{\delta\tilde{\beta}(ke_r - ke_3)}{\beta_0}\right\}$$

$$\quad\quad\quad\quad\quad\quad\quad\quad\quad\quad\quad\quad\quad\quad\quad\quad\quad\quad\quad\quad (4.43a')$$

where the tilde means the Fourier transform in 3-D space as in (4.14). Using the same approach, we get the spherically outgoing scattered waves in the far field and the corresponding scattering amplitude vectors as follows: For P-to-S scattering, using the second term in brackets of (4.41), we get

$$u_i^{1PS}(\mathbf{x},t) = \frac{\delta_{ik} - e_{ri}e_{rk}}{4\pi\rho_0\beta_0^2 r} \int_{-\infty}^{\infty}\int_{-\infty}^{\infty}\int_{-\infty}^{\infty} \delta f_k^P\left(\mathbf{x}',t - \frac{r}{\beta_0} + \frac{e_r\mathbf{x}'}{\beta_0}\right)d\mathbf{x}' = \frac{e^{i(lr - \omega t)}}{r} F_i^{PS} \quad (4.43b)$$

where

$$F_i^{PS} = \frac{l^2}{4\pi}(\delta_{i3} - e_{ri}e_{r3})\left[\left(1 - \frac{2}{\gamma_0}e_{r3}\right)\frac{\delta\tilde{\rho}(le_r - ke_3)}{\rho_0} - \frac{4}{\gamma_0}e_{r3}\frac{\delta\tilde{\beta}(le_r - ke_3)}{\beta_0}\right] \quad (4.43b')$$

For S-to-P scattering

$$u_i^{1SP}(\mathbf{x},t) = \frac{e_{ri}e_{rk}}{4\pi\rho_0\alpha_0^{\,2}r} \int\limits_{-\infty}^{\infty}\int\limits_{-\infty}^{\infty}\int\limits_{-\infty}^{\infty} \delta f_k^{\,S}\!\left(\mathbf{x}',t-\frac{r}{\alpha_0}+\frac{e_r\mathbf{x}'}{\alpha_0}\right)d\mathbf{x}' = \frac{e^{i(kr-\omega t)}}{r}F_i^{SP} \qquad (4.43c)$$

where

$$F_i^{\,SP} = \frac{l^2}{4\pi\gamma_0^{\,2}}e_{ri}e_{r1}\!\left[\left(1-\frac{2}{\gamma_0}e_{r3}\right)\frac{\delta\tilde{\rho}(ke_r-le_3)}{\rho_0} - \frac{4}{\gamma_0}e_{r3}\frac{\delta\tilde{\beta}(ke_r-le_3)}{\beta_0}\right] \qquad (4.43c')$$

For S-to-S scattering

$$u_i^{1SS}(\mathbf{x},t) = \frac{\delta_{ik}-e_{ri}e_{rk}}{4\pi\rho_0\beta_0^{\,2}r} \int\limits_{-\infty}^{\infty}\int\limits_{-\infty}^{\infty}\int\limits_{-\infty}^{\infty} \delta f_k^{\,S}\!\left(\mathbf{x}',t-\frac{r}{\beta_0}+\frac{e_r\mathbf{x}'}{\beta_0}\right)d\mathbf{x}' = \frac{e^{i(kr-\omega t)}}{r}F_i^{SS} \qquad (4.43d)$$

where

$$F_i^{\,SS} = \frac{l^2}{4\pi}\left\{\left[(\delta_{i1}-e_{ri}e_{r1})(1-e_{r3})-(\delta_{i3}-e_{ri}e_{r3})e_{r1}\right]\frac{\delta\tilde{\rho}(le_r-le_3)}{\rho_0}\right.$$
$$\left. -2\left[(\delta_{i1}-e_{ri}e_{r1})e_{r3}+(\delta_{i3}-e_{ri}e_{r3})e_{r1}\right]\frac{\delta\tilde{\beta}(le_r-le_3)}{\beta_0}\right\} \qquad (4.43d')$$

Each argument in the Fourier transform of the fractional fluctuation is the corresponding exchange wavenumber vector defined as the difference between the scattered wavenumber vector and the incident wavenumber vector.

Scattering Amplitudes in the Spherical Coordinate System

Writing the scattering amplitude vectors (4.43a'–d') in spherical coordinates (r,ψ,ζ) gives some insight into the scattering process resulting from the Born approximation. The scattering amplitude for scalar waves is axially symmetric; however, it is more complicated for elastic vector waves. The transformation of the unit base vectors from Cartesian coordinates to spherical coordinates is given by

$$\begin{aligned}
e_r &= \sin\psi\cos\zeta\,e_1 + \sin\psi\sin\zeta\,e_2 + \cos\psi\,e_3 \\
e_\psi &= \cos\psi\cos\zeta\,e_1 + \cos\psi\sin\zeta\,e_2 - \sin\psi\,e_3 \qquad (4.44)\\
e_\zeta &= \qquad -\sin\zeta\,e_1 + \qquad \cos\zeta\,e_2
\end{aligned}$$

where ψ is measured from the positive third axis and ζ from the positive first axis (see Figure 4.1). The P-to-P scattering amplitude in spherical coordinates is given by

$$F^{PP} = \sum_{i=1}^{3} F_i^{PP} \mathbf{e}_i = F_r^{PP} \mathbf{e}_r + F_\psi^{PP} \mathbf{e}_\psi + F_\zeta^{PP} \mathbf{e}_\zeta \qquad (4.45a)$$

In the same way, we define the other terms as

$$F^{PS} = \sum_{i=1}^{3} F_i^{PS} \mathbf{e}_i = F_r^{PS} \mathbf{e}_r + F_\psi^{PS} \mathbf{e}_\psi + F_\zeta^{PS} \mathbf{e}_\zeta$$

$$F^{SS} = \sum_{i=1}^{3} F_i^{SS} \mathbf{e}_i = F_r^{SS} \mathbf{e}_r + F_\psi^{SS} \mathbf{e}_\psi + F_\zeta^{SS} \mathbf{e}_\zeta \qquad (4.45b)$$

$$F^{SP} = \sum_{i=1}^{3} F_i^{SP} \mathbf{e}_i = F_r^{SP} \mathbf{e}_r + F_\psi^{SP} \mathbf{e}_\psi + F_\zeta^{SP} \mathbf{e}_\zeta$$

For an incident P-wave, the scattered P-wave is nonzero only for the radial component

$$F_r^{PP} = \frac{l^2}{4\pi} \frac{1}{\gamma_0^2} \left[\left(-1 + \cos\psi + \frac{2}{\gamma_0^2} \sin^2\psi \right) \frac{\delta\tilde{\rho}(ke_r - ke_3)}{\rho_0} \right.$$

$$\left. -2 \frac{\delta\tilde{\alpha}(ke_r - ke_3)}{\alpha_0} + \frac{4}{\gamma_0^2} \sin^2\psi \frac{\delta\tilde{\beta}(ke_r - ke_3)}{\beta_0} \right] \qquad (4.46a)$$

$$F_\psi^{PP} = F_\zeta^{PP} = 0$$

The scattered S-wave has only transverse components

$$F_\psi^{PS} = \frac{l^2}{4\pi} \sin\psi \left[\left(-1 + \frac{2}{\gamma_0} \cos\psi \right) \frac{\delta\tilde{\rho}(le_r - ke_3)}{\rho_0} + \frac{4}{\gamma_0} \cos\psi \frac{\delta\tilde{\beta}(le_r - ke_3)}{\beta_0} \right] \qquad (4.46b)$$

$$F_r^{PS} = F_\zeta^{PS} = 0$$

The other scattering amplitudes are given by

$$F_r^{SP} = \frac{l^2}{4\pi} \frac{1}{\gamma_0^2} \sin\psi \cos\zeta$$

$$\times \left[\left(1 - \frac{2}{\gamma_0} \cos\psi \right) \frac{\delta\tilde{\rho}(ke_r - le_3)}{\rho_0} - \frac{4}{\gamma_0} \cos\psi \frac{\delta\tilde{\beta}(ke_r - le_3)}{\beta_0} \right] \qquad (4.46c)$$

$$F_\psi^{SP} = F_\zeta^{SP} = 0$$

and

$$F_\psi^{ss} = \frac{l^2}{4\pi}\cos\zeta\left[(\cos\psi - \cos2\psi)\frac{\delta\tilde{\rho}(le_r - le_3)}{\rho_0} - 2\cos2\psi\frac{\delta\tilde{\beta}(le_r - le_3)}{\beta_0}\right]$$

$$F_\zeta^{ss} = \frac{l^2}{4\pi}\sin\zeta\left[(\cos\psi - 1)\frac{\delta\tilde{\rho}(le_r - le_3)}{\rho_0} + 2\cos\psi\frac{\delta\tilde{\beta}(le_r - le_3)}{\beta_0}\right] \qquad (4.46d)$$

$$F_r^{ss} = 0$$

In the forward direction ($\psi = 0$), there are no converted PS- and SP-phases. Also, in the forward direction, PP scattering is caused only by P-wave velocity fluctuation, and SS-scattering is due only to S-wave velocity fluctuation.

4.2.2 Reduction of Independent Medium Fluctuations Using Birch's Law

Well-log data plotted in Figure 2.4b show that P- and S-wave velocities are linearly well correlated. From this result, we assume that P-wave velocity fluctuation is proportional to S-wave velocity fluctuation for rocks that comprise the real earth medium

$$\xi(x) \equiv \frac{\delta\alpha(x)}{\alpha_0} = \frac{\delta\beta(x)}{\beta_0} \qquad (4.47)$$

In addition, Figure 2.4b and the discussion in Section 2.1 show evidence of an empirical linear relationship between wave velocity and mass density for rock. Using the empirical relation, known as Birch's Law [Birch, 1961], $\alpha[km/s] = 3.05\rho[g\cdot cm^{-3}] - 1.87$ as shown in Figure 2.3, we obtain $\delta\rho/\rho_0 = (\alpha_0/(\alpha_0 + 1.87))\delta\alpha/\alpha_0$. Choosing $\alpha_0 = 6.0-8.5$ km/s as typical lithospheric velocities, we get $\delta\rho/\rho_0 = (0.78 - 0.82)\delta\alpha/\alpha_0$. Birch's law along with (4.47) allows us to reduce the number of independent fractional fluctuations in (4.46a–d) from three to one [Sato, 1984a; Malin and Phinney, 1985]:

$$\frac{\delta\rho(x)}{\rho_0} = \nu\xi(x) \qquad (4.48)$$

where we choose the linear coefficient $\nu = 0.8$ for the following equations. Using larger ν values systematically increases the contributions from both velocity and density fluctuations to backward scattering.

We can thus write each nonzero scattering amplitude as a product of the square of the S-wave wavenumber l^2, a basic scattering pattern X_*^{**}, and the Fourier transform of the fractional fluctuation $\tilde{\xi}$:

$$F_r^{PP} = \frac{l^2}{4\pi} X_r^{PP}(\psi,\zeta)\tilde{\xi}(ke_r - ke_3)$$

$$F_\psi^{PS} = \frac{l^2}{4\pi} X_\psi^{PS}(\psi,\zeta)\tilde{\xi}(le_r - ke_3)$$

$$F_r^{SP} = \frac{l^2}{4\pi} X_r^{SP}(\psi,\zeta)\tilde{\xi}(ke_r - le_3) \qquad (4.49)$$

$$F_\psi^{SS} = \frac{l^2}{4\pi} X_\psi^{SS}(\psi,\zeta)\tilde{\xi}(le_r - le_3)$$

$$F_\zeta^{SS} = \frac{l^2}{4\pi} X_\zeta^{SS}(\psi,\zeta)\tilde{\xi}(le_r - le_3)$$

where the basic scattering patterns are given by

$$X_r^{PP}(\psi,\zeta) = \frac{1}{\gamma_0^2}\left[\nu\left(-1+\cos\psi+\frac{2}{\gamma_0^2}\sin^2\psi\right)-2+\frac{4}{\gamma_0^2}\sin^2\psi\right]$$

$$X_\psi^{PS}(\psi,\zeta) = -\sin\psi\left[\nu\left(1-\frac{2}{\gamma_0}\cos\psi\right)-\frac{4}{\gamma_0}\cos\psi\right]$$

$$X_r^{SP}(\psi,\zeta) = \frac{1}{\gamma_0^2}\sin\psi\cos\zeta\left[\nu\left(1-\frac{2}{\gamma_0}\cos\psi\right)-\frac{4}{\gamma_0}\cos\psi\right] \qquad (4.50)$$

$$X_\psi^{SS}(\psi,\zeta) = \cos\zeta[\nu(\cos\psi-\cos2\psi)-2\cos2\psi]$$

$$X_\zeta^{SS}(\psi,\zeta) = \sin\zeta[\nu(\cos\psi-1)+2\cos\psi]$$

Angular dependence of scattering comes from both terms X_*^{**} and $\tilde{\xi}$, but the frequency-dependent nonisotropic scattering comes only from the $\tilde{\xi}$ term. Contrary to the scalar wave case, these scattering patterns have lobes that depend on ψ and ζ. Figure 4.7 shows the ψ-dependence of basic scattering patterns at specific values

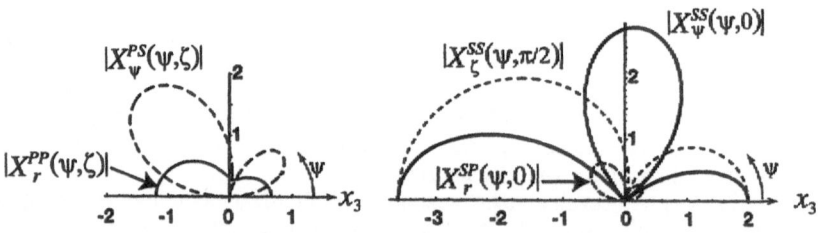

FIGURE 4.7. ψ-dependence of basic scattering patterns for $\gamma_0 = \sqrt{3}$ and $\nu = 0.8$.

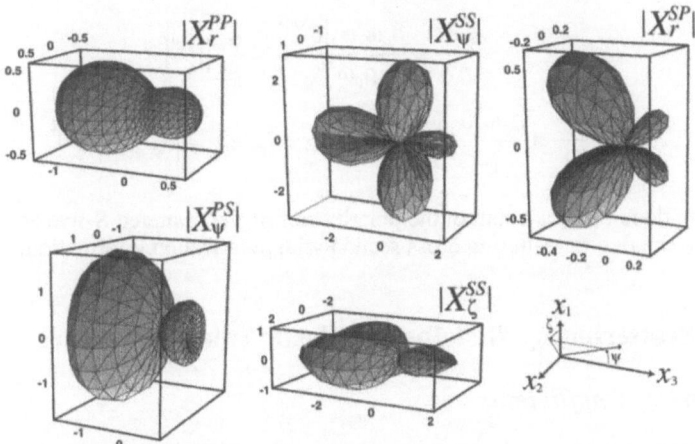

FIGURE 4.8. 3-D views of basic scattering patterns in 3-D elastic random media for $\gamma_0 = \sqrt{3}$ and $\nu = 0.8$. Incident wave is propagating in direction x_3 and S-wave is polarized in direction x_1.

of ζ for given γ_0 and ν. Figure 4.8 shows 3-D views of basic scattering patterns. The angular dependence of basic scattering pattern $X_{\bullet}^{\bullet\bullet}(\psi,\zeta)$ is axially symmetric only for the case of P-wave incidence.

Scattering Cross Sections

We can define the scattering cross section for the block of dimension L for different scattering modes using an extension of (4.18). We note the difference in velocity between incident waves and spherically outgoing scattered waves for conversion scattering. For the stationary process of P-to-P scattering,

$$\frac{d\sigma^{PP}}{d\Omega} \equiv \frac{\rho_0 \omega^2 \alpha_0 |\mathbf{F}^{PP}|^2}{\rho_0 \omega^2 \alpha_0} = \left|F_r^{PP}\right|^2 \tag{4.51a}$$

In the same way,

$$\frac{d\sigma^{PS}}{d\Omega} \equiv \frac{\rho_0 \omega^2 \beta_0 |\mathbf{F}^{PS}|^2}{\rho_0 \omega^2 \alpha_0} = \frac{\beta_0}{\alpha_0}\left|F_\psi^{PS}\right|^2 \tag{4.51b}$$

$$\frac{d\sigma^{SP}}{d\Omega} \equiv \frac{\rho_0 \omega^2 \alpha_0 \left|\mathbf{F}^{SP}\right|^2}{\rho_0 \omega^2 \beta_0} = \frac{\alpha_0}{\beta_0} \left|F_r^{SP}\right|^2 \tag{4.51c}$$

$$\frac{d\sigma^{SS}}{d\Omega} \equiv \frac{\rho_0 \omega^2 \beta_0 \left|\mathbf{F}^{SS}\right|^2}{\rho_0 \omega^2 \beta_0} = \left|F_\psi^{SS} \mathbf{e}_\psi + F_\zeta^{SS} \mathbf{e}_\zeta\right|^2 = \left|F_\psi^{SS}\right|^2 + \left|F_\zeta^{SS}\right|^2 \tag{4.51d}$$

Here, we do not take account of the polarization of the scattered S-waves. If necessary, we can define scattering cross section with polarization information.

4.2.3 Scattering by Distributed Elastic Inhomogeneities

Scattering Coefficients

Following the same procedure as used for scalar waves, we imagine an ensemble of homogeneous and isotropic random media described by fractional fluctuations $\{\xi(\mathbf{x})\}$, where $\langle\xi\rangle = 0$. Then, we can define the statistical scattering coefficients as the scattering power per unit volume of inhomogeneous elastic media. The scattering coefficient for P-to-P scattering is given by

$$g^{PP}(\psi,\zeta;\omega) \equiv \frac{4\pi}{L^3}\left\langle\frac{d\sigma^{PP}}{d\Omega}\right\rangle = \frac{4\pi}{L^3}\left\langle\left|F_r^{PP}\right|^2\right\rangle$$

$$= \frac{l^4}{4\pi}\left|X_r^{PP}\right|^2 \frac{1}{L^3}\int_{-\infty}^{\infty}\int_{-\infty}^{\infty}\int_{-\infty}^{\infty}\int_{-\infty}^{\infty}\int_{-\infty}^{\infty}\langle\xi(\mathbf{x}')\xi(\mathbf{x}'')\rangle\, e^{-i(k\mathbf{e}_r - k\mathbf{e}_3)(\mathbf{x}' - \mathbf{x}'')} d\mathbf{x}'\, d\mathbf{x}'' \tag{4.52a}$$

$$= \frac{l^4}{4\pi}\left|X_r^{PP}\right|^2 P(k\mathbf{e}_r - k\mathbf{e}_3) = \frac{l^4}{4\pi}\left|X_r^{PP}(\psi,\zeta)\right|^2 P\left(\frac{2l}{\gamma_0}\sin\frac{\psi}{2}\right)$$

In the same way,

$$g^{PS}(\psi,\zeta;\omega) \equiv \frac{4\pi}{L^3}\left\langle\frac{d\sigma^{PS}}{d\Omega}\right\rangle = \frac{\beta_0}{\alpha_0}\frac{4\pi}{L^3}\left\langle\left|F_\psi^{PS}\right|^2\right\rangle = \frac{1}{\gamma_0}\frac{l^4}{4\pi}\left|X_\psi^{PS}\right|^2 P(l\mathbf{e}_r - k\mathbf{e}_3)$$

$$= \frac{1}{\gamma_0}\frac{l^4}{4\pi}\left|X_\psi^{PS}(\psi,\zeta)\right|^2 P\left(\frac{l}{\gamma_0}\sqrt{1+\gamma_0^2 - 2\gamma_0\cos\psi}\right) \tag{4.52b}$$

$$g^{SP}(\psi,\zeta;\omega) \equiv \frac{4\pi}{L^3}\left\langle\frac{d\sigma^{SP}}{d\Omega}\right\rangle = \frac{\alpha_0}{\beta_0}\frac{4\pi}{L^3}\left\langle\left|F_r^{SP}\right|^2\right\rangle = \gamma_0\frac{l^4}{4\pi}\left|X_r^{SP}\right|^2 P(k\mathbf{e}_r - l\mathbf{e}_3)$$

$$= \gamma_0\frac{l^4}{4\pi}\left|X_r^{SP}(\psi,\zeta)\right|^2 P\left(\frac{l}{\gamma_0}\sqrt{1+\gamma_0^2 - 2\gamma_0\cos\psi}\right) \tag{4.52c}$$

$$g^{SS}(\psi,\zeta;\omega) \equiv \frac{4\pi}{L^3}\left\langle \frac{d\sigma^{SS}}{d\Omega} \right\rangle = \frac{4\pi}{L^3}\left\langle \left| F_\psi^{SS}\mathbf{e}_\psi + F_\zeta^{SS}\mathbf{e}_\zeta \right|^2 \right\rangle = \frac{4\pi}{L^3}\left\langle \left| F_\psi^{SS} \right|^2 + \left| F_\zeta^{SS} \right|^2 \right\rangle$$

$$= \frac{l^4}{4\pi}\left(\left| X_\psi^{SS} \right|^2 + \left| X_\zeta^{SS} \right|^2 \right) P(le_r - le_3) \qquad (4.52d)$$

$$= \frac{l^4}{4\pi}\left(\left| X_\psi^{SS}(\psi,\zeta) \right|^2 + \left| X_\zeta^{SS}(\psi,\zeta) \right|^2 \right) P\left(2l\sin\frac{\psi}{2} \right)$$

where $l = \omega/\beta_0$ is S-wave wavenumber corresponding to angular frequency ω. The basic scattering patterns defined in (4.50) are independent of angular frequency.

In (4.52a–d), the contribution from the PSDF is symmetric around the axis of the propagation direction. Figure 4.9 illustrates the ψ-dependence of the PSDF terms of the scattering pattern for the case of an exponential ACF for three different angular frequencies. The scattering contribution from the PSDF is nearly isotropic for all four scattering modes for lower angular frequencies; however, at high angular frequencies the contribution of PP- and SS-scattering is larger in a narrow cone around the forward direction compared with those for large scattering angles, since the PSDF term in (4.52a and d) is $8\pi\varepsilon^2 a^3$ for all wavenumbers at $\psi = 0$. The product with the fourth power of wavenumber makes the scattering coefficient much larger at higher angular frequencies. For PS- and SP-conversion scattering at a given angular frequency, the minimum argument of the PSDF in (4.52b and c) is not zero but $(\gamma_0 - 1)l/\gamma_0$, which occurs in the forward direction $\psi = 0$. The PSDF term rapidly decreases with increasing angular frequency even in the forward direction. PSDF contributions to conversion scattering are too small to show in the bottom plot of Figure 4.9. For scalar waves, the angular dependence of scattering coefficient is controlled only by the PSDF term; however, the angular dependence of scattering coefficient for elastic vector waves is a product of the square of the basic scattering patterns $X_\cdot^{\cdot\cdot}$ and the PSDF term, as illustrated in Figures 4.8 and 4.9, respectively. In a narrow cone around the forward direction corresponding to low wavenumbers, the scattering patterns show very small PS- and SP-scattering. This means that the contribution of long wavelength components of random inhomogeneity is very small for conversion scattering.

The average of the scattering cross section over the solid angle gives the total scattering coefficient. For conversion scattering,

$$g_0^{SP} = \frac{1}{2\gamma_0^2} g_0^{PS} \qquad (4.53)$$

from (4.50). This relationship is valid for any kind of inhomogeneity. Aki [1992] derived relationship (4.53) as a consequence of the reciprocal theorem [see also Papanicolaou et al., 1996; Korneev and Johnson, 1996]. This means that P-waves scatter to S-waves more readily than S-waves scatter to P-waves.

S-to-S Backscattering Coefficient

The backscattering coefficient for S-wave to S-wave is important for practical analysis of coda:

$$g_\pi^{SS}(\omega) \equiv g^{SS}(\psi = \pi, \zeta; \omega) = \frac{l^4}{\pi}(1 + \nu)^2 P(2l) \qquad (4.54)$$

As expected from a 1-D reflection study [Aki and Richards, 1980, p. 661], the backscattering coefficient in 3-D is proportional to the PSDF of S-wave impedance:

$$\frac{\delta(\rho\beta)}{\rho_0\beta_0} = \frac{\delta\rho(\mathbf{x})}{\rho_0} + \frac{\delta\beta(\mathbf{x})}{\beta_0} = (1 + \nu)\xi(\mathbf{x}) \qquad (4.55)$$

When the ACF is a von Kármán type, the PSDF is given by (2.12). The S-to-S backscattering coefficient is explicitly written as

$$g_\pi^{SS}(\omega) = \frac{8(1 + \nu)^2 \varepsilon^2 \pi^{1/2} a^3 l^4 \Gamma\left(\kappa + \frac{3}{2}\right)}{\Gamma(\kappa)\left(1 + 4a^2 l^2\right)^{\kappa + \frac{3}{2}}} \qquad (4.56)$$

In the case of an exponential ACF, which is a special case of (4.56) with $\kappa = 0.5$,

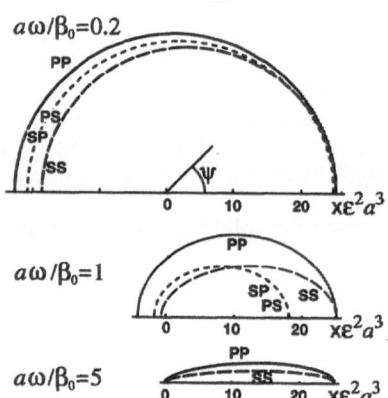

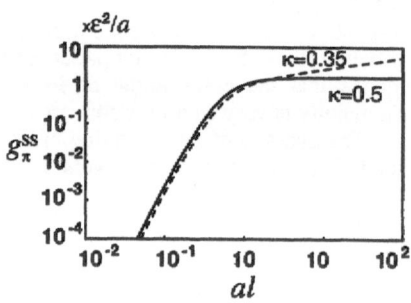

FIGURE 4.9. Angular dependence of the PSDF term in scattering coefficients (4.52a–d) for the case of exponential ACF for three different angular frequencies, where $\nu = 0.8$ and $\gamma_0 = \sqrt{3}$.

FIGURE 4.10. Plots of S-to-S back scattering coefficient against scaled wavenumber al for the exponential ACF ($\kappa = 0.5$, solid) and the von Kármán ACF ($\kappa = 0.35$, broken), where $\nu = 0.8$ and $l = \omega/\beta_0$ is S-wave wavenumber.

$$g_\pi^{ss}(\omega) = \frac{8(1+\nu)^2 \varepsilon^2 a^3 l^4}{\left(1+4a^2 l^2\right)^2}$$

$$\approx \frac{1}{2}(1+\nu)^2 \frac{\varepsilon^2}{a} \qquad \text{for } al \gg 1 \qquad (4.57)$$

$$\approx 1.62 \frac{\varepsilon^2}{a} \qquad (\nu = 0.8)$$

The backscattering coefficient is constant for high frequencies for $\kappa = 0.5$. Figure 4.10 shows the frequency dependence of the S-to-S backscattering coefficient for $\kappa = 0.5$ and 0.35, where $l = \omega / \beta_0$. As κ decreases, the backscattering coefficient increases with increasing frequency for high frequencies because of the increasing richness of the PSDF in short-wavelength components.

CHAPTER 5

Attenuation of High-Frequency Seismic Waves

We will now discuss the attenuation with propagation distance of seismic wave amplitude in the lithosphere for frequencies mostly higher than 1 Hz. First we review the frequency dependence of observed amplitude attenuation in the earth's lithosphere. We discuss various proposed mechanisms of intrinsic attenuation and describe their frequency characteristics. We have already discussed the scattering of seismic waves caused by random heterogeneities as a mechanism to explain the excitation of incoherent S-coda waves. The amplitude decay with travel distance will now be derived as a natural consequence of the application of energy conservation to the scattering model; scattering attenuates the direct wave amplitude and excites coda waves. Taking scalar waves as an example, we introduce an approach for calculating the amount of scattering attenuation in a manner consistent with conventional seismological attenuation measurements. Then, extending the method to elastic waves, we calculate the scattering attenuation of P- and S-waves in inhomogeneous elastic media. The randomness of the lithosphere will then be quantitatively estimated from S-wave attenuation and S-coda excitation measurements.

5.1 ATTENUATION IN THE LITHOSPHERE

Seismic wave amplitude generally decreases with increasing travel distance through the earth. Except where wave interference occurs, the observed change in amplitude is usually exponentially related to travel distance, and decay rates are proportional to Q_P^{-1} and Q_S^{-1} which characterize the spatial attenuation for P- and S-waves, respectively. For plane waves of frequency f, the exponential decay with travel distance r is given by $-\pi r f Q_P^{-1}/\alpha_0$ for P-waves and $-\pi r f Q_S^{-1}/\beta_0$ for S-waves. For spherically outgoing body waves in a uniform velocity structure in 3-D space, there is an additional geometrical spreading factor r^{-1}, so the spectral amplitudes of P- and S-waves, u^P and u^S, go roughly as

$$u^P(r;f) \propto \frac{e^{-\pi r f Q_P^{-1}/\alpha_0}}{r} \quad \text{and} \quad u^S(r;f) \propto \frac{e^{-\pi r f Q_S^{-1}/\beta_0}}{r} \qquad (5.1)$$

A popular method of making attenuation measurements is the spectral decay

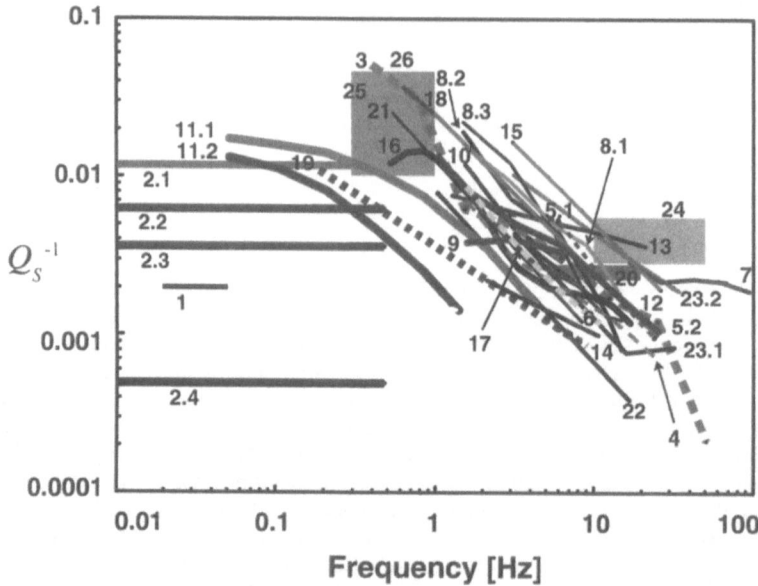

FIGURE 5.1. Reported values of Q_s^{-1} for the lithosphere. **Surface wave analysis**: 1, depth < 45 km, SL8 global model [Anderson and Hart, 1978]; 2.1, upper crust (depth < 18 km) in the Basin and Range Province; 2.2, upper crust in the Colorado plateau; 2.3, upper crust in the eastern U. S. A.; 2.4, lower crust (depth>18 km) in the U. S. A. [Cheng and Mitchell, 1981]. **Coda-normalization method** (see Section 3.4.3): 3, Hindu–Kush [Roecker et al., 1982]; 4, Kanto, Japan [Aki, 1980a]; 5.1, eastern Kanto, Japan [Sato and Matsumura, 1980]; 5.2, Kanto, Japan [Yoshimoto et al., 1993]; 6, northern Greece [Hatzidimitriou, 1995]; 7, shallow crust at western Nagano, Japan [Yoshimoto et al., 1994]. **Multiple lapse-time window analysis** (see Section 7.2): 8.1, Central California; 8.2, Hawaii; 8.3, Long Valley in California, U. S. A. [Mayeda et al. 1992]; 9, Kanto-Tokai, Japan [Fehler et al., 1992]; 10, Kyushu, Japan [Hoshiba, 1993]. **Spectral decay analysis**: 11.1, Basin and Range Province; 11.2, U. S. Shield [Taylor et al., 1986]; 12, Sg and Lg, Utah, U. S. A. [Brockman and Bollinger, 1992]; 13, depth 5–25 km, southern Kurils [Fedotov and Boldyrev, 1969]; 14, Lg, France [Campillo and Plantet, 1991]; 15, Imperial fault, California [Singh et al., 1982]; 16, depth < 50 km, southern Kanto, Japan [Kinoshita, 1994]; 17, Pacific coast of Kanto, Japan [Takemura et al., 1991]; 18, Montenegro, Yugoslavia [Rovelli, 1984]; 19, Mexico [Ordaz and Singh, 1992]; 20, depth < 40 km, northern Caribbean [Frankel, 1982]; 21, northern Italy [Console and Rovelli, 1981]; 22, southern Norway [Kvamme and Havskov, 1989]; 23.1, New York State, U. S. A.; 23.2, southern California [Frankel et al., 1990]; 24, depth < 10 km, Arette, Pyrénées [Modiano and Hatzfeld, 1982]; 25, San Andreas Fault, California [Kurita, 1975]; 26, southern Italy [Rovelli, 1983].

method for body waves or surface waves. The spectral decay method uses measurements of spectral amplitudes vs. frequency for at least two propagation distances. If we know $u^P(r_1; f_1)$ and $u^P(r_2; f)$,

$$\ln \frac{r_2\, u^P(r_2; f)}{r_1\, u^P(r_1; f)} = -\pi(r_2 - r_1) f Q_P^{-1}/\alpha_0 + \text{Const.} \tag{5.2}$$

If Q_P^{-1} is assumed to be frequency independent, its value can be determined from the slope of the left-hand side of (5.2) vs. f from data at a single station. Other measurements have been based on observations of the change in direct-wave amplitude with distance using the coda-normalization method with data from a single station as mentioned in Section 3.4.3. More recent Q_S^{-1} measurements have been based on the multiple lapse-time window analysis of whole S-wave seismograms (see Section 7.2).

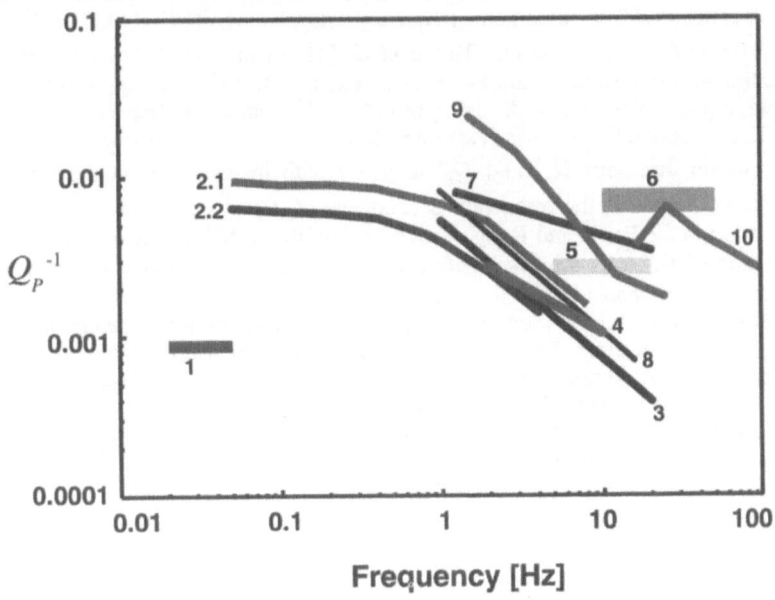

FIGURE 5.2. Reported values of Q_P^{-1} for the lithosphere. **Surface-wave analysis:** 1, depth < 45 km of the SL8 global model [Anderson and Hart, 1978]. **Spectral decay analysis:** 2.1, Basin and Range Province; 2.2, U. S. Shield [Taylor et al., 1986]; 3, Pn, eastern Canada [Zhu et al., 1991]; 4, Pg, France [Campillo and Plantet, 1991]; 5, depth < 40 km, northern Caribbean [Frankel, 1982]; 6, depth < 10 km, Arette, Pyrénées [Modiano and Hatzfeld, 1982]; 7, depth 5–25 km, southern Kurils [Fedotov and Boldyrev, 1969]; 8, southern Norway [Kvamme and Havskov, 1989]. **Extended coda-normalization method** (see Section 3.4.3): 9, Kanto, Japan [Yoshimoto et al., 1993]; 10, upper crust of western Nagano, Japan [Yoshimoto et al., 1994].

We will first discuss reported values of Q_S^{-1} and Q_P^{-1} for the lithosphere. We focus our attention on Q_S^{-1} and Q_P^{-1} as opposed to coda Q_C^{-1}, which is a phenomenological parameter discussed in Chapter 3. We will briefly enumerate the measurements; plots of reported Q_S^{-1}, Q_P^{-1} and the ratio Q_P^{-1}/Q_S^{-1} are shown in Figures 5.1, 5.2 and 5.3, respectively.

For low frequencies, the ratio $Q_P^{-1}/Q_S^{-1} \approx 0.4$–$0.47$ in global model MM8 of Anderson et al. [1965] derived from analysis of surface wave data. Anderson and Hart [1978] proposed Q models of the earth having $Q_S^{-1} \approx 0.002$ and $Q_P^{-1} \approx 0.0009$ and ratio $Q_P^{-1}/Q_S^{-1} \approx 0.5$ for frequencies < 0.05 Hz over a depth range from the surface to 45 km. Analyzing higher mode surface waves, Cheng and Mitchell [1981] determined crustal attenuation in the United States for 0.01–0.5 Hz and found that Q_S^{-1} in the upper crust (depth < 18 km) is smallest in the eastern U. S. A., larger in the Colorado Plateau region, and largest in the Basin and Range Province. Zhu et al. [1991] obtained $Q_P^{-1} \approx 0.0053 f^{-0.87}$ for 1–20 Hz from the analysis of Pn waves in eastern Canada. Analyzing Pg- and Lg-waves in France, Campillo and Plantet [1991] obtained frequency dependent $Q_S^{-1} \approx 0.0031 f^{-0.5}$ and $Q_P^{-1} \approx 0.042 f^{-0.6}$ for 2–10 Hz. Taylor et al. [1986] measured the differential attenuation of teleseismic P- and S-waves across the Basin and Range Province and Shield regions of the U. S. A. using broad-band seismic data between 0.05 and 5 Hz. They reported that the depth sampled by the waves reached a few hundred km. They found that both Q_S^{-1} and Q_P^{-1} decrease with increasing frequency for both provinces, however, the frequency dependence of Q_S^{-1} is more pronounced. Q_S^{-1} was larger in the Basin and Range Province than in the Shield region. Their measured ratio of Q_P^{-1}/Q_S^{-1} increases with frequency and becomes larger than 1 for frequencies higher than 1 Hz in both regions.

There are several reported measurements made near subduction zones and other seismically active areas. Fedotov and Boldyrev [1969] measured values for several layers in the crust to the upper mantle in the southern Kurils. From spectral analysis of P- and S-waves near Garm, Tadjikistan in Central Asia, Rautian et al. [1978] reported that the ratio Q_P^{-1}/Q_S^{-1} is equal to the velocity ratio of P- to S-waves for 2–12 Hz. In Mexico, Ordaz and Singh [1992] estimated $Q_S^{-1} \approx 0.037 f^{-0.66}$ for 0.2–10 Hz.

Analyzing records of microearthquakes in Kanto, Japan using the coda normalization method as discussed in Section 3.4.3, Aki [1980a] found that Q_S^{-1} decreases with increasing frequency as a power law $Q_S^{-1} \propto f^{-(0.6-0.8)}$ for frequencies 1–25 Hz. Combining the low Q_S^{-1} estimated around 0.05 Hz from surface wave analysis with measurements of Q_S^{-1} at frequencies higher than 1 Hz, Aki [1980a] conjectured that Q_S^{-1} has a peak around 0.5 Hz and decreases for both lower and higher frequencies (Figure 3.32b). His conjecture was confirmed observationally by Kinoshita [1994] from spectral decay analysis of strong motion records of earthquakes having focal depths < 50 km in southern Kanto, Japan. Kinoshita [1994] found a decrease in attenuation with decreasing frequency for frequencies less than about 0.8 Hz as shown by curve 16 in Figure 5.1. At higher frequencies, he found $Q_S^{-1} = 0.0077\, f^{-0.7}$ for 2–16 Hz. He estimated that the uncertainty in the measurements of Q_S^{-1} is about a factor of two. The power of frequency dependence of about 0.7 found by Kinoshita [1994] for this area agrees with previous measurements [Aki, 1980a; Sato, 1984b, 1990]. Extending the coda-normalization method (see Section 3.4.3), Yoshimoto et al. [1993] estimated attenuation in Kanto, Japan for depth < 100 km and found $Q_S^{-1} \approx 0.012\, f^{-0.73}$ and $Q_P^{-1} \approx 0.031\, f^{-0.5}$ for 1–32 Hz and the resultant ratio Q_P^{-1}/Q_S^{-1} is larger than 1. Fixing the power of the frequency at 1, since this form simply makes the amplitude attenuation independent of frequency, Sato [1984a] estimated $Q_S^{-1} \approx 0.014\, f^{-1}$ for 2–30 Hz in Kanto.

Attenuation measurements in the upper crust and/or near active faults include

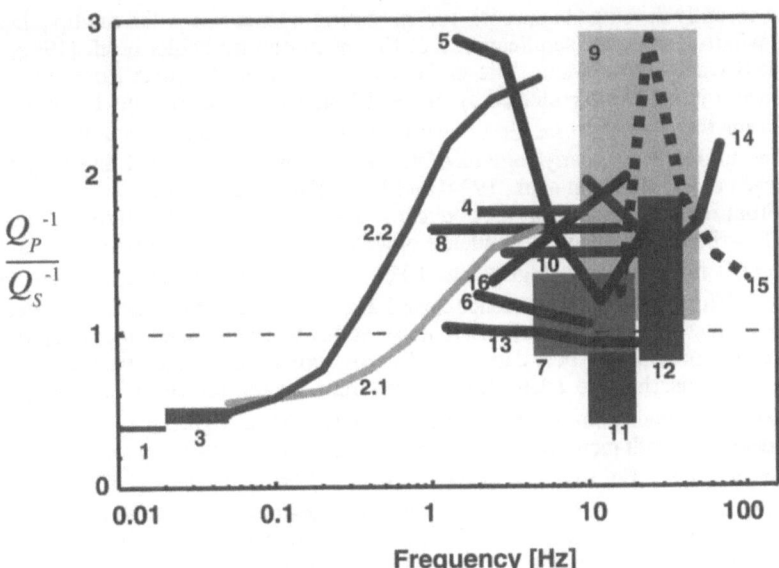

FIGURE 5.3. Ratio Q_P^{-1}/Q_S^{-1} for the lithosphere based on measurements of P- and S-wave attenuation: 1, MM8 global model [Anderson et al., 1965]; 2.1, Basin and Range Province; 2.2, U. S. Shield [Taylor et al., 1986]; 3, depth < 45 km of the SL8 global model [Anderson and Hart, 1978]; 4, Garm, Central Asia [Rautian et al., 1978]; 5, Kanto, Japan [Yoshimoto et al., 1993]; 6, Pg and Lg, France [Campillo and Plantet, 1991]; 7, depth < 40 km, northern Caribbean [Frankel, 1982]; 8, southern Norway [Kvamme and Havskov, 1989]; 9, depth < 10 km, Arette, Pyrénées [Modiano and Hatzfeld, 1982]; 10, upper crust of the Rio Grande Rift, U. S. A. [Carpenter and Sanford, 1985]; 11, depth < 7 km, San Andreas Fault [Bakun et al., 1976]; 12, depth < 7 km, Swabian Jura, Germany [Hoang-Trong, 1983]; 13, southern Kurils [Fedotov and Boldyrev, 1969]; 14, Anza, California [Hough et al., 1988]; 15, upper crust of western Nagano, Japan [Yoshimoto et al., 1994]; 16, western Pacific [Butler et al., 1987].

those made by Kurita [1975], who reported $Q_S^{-1} \approx 0.01 - 0.05$ for 0.3–1 Hz in the shallow crust along the San Andreas Fault. From measurements of SH waves along the Imperial fault in California for depth > 4 km, Singh et al. [1982] reported strong attenuation $Q_S^{-1} \approx 0.05 f^{-1}$ for 3–25 Hz. Modiano and Hatzfeld [1982] analyzed seismograms of microearthquakes having depth < 10 km in the Arette region in the Pyrénées and obtained both Q_S^{-1} and Q_P^{-1} for 10–50 Hz and the resultant ratio $Q_P^{-1}/Q_S^{-1} \approx 1.64$. Carpenter and Sanford [1985] measured apparent attenuation in the upper crustal regions of the central Rio Grande Rift, U. S. A. and found $Q_P^{-1}/Q_S^{-1} \approx 1.5$ for 3–30 Hz. Yoshimoto et al. [1994] applied the extended coda-normalization method to waveforms from shallow microearthquakes having S-P time less than 1 s in western Nagano, central Japan. They found rather strong attenuation for both Q_S^{-1} and Q_P^{-1} for frequencies from 13 to 128 Hz: $Q_S^{-1} \approx 0.0047 f^{-0.20}$ and $Q_P^{-1} \approx 0.012 f^{-0.3}$.

A recently developed approach for measuring attenuation is the multiple lapse-time window analysis (see Section 7.2) first proposed by Fehler et al. [1992] for whole S-wave seismograms. The method was first applied to data from epicentral distances up to 250 km collected by the NIED network in the Kanto-Tokai region, Japan for the estimation of Q_S^{-1}. Since then, this method has become popular and applied to data from many regions of the world including Hawaii and California [Mayeda et al., 1992; Jin et al., 1994] and Japan [Hoshiba, 1993].

From results presented in the above discussion and Figure 5.1, it seems reasonable to write the frequency dependence of attenuation in the form of a power law as $Q_S^{-1} \propto f^{-n}$ for frequencies higher than 1 Hz. Values of the exponent n range from 0.5 to 1. The frequency dependence near 1 Hz remains poorly understood because seismic measurements are difficult to make at this frequency and, since most of the events that provide the best data have S-wave corner frequencies near 1 Hz, it is difficult to discriminate attenuation effects from source effects. Even though P-wave data are poor compared to S-wave data, results in Figure 5.2 show that Q_P^{-1} also decreases with increasing frequency according to a power law for frequencies higher than 1 Hz. For frequencies lower than 0.05 Hz, the ratio Q_P^{-1}/Q_S^{-1} has been taken to be constant at 0.4–0.47 by many investigators. Many have assumed that the ratio for higher frequencies is the same as for low frequencies. However, recent observations have clearly shown that the ratio ranges between 1 and 2, and sometimes up to 3, for frequencies higher than 1 Hz, as shown in Figure 5.3.

We may summarize the characteristics of observed attenuation in the lithosphere as follows: Q_S^{-1} is of the order of 10^{-2} at 1 Hz and decreases to the order of 10^{-3} at 20 Hz. Considering Q_S^{-1} to be of the order of 10^{-3} at 0.01 Hz from surface wave analysis, we may expect that Q_S^{-1} has a peak of the order of 10^{-2} around 0.5 Hz and decays for both increasing and decreasing frequencies as conjectured by Aki [1980a]. The ratio Q_P^{-1}/Q_S^{-1} is smaller than 1 for frequencies lower than 1 Hz; however, the ratio increases and becomes larger than 1 for frequencies higher than 1 Hz.

5.2 INTRINSIC ATTENUATION MECHANISMS

The mechanism of seismic wave attenuation has been a topic of interest among seismologists and rock physicists for many years and numerous physical mechanisms to explain the cause of seismic wave attenuation have been proposed. Seismic attenuation is usually considered to be caused by two mechanisms, scattering and intrinsic mechanisms, so that total attenuation is the sum of the two types:

$$Q_P^{-1} = {}^{Sc}Q_P^{-1} + {}^{I}Q_P^{-1} \text{ and } Q_S^{-1} = {}^{Sc}Q_S^{-1} + {}^{I}Q_S^{-1} \tag{5.3}$$

As we have seen, scattering redistributes wave energy within the medium but does not remove energy from the overall wavefield. Conversely, intrinsic attenuation refers to various mechanisms that convert vibration energy into heat through friction, viscosity, and thermal relaxation processes. Measurements of attenuation of direct seismic waves give values for total attenuation. There has been considerable speculation about which process, intrinsic or scattering, dominates attenuation and several methods have been proposed to determine the amounts of both scattering and intrinsic attenuation [Jacobson, 1987; Fehler et al., 1992].

Models of seismic attenuation were initially developed to explain an apparently observed frequency-independence of Q^{-1} at low frequencies. There are several review papers that discuss proposed mechanisms for intrinsic attenuation that lead to frequency-independent Q_P^{-1} and Q_S^{-1} [Knopoff, 1964; Jackson and Anderson, 1970; Dziewonski, 1979]. Attenuation has been considered an important parameter to measure and characterize in sedimentary rocks for petroleum exploration and this has led to an effort to develop models explaining the observed attenuation in sedimentary rocks [Mavko et al., 1979; Toksöz and Johnston, 1981]. Many proposed intrinsic attenuation models are relaxation mechanisms having characteristic relaxation times that depend on the physical dimensions of the elements in the rock. The characteristic time leads to a Q^{-1} that peaks at some frequency and decreases rapidly away from that frequency. By assuming that rocks are composed of elements with a range of dimensions, the attenuation caused by the mechanism can be made frequency-independent over some frequency range. For seismic waves to remain causal in the presence of attenuation, there must be frequency-dependent amplitude and phase changes [Aki and Richards, 1980, p 173]. The relationship between frequency-dependent attenuation and velocity dispersion was discussed by Liu et al. [1976].

Although we will not describe all proposed mechanisms of intrinsic attenuation, we will briefly examine some and give their predicted relation between physical dimensions and characteristic frequencies. Our discussion follows closely that of Aki [1980a]. Many of the papers we will refer to have been reprinted in Toksöz and Johnston [1981]. Models whose characteristic frequencies are well removed from the frequency band of observed regional seismic phases cannot be considered as the dominant attenuation mechanisms in that band. Any viable model must be consistent with the observed and partially conjectured frequency-dependence of Q_S^{-1} having a peak on the order of 0.01 around 0.5 Hz.

Many proposed mechanisms of intrinsic attenuation are based on the observation that crustal rocks have microscopic cracks and pores which may contain fluids. These features have dimensions much smaller than the wavelengths of regional seismic phases. As discussed in Section 2.1, these cracks can have a profound influence on the propagation velocity of P- and S-waves through rocks (Figure 2.1). It is well-known that static stress-strain curves of rocks show hysteresis [McCall and Guyer, 1994]. The area enclosed by the hysteretic loop is the energy lost from the elastic field during the stress-strain cycle. McCall and Guyer [1994] discuss the effects of hysteresis in dynamic behavior of rocks and show that attenuation caused by hysteresis is frequency-independent. This model is consistent with the presence of cracks that open, close, and slip during elastic loading. Although they do not model the physical process that causes the hysteresis, they successfully account for the difference between the static and dynamic behavior of rocks, the observed hysteresis in stress-strain measurements, and the nonlinear behavior of rocks.

Crack aspect ratio d, which is the ratio of width to length of a crack, is one of the dominant parameters controlling the frequency-dependence of many attenuation models. Hadley [1976] used a scanning electron microscope to measure crack lengths and aspect ratios of virgin and prestressed samples of Westerly granite. She found crack lengths up to 150 microns and aspect ratios of 10^{-4} to 10^{-1}.

Walsh [1966] proposed frictional sliding on dry surfaces of thin cracks as an intrinsic attenuation mechanism. The frictional model predicts that $^{1}Q^{-1}$ is frequency-independent over the frequency range of regional seismic phases. Walsh [1969] proposed viscous dissipation of energy due to liquid movement through cracks as another attenuation mechanism. This model predicts a peak in attenuation at frequency $d\mu/2\pi\eta$, where μ is the rigidity of the surrounding rock and η is the viscosity of the fluid. If water fills the pores, the viscosity $\eta \approx 10^{-2}$ poise at 20 °C decreases with increasing temperature and increases with increasing pressure [Keenan et al., 1969]. Using $\mu \approx 10^{12}$ g/cm·s^2 for rocks and the range of aspect ratios for rocks found by Hadley [1976], we find that predicted attenuation peaks at 10^9 to 10^{12} Hz for this mechanism. To get a peak frequency at 0.5 Hz as conjectured by Aki [1980a] would require aspect ratios of $d \approx 3 \times 10^{-14}$ which is inconsistent with Hadley's [1976] measurements. Nur [1971] proposed viscous dissipation in a zone of partially molten rock to explain the low velocity/high attenuation zone at the base of the lithosphere. The addition of water reduces the melting temperature of rocks. However, the melting temperature of granite at 15 kb is 600 °C, and it is 800 °C for peridotite [Boettcher, 1977]. At Moho depths beneath Kanto, Japan, the temperature is estimated to be around 200–300 °C [Uyeda and Horai, 1964]. Therefore, it is unlikely that melted rock exists in most regions of the lithosphere.

Biot [1956a, b] analyzed wave propagation in isotropic porous solids where the coupling of motion between the fluid and the solid matrix was considered. He arrives at expressions for attenuation due to the flow of fluids within nonconnecting pores initiated by elastic waves. White [1965, p.131] discusses Biot's models and concludes that the attenuation predicted by this model is extremely small for frequencies less than 100 Hz. He shows that the model includes the loss of elastic energy only through viscous drag on the fluid at the crack walls and that this loss is too small to be consistent with seismic measurements. Mavko and Nur [1979] examined the effect of partial saturation of cracks on attenuation. In their model, fluid movement within cracks is enhanced by the presence of gas bubbles, and predicted attenuation is larger than that in Biot's [1956a, b] models. The partial saturation model has a peak attenuation at frequency $\sqrt{K_f/\rho_f}/2\pi a_f$, where K_f is the fluid bulk modulus, ρ_f the fluid density and a_f the half-length of the fluid drop in the crack: $^IQ^{-1}$ is proportional to $\omega^{3/2}$ for lower frequencies and to $\omega^{-3/2}$ at higher frequencies. For water, $K_f \approx 10^{12}$ g/cm·s^2 and $\rho_f \approx 1$ g/cm^3. For attenuation to peak in the regional seismic frequency band, $a_f \approx 10^5$ cm, which is too large.

As an alternative to considering just the effects of fluid movement within one crack, O'Connell and Budiansky [1977] proposed a model in which fluid moves between closely spaced adjacent cracks. There is a characteristic frequency corresponding to the transition from saturated isolated to saturated isobaric behaviors: $f \approx Kd^3/2\pi\eta$, where K is the bulk modulus of the rock. This frequency is lower than the peak frequency predicted by the Walsh [1969] viscous dissipation model. When $\eta \approx 10^{-2}$ poise for water and $K \approx 10^{12}$ g/cm·s^2, a 0.5 Hz attenuation peak in rock requires aspect ratio $d \approx 10^{-5}$, which is close to a range consistent with Hadley's [1976] measurements. However, numerical simulation by O'Connell and Budiansky [1977] predicts $Q_P^{-1} < Q_S^{-1}$, which contradicts observations discussed in Section 5.1.

After drying an olivine basalt sample in a moderately heated high vacuum, Tittmann [1977] found that Q_P^{-1} decreased from 2×10^{-2} to 0.9×10^{-3} at 56 Hz. This measured low attenuation for a dry rock is consistent with the very low attenuation values measured on lunar rock samples that contain little water [Tittmann et al., 1976]. Gradually adding a small amount of volatile to a dry rock, Tittmann et al. [1980] measured an increase of Q_S^{-1} and a change in electric dipole moment which indicated adsorption of the volatile. They found that the rapid increase of Q_S^{-1} was not due to the classical viscous fluid movement through fractures but due to an interaction between adsorbed water film on the solid surface by thermally activated motions. This is due to relaxation involving liquid molecules. Controlling the amount of water, Spencer [1981] identified individual relaxation peaks in rocks. He found a peak in Q_E^{-1} at frequencies as low as 17 Hz in limestone, where E is Young's modulus. However, the peak frequency is of the order of kHz for other kinds of rocks. He argued that most rocks have a range of relaxation frequencies and that the dominant mechanism of attenuation observed in his measurements is a frequency-dependent softening of the rock due to the bonding of fluid molecules to crack surfaces.

Thermally activated processes at grain boundaries have been proposed as an attenuation mechanism for the upper mantle [Anderson and Hart, 1978; Lundquist and Cormier, 1980]. In polycrystalline rocks, temperature-activated relaxation thought to be stress relaxation at a viscous boundary layer between grains took place [Jackson and Anderson, 1970]. In this case attenuation has a peak whose amplitude increases with increasing temperature. Dislocation motion in rock materials has been proposed as an attenuation mechanism [Mason, 1969]. However, the peak frequency of Q^{-1} is in the MHz range [Mason et al., 1978]. Aki [1980a] excluded dislocation glide as a factor in attenuation at low temperature in the lithosphere.

Spatial temperature differences induced by a passing wave due to adiabatic compression are reduced by thermal diffusion [Zener, 1948; Savage, 1966a]. This thermoelastic effect removes vibrational energy from a wavefield. Grain-sized heterogeneities in a rock increase the amount of predicted attenuation. Thermoelastic attenuation peaks at frequency D_T / a_g^2, where a_g is the grain size and D_T is the thermal diffusivity. For $D_T \approx 5 \times 10^{-2}$ cm^2/s for quartz and a peak frequency of 0.5 Hz, $a_g \approx 0.3$ cm, which is reasonable for rocks. The exchange of heat between adjacent grains plays an important role for Q_P^{-1}. Because of rock heterogeneity, thermoelasticity causes S-wave attenuation, but it causes more attenuation for compressional waves. Therefore, the model predicts $Q_P^{-1} > Q_S^{-1}$. Savage [1966b] investigated thermoelasticity caused by stress concentrations induced by the presence of empty cracks having the shape of elliptic cylinders. This model predicts a peak in attenuation at a frequency given by D_T / a_C^2, where a_C is the half-length of the crack, which yields crack sizes similar to grain sizes predicted by Zener's [1948] model. For ordinary materials containing cracks, the theory predicts $Q_P^{-1} > Q_S^{-1}$, which is consistent with the measurements illustrated in Figure 5.3.

Most of the mechanisms discussed above can predict Q_S^{-1} having values in the range of 10^{-3}; however, the importance of various mechanisms varies with depth, temperature, fracture content, fracture aspect ratios, pressure, and the presence of fluids. Aki [1980a] preferred thermoelasticity as the most viable model to explain intrinsic attenuation at lithospheric temperatures because the required scales for rock grains and cracks along with the amount of attenuation caused by thermoelasticity are in closest agreement with observations.

5.3 SCATTERING ATTENUATION DUE TO DISTRIBUTED RANDOM INHOMOGENEITIES

A first step in making a model of attenuation is to determine whether it is controlled by some characteristic scale in time or space. In Figures 5.1–5.3 we took frequency as the abscissa, which allows us to look at characteristic time scales. Choosing wavenumber as the abscissa allows us to investigate the spatial scale of attenuation. Figure 5.4 shows direct wave attenuation Q_S^{-1} and Q_P^{-1} in Kanto, Japan, measured using an extension of the coda-normalization method described in Section 3.4.3 [Yoshimoto et al., 1993], plotted against wavenumber, where frequency 0.5 Hz corresponds to S-wave wavenumber of 0.8 km^{-1}. The results show good coincidence between Q_S^{-1} and Q_P^{-1}. This coincidence implicitly suggests that attenuation is characterized by a spatial scale. As described in Section 5.1 and Figure 3.32b, attenuation for S-waves Q_S^{-1} is conjectured to have a peak of amplitude about 10^{-2} at about 0.5 Hz and to decrease for both increasing and decreasing frequency away from 0.5 Hz. Figure 5.4 shows that attenuation per travel distance

$2\pi f Q_S^{-1}/\beta_0 = l\,Q_S^{-1}$ is approximately constant for 1 to 20 Hz and has a value on the order of 10^{-2} km^{-1}, which is nearly the same order as the total scattering coefficient g_0 of S-waves as shown in Figure 3.10. The coincidence between g_0 and $l\,Q_S^{-1}$ leads to the idea that scattering attenuation may be the dominant mechanism for amplitude attenuation of seismic waves in the lithosphere [Aki, 1980a, 1981, 1982]. As schematically illustrated in Figure 2.28, we may expect that scattering attenuates direct wave amplitude and excites coda waves. However, we will show that the ordinary derivation of amplitude attenuation using the Born approximation to estimate scattering attenuation leads to a prediction that Q_S^{-1} increases with frequency, as shown in Figure 5.5. There have been two attempts to resolve the discrepancy between observations that Q_S^{-1} decreases with frequency above 0.5 Hz and the scattering theory, which predicts that Q_S^{-1} increases with frequency. One improves the statistical averaging procedure by isolating the effect of the travel-time fluctuation caused by slowly varying velocity fluctuation from other scattering phenomena that are caused by more rapidly varying velocity inhomogeneities [Sato, 1982a, b]; the second attempt neglects scattering in the forward direction during calculation of the attenuation [Wu, 1982a]. Using scalar wave propagation as an example, we will demonstrate the discrepancy between attenuation observations and the theory based on the ordinary Born approximation and show how the two proposals to resolve the discrepancy are implemented. We will show that the two approaches are equivalent, and then we will extend the analysis to elastic waves.

5.3.1 Use of the Born Approximation for Estimating Scattering Attenuation of Scalar Waves

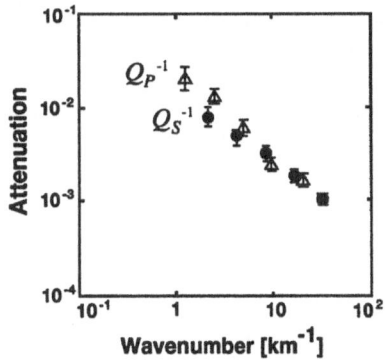

FIGURE 5.4. Q_s^{-1} (closed circle) and Q_p^{-1} (triangle) vs. wavenumber measured in Kanto, Japan. [From Yoshimoto et al., 1993, with permission from Blackwell Science, United Kingdom.]

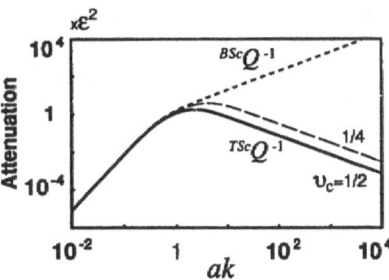

FIGURE 5.5. Scattering attenuation vs. normalized wavenumber for scalar waves: dotted, ordinary Born approximation; solid (υ_c=1/2) and broken (υ_c=1/4), travel-time corrected Born approximation, where $k = \omega / V_0$.

Here, we study the scattering attenuation of scalar waves that travel through randomly inhomogeneous media. Using an ensemble of random media whose spatial velocity distributions are described by $\{\xi(\mathbf{x})\}$, where $\langle\xi\rangle = 0$, we can calculate the statistical scattering coefficient g from the PSDF using the Born approximation derived in Section 4.1.2. The integral over the solid angle of the average of the square scattering amplitude over an ensemble of inhomogeneous media is identified as the scattering energy loss from the incident plane wave. Taking the ensemble average of (4.17) with (4.21), the scattered wave energy generated per unit time by a cube of inhomogeneity having volume L^3 is given by $\rho_0\omega^2 V_0 \oint d\Omega r^2 \left(gL^3 / 4\pi r^2\right) = \rho_0\omega^2 V_0 g_0 L^3$, where the incident energy-flux having unit amplitude passing through an area L^2 is $\rho_0\omega^2 V_0 L^2$. The fractional scattering attenuation of the incident-wave energy per unit travel distance is thus equal to g_0. Dividing g_0 by k, the scattering attenuation based on the ordinary Born approximation [Aki and Richards, p. 742; Chernov, 1960] is given by

$$^{BSc}Q^{-1}(\omega) \equiv \frac{1}{k}g_0 \equiv \frac{1}{4\pi k}\oint g\,d\Omega = \frac{1}{k}\oint \frac{1}{L^3}\left\langle\frac{d\sigma}{d\Omega}\right\rangle d\Omega$$

$$= \frac{1}{k}\oint \frac{1}{L^3}\left\langle|F|^2\right\rangle d\Omega = \frac{k^3}{4\pi^2}\oint P\left(2k\sin\frac{\psi}{2}\right)d\Omega(\psi,\zeta) \qquad (5.4)$$

$$= \frac{k^3}{2\pi}\int_0^\pi P\left(2k\sin\frac{\psi}{2}\right)\sin\psi\,d\psi$$

where scattering is axially symmetric, ψ is the scattering angle, and the prefix "BSc" explicitly means the attenuation due to scattering by distributed random inhomogeneities based on the ordinary Born approximation. The resultant representation of scattering attenuation is independent of dimension L.

In the case where the exponential ACF describes the random media, substituting (2.10) for the PSDF, we may write the above integral as

$$^{BSc}Q^{-1}(\omega) = 4\varepsilon^2 a^3 k^3 \int_0^\pi \frac{2\sin\frac{\psi}{2}\cos\frac{\psi}{2}}{\left(1+4a^2k^2\sin^2\frac{\psi}{2}\right)^2}d\psi = 4\varepsilon^2 a^3 k^3 \int_0^2 \frac{\upsilon}{\left(1+a^2k^2\upsilon^2\right)^2}d\upsilon$$

$$(5.5)$$

$$= \frac{-2\varepsilon^2 ak}{1+a^2k^2\upsilon^2}\bigg|_0^2 = \frac{8\varepsilon^2 a^3 k^3}{1+4a^2k^2} \approx \begin{cases} 8\varepsilon^2 a^3 k^3 & \text{for } ak \ll 1 \\ 2\varepsilon^2 ak & \text{for } ak \gg 1 \end{cases}$$

where $\upsilon = 2\sin(\psi/2)$. The scattering attenuation is proportional to the mean square (MS) fractional fluctuation of velocity. The dotted curve in Figure 5.5 shows the predicted scattering attenuation against ak. It is proportional to the cube of the wavenumber or frequency for low frequencies, and it increases linearly with frequency for high frequencies. Even if the MS fractional fluctuation is small, (5.5) predicts a larger attenuation for large ak compared to small ak which does not agree with observations like those shown in Figure 5.4. The large theoretically predicted scattering attenuation for high frequencies is caused by strong forward scattering. As discussed in Section 4.1, the Born approximation is valid only when the energy loss per distance L is small [Aki and Richards, 1980, p. 742; Hudson and Heritage, 1981]: $^{BSc}Q^{-1}kL \ll 1$. Replacing L with a, we get the least restrictive condition for the applicability of the Born approximation as

$$^{BSc}Q^{-1}ak \ll 1 \qquad (5.6)$$

To better understand the effects of the slowly varying velocity inhomogeneity on the prediction of scattering attenuation, consider an ensemble of wave propagation experiments through 1-D random media whose wave velocities vary slowly. The experiments are done for high frequencies so we choose a wavelength for the incident seismic wave that is much shorter than the scale length of the velocity inhomogeneity. Figure 5.6a is a schematic diagram showing the time traces (bold curves) u obtained from these experiments for different random media. We expect good resemblance in waveform between differing traces; however, first arrival travel-times are expected to vary considerably from trace to trace. The bottom trace is the average over the ensemble of the traces corresponding to the mean wavefield $\langle u \rangle$. It differs greatly from all measured traces because of travel-time fluctuations. The amplitude of the ensemble average trace is much smaller than that of individual traces. The wave trace next to the bottom shows the wave trace u^0 in the homogenous medium. Each fine broken curve in Figure 5.6a is the difference between the measured (bold) trace and the wave trace (next to the bottom) in the homogenous

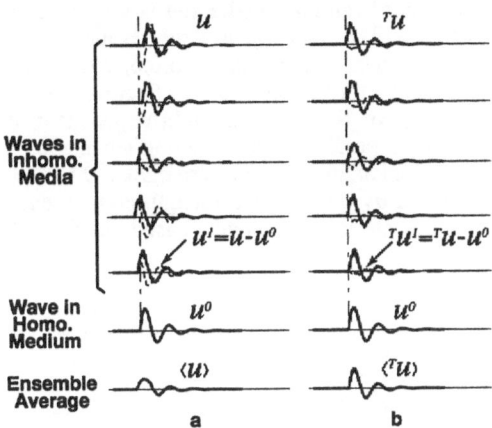

FIGURE 5.6. Time traces (bold curves) after traveling through random media containing inhomogeneities with wavelengths longer than the wavelength of the propagating wave. Fine broken traces show the differences from the time trace in the homogenous medium u^0, which is shown next to the bottom. The bottom trace is the ensemble average: **(a)** raw traces; **(b)** travel-time corrected traces.

medium, which corresponds to scattered waves $u^1 = u - u^0$. The ensemble average of the square of the fine broken traces (not shown), which is used to predict scattering attenuation by the ordinary Born approximation, is large. A blind application of the Born approximation thus predicts a large attenuation because of the relative travel-time shift due to the long wavelength structure. Thus, we find a link between the predicted large attenuation caused by large forward scattering for high frequencies and the travel-time fluctuation caused by the velocity inhomogeneity.

Seismological attenuation measurements are done by measuring amplitudes of pulse-like direct waves irrespective of travel times because travel-time fluctuations are unobservable on individual seismograms. Ignoring travel-time fluctuations is similar to correcting for them, so that waveforms $^T u$ appear aligned, as illustrated in Figure 5.6b. The ensemble average trace after travel-time correction $\langle ^T u \rangle$ is shown as the bottom trace in Figure 5.6b. The difference between each observed trace and the wave trace in the homogenous medium, $^T u^1 = ^T u - u^0$, has a small amplitude, as shown by a fine broken curve. Since the difference is small, we find that predicted scattering attenuation is small. We may say that the amplitude decay of $\langle ^T u \rangle$ corresponds to the conventional attenuation measurement [Sato, 1982a, b, 1984a, b].

The stochastic treatment of wave propagation through random media has been extensively studied using the mean wavefield theory and the smooth perturbation method [Karal and Keller, 1964; Frisch, 1968; Beaudet, 1970; Howe, 1971; Sato, 1979]. Wu [1982b] showed that the mean wavefield $\langle u \rangle$ decays exponentially due to a loss of coherency at a more rapid rate than predicted by point measurements made for a single realization of the random medium. Wu [1982a, b] pointed out that the predicted attenuation of the mean wavefield is related only to the statistical treatment of the ensemble of random media and is unrelated to attenuation measurements in a real medium. The relationship between the stochastic averaging procedure and the attenuation measurement in seismology has been made clear based on these studies. This will be further discussed in Section 5.3.2.

We can use the results of the above thought experiment to modify the Born scattering theory to make a prediction of scattering attenuation consistent with the manner in which seismological observations are made. Our approach is to subtract the travel-time shift caused by the long wavelength component of velocity fluctuation and then calculate scattering amplitude based on the Born approximation.

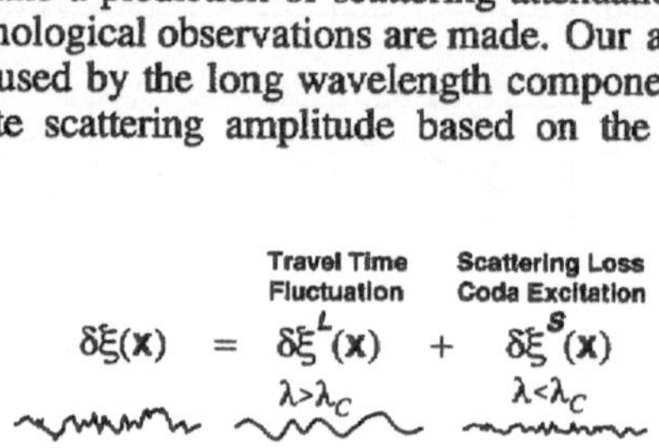

FIGURE 5.7. Decomposition of the fractional fluctuation of wave velocity into two components in the case of cutoff-wavelength $\lambda_c = 2\lambda_w$, where λ_w is the dominant wavelength.

Scattering attenuation will then be given by an ensemble average of the integral over the solid angle of the square of travel-time corrected scattering amplitude.

For an incident wave of dominant wavelength λ_w, we first decompose the fractional fluctuation of wave velocity $\xi(\mathbf{x})$ into long-wavelength and short-wavelength components by choosing a cutoff wavelength $\lambda_c = \lambda_w / \upsilon_c$:

$$\xi(\mathbf{x}) = \xi^L(\mathbf{x}) + \xi^s(\mathbf{x}) \tag{5.7}$$

Figure 5.7 shows the concept of decomposition in the case of $\upsilon_c = 1/2$. The decomposition is accomplished using the Fourier transform:

$$\xi^L(\mathbf{x}) \equiv \frac{1}{(2\pi)^3} \int\int\int_{-\infty}^{\infty} H(\upsilon_c k - m)\tilde{\xi}(\mathbf{m})\; e^{i\mathbf{m}\mathbf{x}} d\mathbf{m}$$

$$\xi^s(\mathbf{x}) \equiv \frac{1}{(2\pi)^3} \int\int\int_{-\infty}^{\infty} H(m - \upsilon_c k)\tilde{\xi}(\mathbf{m})\; e^{i\mathbf{m}\mathbf{x}} d\mathbf{m} \tag{5.8}$$

where $k = 2\pi/\lambda_w$ is the wavenumber of the incident wave and $\upsilon_c k$ is the cutoff wavenumber for the velocity fluctuation. The corresponding power spectral densities are given by

$$P^L(m) = P(m)H(\upsilon_c k - m) \quad \text{and} \quad P^s(m) = P(m)H(m - \upsilon_c k) \tag{5.9}$$

Then the long-wavelength component of velocity fluctuation causes a travel-time fluctuation whose size is given by a 1-D integral along the incident ray path. For plane wave incident along the third axis,

$$\delta t(\mathbf{x}) = \int_{\text{Ray}}^{x_3} \left[\frac{1}{V_0} - \frac{1}{V(\mathbf{x}')} \right]_{\substack{\text{Long-}\\\text{Wavelength}\\\text{Component}}} dx_3' \approx \frac{1}{V_0} \int_{\text{Ray}}^{x_3} \xi^L(\mathbf{x}')\, dx_3' \tag{5.10}$$

In differential form,

$$\partial_i \delta t = \frac{1}{V_0}\delta_{i3}\,\xi^L \quad \text{and} \quad \partial_i \partial_j \delta t = 0 \tag{5.11}$$

where the latter condition is added [Yoshimoto, 1995] to ensure that the travel-time correction term is locally constant on a plane normal to the incident ray and its second derivative with respect to the propagation direction is also zero since the spatial variation of the long-wavelength component is small. Subtracting the travel-time fluctuation $\delta t(\mathbf{x})$, we can define the travel-time corrected wavefield as

$$u(\mathbf{x},t) = {}^T u[\mathbf{x}, t + \delta t(\mathbf{x})] \tag{5.12}$$

where prefix "T" denotes the travel-time correction. Substituting (5.12) in (4.4) and neglecting second-order quantities, we get the wave equation

$$\left(\Delta - \frac{1}{V_0^2}\partial_t^2\right){}^T u = -\left(\frac{2}{V_0^2}\xi\partial_t^2 + 2\partial_i\delta t \cdot \partial_i\partial_t + \Delta\delta t \cdot \partial_t\right){}^T u \qquad (5.13)$$

We decompose the wavefield into the incident plane wave u^0 which satisfies the homogenous wave equation (4.6) and the first-order perturbation term ${}^T u^1$:

$$ {}^T u = u^0 + {}^T u^1 \qquad (5.14)$$

where $|{}^T u^1| << |u^0|$. Substituting $u^0(\mathbf{x},t) = e^{i(k\mathbf{e}_3\mathbf{x}-\omega t)}$ as the incident wave propagating to the third direction in (5.13) and using (5.11),

$$\left(\Delta - \frac{1}{V_0^2}\partial_t^2\right){}^T u^1 = 2k^2\xi^s(\mathbf{x})e^{i(k\mathbf{e}_3\mathbf{x}-\omega t)} \qquad (5.15)$$

where $\omega = V_0 k$. Accounting for the travel-time correction gives the result that waves are scattered only by the short-wavelength components of the inhomogeneity. As done in Section 4.1, we solve (5.15) under the condition that the inhomogeneity is localized in a volume having dimension L around the origin, where $L > a$. Using the retarded Green function given by (4.10) and following a procedure like the one that leads to (4.13), spherically outgoing scattered waves in the far field are given by

$$ {}^T u^1(\mathbf{x},t) = \frac{e^{i(kr-\omega t)}}{r}\left(\frac{-k^2}{2\pi}\right)\tilde{\xi}^s(k\mathbf{e}_r - k\mathbf{e}_3) = \frac{e^{i(kr-\omega t)}}{r}{}^T F \qquad (5.16)$$

The longer wavelength component $\xi^L(\mathbf{x})$ causes travel-time fluctuation; scattering due to shorter wavelength component $\xi^s(\mathbf{x})$ excites coda waves. The travel-time corrected scattering amplitude can now be written using the Fourier transform of the short-wavelength component of the fractional fluctuation:

$$ {}^T F = \left(\frac{-k^2}{2\pi}\right)\tilde{\xi}^s(k\mathbf{e}_r - k\mathbf{e}_3) \qquad (5.17)$$

Substituting (5.17) in (4.20), we get the travel-time corrected scattering coefficient

$$\begin{aligned}
{}^T g(\psi,\zeta;\omega) &\equiv \frac{4\pi}{L^3}\langle|{}^T F|^2\rangle = \frac{k^4}{\pi}P^s(k\mathbf{e}_r - k\mathbf{e}_3) \\
&= \frac{k^4}{\pi}P(k\mathbf{e}_r - k\mathbf{e}_3)H(|k\mathbf{e}_r - k\mathbf{e}_3| - \upsilon_c k) \\
&= \frac{k^4}{\pi}P\left(2k\sin\frac{\psi}{2}\right)H(\psi - \psi_c)
\end{aligned} \qquad (5.18)$$

where ψ_C is the cutoff scattering angle corresponding to the cutoff wavenumber in (5.8):

$$\psi_C \equiv 2\sin^{-1}\frac{\upsilon_C}{2} \tag{5.19}$$

Integrating (5.18) over the solid angle as in (5.4), we get the scattering attenuation as

$$^{TSc}Q^{-1}(\omega) = \frac{1}{4\pi k}\oint {}^Tg\,d\Omega = \frac{k^3}{2\pi}\int_0^{\pi} P^S\left(2k\sin\frac{\psi}{2}\right)\sin\psi\,d\psi$$

$$= \frac{k^3}{2\pi}\int_{\psi_C}^{\pi} P\left(2k\sin\frac{\psi}{2}\right)\sin\psi\,d\psi = \frac{k^3}{2\pi}\int_{\upsilon_C}^{2} P(k\upsilon)\upsilon\,d\upsilon \tag{5.20}$$

where prefix "*TSc*" denotes scattering attenuation based on the travel-time corrected Born approximation. Adjusting for the travel-time fluctuation resulted in the introduction of a lower bound for the integral in (5.4) that reduces the scattering attenuation for large wavenumbers, that is, the travel-time correction is equivalent to neglecting the contribution of large forward scattering within a cutoff scattering angle when calculating scattering attenuation.

When the ACF is exponential,

$$^{TSc}Q^{-1}(\omega) = 4\varepsilon^2 a^3 k^3 \int_{\upsilon_C}^{2} \frac{\upsilon}{\left(1+a^2k^2\upsilon^2\right)^2}d\upsilon = \frac{2\varepsilon^2 a^3 k^3\left(4-\upsilon_C^2\right)}{\left(1+\upsilon_C^2 a^2 k^2\right)\left(1+4a^2k^2\right)}$$

$$\approx \begin{cases} 2\left(4-\upsilon_C^2\right)\varepsilon^2 a^3 k^3 & \text{for } ak \ll 1 \\ \dfrac{\left(4-\upsilon_C^2\right)}{2\upsilon_C^2}\dfrac{\varepsilon^2}{ak} & \text{for } ak \gg 1 \end{cases} \tag{5.21}$$

The resultant scattering attenuation decreases with increasing frequency for high frequencies. We plot the travel-time corrected scattering attenuation for $\upsilon_C=1/2$ (solid) and 1/4 (broken) in Figure 5.5. As υ_C decreases, the travel-time correction becomes weaker and scattering attenuation increases, particularly for larger wavenumbers. The minimum fractional velocity fluctuation cutoff wavelength λ_C, for which a wave having wavelength λ_w will have the same sign of travel-time fluctuation over its wavelength, is $\lambda_C = 2\lambda_w$. This corresponds to $\upsilon_C=1/2$ and $\psi_C \equiv 2\sin^{-1}\left(\upsilon_C/2\right) \approx 29°$ [Sato, 1982a, b]. We will use this value in the following. Then, (5.21) becomes

$$^{TSc}Q^{-1}(\omega) = \frac{15\varepsilon^2 a^3 k^3}{2\left(1+\frac{1}{4}a^2k^2\right)\left(1+4a^2k^2\right)} \approx \begin{cases} \dfrac{15}{2}\varepsilon^2 a^3 k^3 & \text{for } ak \ll 1 \\ \dfrac{15}{2}\dfrac{\varepsilon^2}{ak} & \text{for } ak \gg 1 \end{cases} \tag{5.22}$$

and

$$^{TSc}Q^{-1}{}_{Max} \approx 1.8\varepsilon^2 \quad \text{at} \quad ak \approx 2.2 \tag{5.23}$$

Thus, correcting for the travel-time fluctuation, we get scattering attenuation that has a peak whose amplitude is of the order of the MS fractional fluctuation and that decreases with the reciprocal of wavenumber for large wavenumbers. The resultant scattering attenuation satisfies condition (5.6) $^{TSc}Q^{-1}ak \ll 1$ for high frequencies if $\varepsilon^2 \ll 1$. For the calculation of attenuation, Chernov [1960, p.56] proposed to integrate outside of angle $\psi_C = 1/ak$, by arguing that forward scattering causes only phase fluctuations. Wu [1982a] proposed a method to calculate the scattering attenuation specifying $\psi_C = 90°$ in (5.20) by arguing that this is the back-scattered energy, which is lost, and that forward scattered energy is not lost. Wu's [1982a] proposal corresponds to $\upsilon_c = \sqrt{2}$, which gives a smaller peak attenuation for the same fractional velocity fluctuation. Dainty [1984] and Menke [1984b] discussed the general relationship between the spectra of inhomogeneity and corresponding scattering attenuation in the frequency domain.

Taking the third axis as the incident ray direction and writing $\mathbf{x} = (\mathbf{x}_\perp, z)$ in (5.10), we get the MS travel-time fluctuation for travel distance $Z \gg a$ as

$$\left\langle \delta t(Z)^2 \right\rangle = \frac{1}{V_0^2} \int_0^Z \int_0^Z \left\langle \xi^L(\mathbf{x}_\perp' = 0, z') \xi^L(\mathbf{x}_\perp'' = 0, z'') \right\rangle dz' \, dz''$$

$$= \frac{1}{V_0^2} \int_0^Z dz_d \int_{\frac{z_d}{2}}^{z - \frac{z_d}{2}} dz_c R^L(\mathbf{x}_\perp = 0, z_d) + \frac{1}{V_0^2} \int_{-z}^0 dz_d \int_{-\frac{z_d}{2}}^{z + \frac{z_d}{2}} dz_c R^L(\mathbf{x}_\perp = 0, z_d)$$

$$= \frac{2}{V_0^2} \int_0^Z dz_d R^L(\mathbf{x}_\perp = 0, z_d)(Z - z_d) \approx \frac{Z}{V_0^2} \int_{-\infty}^{\infty} dz_d R^L(\mathbf{x}_\perp = 0, z_d)$$

$$= \frac{Z}{2\pi V_0^2} \int_0^{\infty} P^L(\mathbf{m}_\perp, m_z = 0) m_\perp dm_\perp = \frac{Z}{2\pi V_0^2} \int_0^{\upsilon_c k} P(m) m \, dm \tag{5.24}$$

where $z' = z_c + z_d/2$ and $z'' = z_c - z_d/2$. This travel-time fluctuation does not contain diffraction effects. For the exponential ACF,

$$\left\langle \delta t(Z)^2 \right\rangle = \frac{Z}{V_0^2} \frac{1}{2\pi} \int_0^{\upsilon_c k} \frac{8\pi \varepsilon^2 a^3}{\left(1 + a^2 m^2\right)^2} m \, dm = \frac{2\varepsilon^2 aZ}{V_0^2} \frac{a^2 k^2 \upsilon_C^2}{1 + a^2 k^2 \upsilon_C^2}$$

$$\approx \frac{2\varepsilon^2 a}{V_0^2} Z \qquad \text{for} \quad ak \gg 1 \tag{5.25}$$

where we used (2.10). For the Gaussian ACF,

$$\langle \delta t(Z)^2 \rangle = \frac{Z}{V_0^2} \frac{1}{2\pi} \int_0^{v_c k} \varepsilon^2 a^3 \sqrt{\pi^3} e^{-a^2 m^2/4} m \, dm = \frac{\sqrt{\pi} \varepsilon^2 a Z}{V_0^2} \left(1 - e^{-v_c^2 a^2 k^2/4} \right)$$

$$\approx \frac{\sqrt{\pi} \varepsilon^2 a Z}{V_0^2} \quad \text{for} \quad ak \gg 1 \tag{5.26}$$

where we used (2.8). The MS travel-time fluctuation increases with increasing travel distance. The travel-time fluctuation will be discussed in relation to the phase fluctuation in the parabolic approximation in Chapter 8.

5.3.2 Use of the Born Approximation for Estimating Scattering Attenuation of Elastic Vector Waves

Following the procedure described in Section 5.3.1 for scalar waves, we will now describe the procedure to correct for travel-time fluctuation due to long wavelength velocity structure to estimate scattering attenuation for vector waves that is consistent with seismological observation methods. We define the travel-time corrected vector wavefield $^T\mathbf{u}$, which is related to vector wavefield $\mathbf{u}$ as

$$\mathbf{u}(\mathbf{x}, t) = {}^T\mathbf{u}\left[\mathbf{x}, t + \delta t(\mathbf{x})\right] \tag{5.27}$$

where travel-time fluctuation $\delta t = \delta t^P$ or δt^S for incident P- or S-waves propagating to the third direction, respectively. Substituting (5.27) in (4.27), we get the wave equation for $^T\mathbf{u}$ as

$$
\begin{aligned}
\rho_0 {}^T\ddot{u}_i = & \left[\lambda_0 \partial_i \partial_j {}^T u_j + \mu_0 \partial_j \left(\partial_i {}^T u_j + \partial_j {}^T u_i \right) \right] \\
& - \delta\rho \, {}^T\ddot{u}_i + \partial_i \delta\lambda \partial_j {}^T u_j + \partial_j \delta\mu \left(\partial_i {}^T u_j + \partial_j {}^T u_i \right) \\
& + \delta\lambda \partial_i \partial_j {}^T u_j + \delta\mu \partial_j \left(\partial_i {}^T u_j + \partial_j {}^T u_i \right) \\
& + \left(\lambda_0 + \mu_0 \right) \left(\partial_i \delta t \cdot \partial_j {}^T \dot{u}_j + \partial_j \delta t \cdot \partial_i {}^T \dot{u}_j \right) + 2\mu_0 \partial_j \delta t \cdot \partial_j {}^T \dot{u}_i \\
& + \left(\lambda_0 + \mu_0 \right) \partial_i \partial_j \delta t \cdot {}^T \dot{u}_j + \mu_0 \Delta \delta t \cdot {}^T \dot{u}_i
\end{aligned}
\tag{5.28}
$$

where terms of the second power of δt or higher order and cross terms of δt and fluctuations of elastic coefficients are neglected. We solve (5.28) using the first-order perturbation method. We decompose the vector wave into the incident plane wave that satisfies the homogeneous equation (4.33) and the scattered wave having small amplitude:

$$^T\mathbf{u} = \mathbf{u}^0 + {}^T\mathbf{u}^1 \tag{5.29}$$

where $\left|{}^T\mathbf{u}^1\right| << \left|\mathbf{u}^0\right|$. The perturbation term satisfies

$$\rho_0\ {}^T\ddot{u}_i^{\ 1} - \partial_j T_{ij}\left(\lambda_0, \mu_0;\ {}^T u_i^{\ 1}\right) = \delta f_i(\mathbf{x},t) + {}^C \delta f_i(\mathbf{x},t) \tag{5.30}$$

where T_{ij} is the stress tensor defined by (4.28), the first term on the right-hand side is the equivalent body force due to the inhomogeneity given by (4.35), and the second term is the equivalent body force corresponding to the travel-time correction:

$$\begin{aligned}
{}^C\delta f_i(\mathbf{x},t) = &\left(\lambda_0 + \mu_0\right)\!\left(\partial_i \delta t \cdot \partial_j \dot{u}_j^0 + \partial_j \delta t \cdot \partial_i \dot{u}_j^0\right) + 2\mu_0 \partial_j \delta t \cdot \partial_j \dot{u}_i^0 \\
&+ \left(\lambda_0 + \mu_0\right)\partial_i \partial_j \delta t \cdot \dot{u}_j^0 + \mu_0 \Delta \delta t \cdot \dot{u}_i^0
\end{aligned} \tag{5.31}$$

We first decompose the fluctuation of P-wave velocity following (5.7):

$$\delta\alpha(\mathbf{x}) = \delta\alpha^L(\mathbf{x}) + \delta\alpha^S(\mathbf{x}) \tag{5.32}$$

where

$$\begin{aligned}
\delta\alpha^L(\mathbf{x}) &\equiv \frac{1}{(2\pi)^3} \int\!\!\!\int\!\!\!\int_{-\infty}^{\ \ \infty\ \infty\ \infty} H(\upsilon_c k - m)\delta\tilde{\alpha}(\mathbf{m})\ e^{im\mathbf{x}}dm \\
\delta\alpha^S(\mathbf{x}) &\equiv \frac{1}{(2\pi)^3} \int\!\!\!\int\!\!\!\int_{-\infty}^{\ \ \infty\ \infty\ \infty} H(m - \upsilon_c k)\delta\tilde{\alpha}(\mathbf{m})\ e^{im\mathbf{x}}dm
\end{aligned} \tag{5.33}$$

where $\upsilon_c k$ is the cutoff wavenumber for a given angular frequency. For an incident plane P-wave propagating along the third axis, the travel-time fluctuation satisfies (5.11):

$$\partial_i \delta t^P = \delta_{i3} \frac{\delta\alpha^L}{\alpha_0^{\ 2}} \quad \text{and} \quad \partial_i \partial_j \delta t^P = 0 \tag{5.34}$$

The latter condition is according to Yoshimoto et al. [1997a] as discussed in relation to (5.11). This condition allows us to neglect the second line of eq. (5.31). Combining (5.31) and (5.34) for an incident plane P-wave of unit amplitude propagating along the third direction (4.36),

$$^C\delta f_i^P(\mathbf{x},t) = 2k\omega\rho_0 \delta\alpha^L(\mathbf{x})\delta_{i3} e^{i(k e_3 \mathbf{x} - \omega t)} \tag{5.35}$$

Solving (5.30) for body forces (5.35) and (4.37), we get scattered waves as outgoing spherical waves from the inhomogeneity. Then, the travel-time corrected PP-scattering amplitude having prefix "T" is given as a sum of terms given by (4.46) and correction terms:

$$^T F^{PP}_{r,\psi,\zeta} = F^{PP}_{r,\psi,\zeta} + {}^C F^{PP}_{r,\psi,\zeta} \tag{5.36}$$

where the correction terms having prefix "C" are

$$
\begin{aligned}
{}^{C}F_r{}^{PP} &= \frac{l^2}{4\pi}\frac{2}{\gamma_0{}^2}\cos\psi\,\frac{\delta\tilde{\alpha}^{L}(ke_r - ke_3)}{\alpha_0} \\
&= \frac{l^2}{4\pi}\left[\frac{2}{\gamma_0{}^2}\cos\psi\, H(\psi_C - \psi)\frac{\delta\tilde{\alpha}(ke_r - ke_3)}{\alpha_0}\right] \\
{}^{C}F_{\mathbf{v}}{}^{PP} &= {}^{C}F_{\zeta}{}^{PP} = 0
\end{aligned}
\tag{5.37}
$$

In the same manner, the fluctuation of S-wave velocity is decomposed to

$$
\delta\beta(\mathbf{x}) = \delta\beta^{L}(\mathbf{x}) + \delta\beta^{S}(\mathbf{x})
\tag{5.38}
$$

where

$$
\begin{aligned}
\delta\beta^{L}(\mathbf{x}) &\equiv \frac{1}{(2\pi)^3}\int_{-\infty}^{\infty}\int_{-\infty}^{\infty}\int_{-\infty}^{\infty} H(\upsilon_c l - m)\,\delta\tilde{\beta}(\mathbf{m})\,e^{im\mathbf{x}}\,d\mathbf{m} \\
\delta\beta^{S}(\mathbf{x}) &\equiv \frac{1}{(2\pi)^3}\int_{-\infty}^{\infty}\int_{-\infty}^{\infty}\int_{-\infty}^{\infty} H(m - \upsilon_c l)\,\delta\tilde{\beta}(\mathbf{m})\,e^{im\mathbf{x}}\,d\mathbf{m}
\end{aligned}
\tag{5.39}
$$

where $\upsilon_c l$ is the cutoff wavenumber for a given angular frequency. For an incident plane S-wave propagating along the third axis, the travel-time fluctuation satisfies

$$
\partial_i \delta t^s = \delta_{i3}\frac{\delta\beta^{L}(\mathbf{x})}{\beta_0{}^2} \quad \text{and} \quad \partial_i\partial_j\delta t^s = 0
\tag{5.40}
$$

For the incidence of a plane S-wave (4.38) propagating in the third direction, having unit amplitude, and polarized in the first direction, we combine (5.31) and (5.40) to find the equivalent body force term

$$
{}^{C}\delta f_i^{s}(\mathbf{x},t) = 2l\omega\rho_0\delta\beta^{L}(\mathbf{x})\delta_{i1}e^{i(le_3\mathbf{x}-\omega t)}
\tag{5.41}
$$

Solving (5.30) for body forces (5.41) and (4.39), we get the scattered waves as outgoing spherical waves from the inhomogeneity. The travel-time corrected SS-scattering amplitude is given as a sum of terms given by (4.46) and correction terms:

$$
{}^{T}F_{r,\psi,\zeta}^{SS} = F_{r,\psi,\zeta}^{SS} + {}^{C}F_{r,\psi,\zeta}^{SS}
\tag{5.42}
$$

where the correction terms are

$$^{C}F_\psi^{SS} = \frac{l^2}{4\pi} 2\cos\zeta\cos\psi\, H(\psi_C - \psi)\frac{\delta\tilde{\beta}(le_r - le_3)}{\beta_0}$$

$$^{C}F_\zeta^{SS} = \frac{l^2}{4\pi}(-2\sin\zeta)H(\psi_C - \psi)\frac{\delta\tilde{\beta}(le_r - le_3)}{\beta_0} \qquad (5.43)$$

$$^{C}F_r^{SS} = 0$$

The correction terms for both P- and S-waves are nonzero only within a cone in the forward direction satisfying $\psi < \psi_C$. Correction terms (5.37) and (5.43) are slightly different from the corresponding correction terms in Sato [1984a] because of the second condition of (5.34) and (5.40), but the following results are quantitatively similar to those given in Sato [1984a, 1990].

As discussed in Section 4.2.2, we assume that the fractional fluctuations for α and β are given by one isotropic and homogeneous random function $\xi(\mathbf{x})$ as in (4.47). Using Birch's law, the fractional fluctuation of density is taken to be proportional to $\xi(\mathbf{x})$ as given by (4.48). Then, combining (4.49) with (5.36), (5.37), (5.42), and (5.43), the scattering amplitudes are written by using the Fourier transform of $\xi(\mathbf{x})$, where the argument is the exchange wavenumber vector corresponding to each scattering mode:

$$^{T}F_r^{PP} = \frac{l^2}{4\pi}\,^{T}X_r^{PP}(\psi,\zeta)\,\tilde{\xi}(ke_r - ke_3)$$

$$F_\psi^{PS} = \frac{l^2}{4\pi}X_\psi^{PS}(\psi,\zeta)\,\tilde{\xi}(le_r - ke_3)$$

$$F_r^{SP} = \frac{l^2}{4\pi}X_r^{SP}(\psi,\zeta)\,\tilde{\xi}(ke_r - le_3) \qquad (5.44)$$

$$^{T}F_\psi^{SS} = \frac{l^2}{4\pi}\,^{T}X_\psi^{SS}(\psi,\zeta)\,\tilde{\xi}(le_r - le_3)$$

$$^{T}F_\zeta^{SS} = \frac{l^2}{4\pi}\,^{T}X_\zeta^{SS}(\psi,\zeta)\,\tilde{\xi}(le_r - le_3)$$

Travel-time correction has been done only for PP- and SS-scattering since the time correction is necessary only when the scattered wave and the incident wave are the same wave type. As shown in Figure 4.9, basic scattering patterns $X_\bullet^{\bullet\bullet}$ for conversion scattering have no lobes in the forward direction. Here, $^{T}X_\bullet^{\bullet\bullet}$ is a function of angle (ψ,ζ) representing the basic scattering pattern including the effect of travel-time correction:

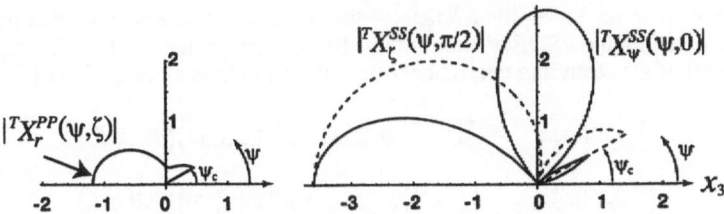

FIGURE 5.8. ψ dependence of basic scattering patterns for the travel-time corrected Born approximation for $\gamma_0 = \sqrt{3}$, $v = 0.8$, and $\upsilon_c = 1/2$ ($\psi_c \approx 29°$). Compare with Figure 4.7.

$$^T X_r^{PP}(\psi,\zeta) = \frac{1}{\gamma_0^2}\left[v\left(-1+\cos\psi+\frac{2}{\gamma_0^2}\sin^2\psi\right)-2\right.$$

$$\left.+\frac{4}{\gamma_0^2}\sin^2\psi+2\cos\psi H(\psi_C-\psi)\right] \qquad (5.45)$$

$$^T X_\psi^{SS}(\psi,\zeta) = \cos\zeta\left[v(\cos\psi-\cos 2\psi)-2\cos 2\psi+2\cos\psi H(\psi_C-\psi)\right]$$

$$^T X_\zeta^{SS}(\psi,\zeta) = \sin\zeta\left[v(\cos\psi-1)+2\cos\psi-2H(\psi_C-\psi)\right]$$

Contrary to the scalar wave case, the travel-time correction for vector waves does not completely eliminate the contribution of scattering within a cone defined by cutoff scattering angle ψ_C around the forward direction. However, for angles smaller than ψ_C, the travel-time correction makes the scattering amplitude very small. In addition, $^T X_r^{PP}(0,\zeta)=^T X_\psi^{SS}(0,\zeta)=^T X_\zeta^{SS}(0,\zeta)=0$. This means that the correction for the travel-time fluctuation is almost the same as neglecting scattering loss within a cone around the forward direction. Figure 5.8 shows the ψ dependence of basic scattering patterns for the travel-time corrected Born approximation, where $\gamma_0=\sqrt{3}$, $v=0.8$ and $\upsilon_c=1/2$. The backward scattering coefficient for the S-wave is the same as (4.54) since (5.43) shows that the travel-time correction does not effect scattering for angles larger than the cutoff scattering angle ψ_C.

We imagine an ensemble of media having fluctuations described by $\{\xi(\mathbf{x})\}$. Then, we define scattering coefficients with travel-time correction as in (4.52):

$$^T g^{PP}(\psi,\zeta;\omega) \equiv \frac{4\pi}{L^3}\left\langle\left|^T F_r^{PP}\right|^2\right\rangle = \frac{l^4}{4\pi}\left|^T X_r^{PP}\right|^2 P\left(\frac{2l}{\gamma_0}\sin\frac{\psi}{2}\right)$$

$$(5.46)$$

$$^T g^{SS}(\psi,\zeta;\omega) \equiv \frac{4\pi}{L^3}\left\langle\left|^T F_\psi^{SS}\right|^2+\left|^T F_\zeta^{SS}\right|^2\right\rangle = \frac{l^4}{4\pi}\left(\left|^T X_\psi^{SS}\right|^2+\left|^T X_\zeta^{SS}\right|^2\right)P\left(2l\sin\frac{\psi}{2}\right)$$

where we write the ensemble average of the squared scattering amplitude per unit volume by using the PSDF of the random fluctuation. Scattering loss is written as an integral of the scattering coefficient over a solid angle as an extension of (5.4):

$$^{TSc}Q_P^{-1}(\omega) \equiv \frac{1}{4\pi k} \oint \left[{}^T g^{PP}(\psi,\zeta;\omega) + g^{PS}(\psi,\zeta;\omega) \right] d\Omega(\psi,\zeta)$$

$$^{TSc}Q_S^{-1}(\omega) \equiv \frac{1}{4\pi l} \oint \left[{}^T g^{SS}(\psi,\zeta;\omega) + g^{SP}(\psi,\zeta;\omega) \right] d\Omega(\psi,\zeta)$$

(5.47)

Substituting (5.46) and (4.52b and c) in (5.47), we finally get

$$^{TSc}Q_P^{-1}(\omega) = \frac{\gamma_0 l^3}{(4\pi)^2} \oint \left[\left| {}^T X_r^{PP}(\psi,\zeta) \right|^2 P\left(\frac{2l}{\gamma_0} \sin\frac{\psi}{2} \right) \right. $$

$$\left. + \frac{1}{\gamma_0} \left| X_\psi^{PS}(\psi,\zeta) \right|^2 P\left(\frac{l}{\gamma_0} \sqrt{1+\gamma_0^2 - 2\gamma_0 \cos\psi} \right) \right] d\Omega(\psi,\zeta)$$

$$^{TSc}Q_S^{-1}(\omega) = \frac{l^3}{(4\pi)^2} \oint \left\{ \left[\left| {}^T X_\psi^{SS}(\psi,\zeta) \right|^2 + \left| {}^T X_\zeta^{SS}(\psi,\zeta) \right|^2 \right] P\left(2l \sin\frac{\psi}{2} \right) \right. $$

$$\left. + \gamma_0 \left| X_r^{SP}(\psi,\zeta) \right|^2 P\left(\frac{l}{\gamma_0} \sqrt{1+\gamma_0^2 - 2\gamma_0 \cos\psi} \right) \right\} d\Omega(\psi,\zeta)$$

(5.48)

For PS- and SP-conversion scattering, the argument of the PSDF in (5.48) cannot take a value smaller than $(\gamma_0 - 1)l/\gamma_0$ for a given angular frequency, that is, only the short wavelength components of the random inhomogeneity contribute to scattering loss.

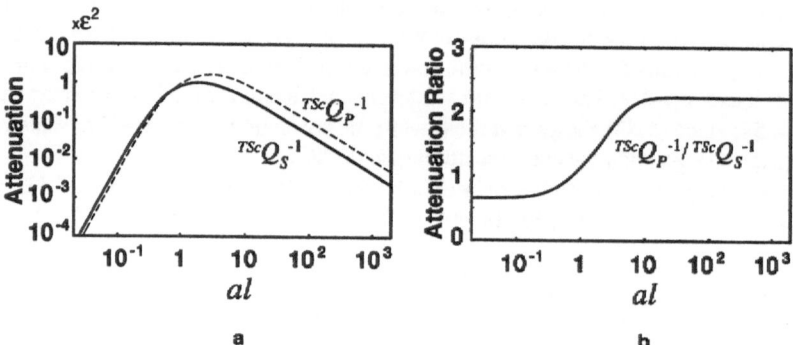

FIGURE 5.9. (a) Frequency dependence of scattering attenuation $^{TSc}Q_S^{-1}$ (solid curve) and $^{TSc}Q_P^{-1}$ (broken curve) and (b) ratio of $^{TSc}Q_P^{-1}$ to $^{TSc}Q_S^{-1}$ theoretically predicted by the travel-time corrected Born approximation for the exponential ACF ($\kappa = 0.5$) for $\gamma_0 = \sqrt{3}$, $\nu = 0.8$, and $\upsilon_c = 1/2$ ($\psi_c \approx 29°$), where $l = \omega/\beta_0$.

Exponential ACF

For the case of the exponential ACF, substituting the PSDF (2.10) in (5.48) and taking the grid size to be $0.5° \otimes 0.5°$ for $\cos^{-1} w \otimes \phi$, we numerically integrate to get

$$^{TSc}Q_P^{-1}{}_{Max} \approx 1.6\,\varepsilon^2 \quad \text{at } a\omega/\beta_0 \approx 3.0$$
$$^{TSc}Q_S^{-1}{}_{Max} \approx 1.0\,\varepsilon^2 \quad \text{at } a\omega/\beta_0 \approx 2.0 \tag{5.49}$$

For low frequency, $a\omega/\beta_0 \ll 1$,

$$^{TSc}Q_P^{-1} \approx 4.7\varepsilon^2 \left(\frac{a\omega}{\beta_0}\right)^3 \quad \text{and} \quad ^{TSc}Q_S^{-1} \approx 7.2\varepsilon^2 \left(\frac{a\omega}{\beta_0}\right)^3 \tag{5.50}$$

where $^{TSc}Q_P^{-1}/^{TSc}Q_S^{-1} \approx 0.66$. For high frequency, $a\omega/\beta_0 \gg 1$,

$$^{TSc}Q_P^{-1} \approx 10\,\varepsilon^2 \left(\frac{\beta_0}{a\omega}\right) \quad \text{and} \quad ^{TSc}Q_S^{-1} \approx 4.6\,\varepsilon^2 \left(\frac{\beta_0}{a\omega}\right) \tag{5.51}$$

where $^{TSc}Q_P^{-1}/^{TSc}Q_S^{-1} \approx 2.2$. Thus, scattering attenuation decreases according to the reciprocal of frequency for both P- and S-waves.

Figure 5.9 shows the frequency dependence of scattering attenuation for P-waves, S-waves, and their ratio, where the abscissa is scaled S-wave wavenumber al. Ratio $^{TSc}Q_P^{-1}/^{TSc}Q_S^{-1}$ is smaller than 1 for lower frequencies; however, it be-

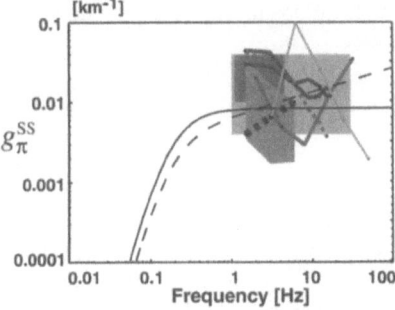

FIGURE 5.10. Theoretical prediction of S-to-S backscattering coefficient g_π^S vs. frequency for vector waves ($\beta_0 = 4\text{km/s}$, $\gamma_0 = \sqrt{3}$, $\nu = 0.8$): solid curve for the exponential ACF ($\kappa = 0.5, \varepsilon^2 = 0.01$ and $a = 2$ km); broken curve for the von Kármán ACF ($\kappa = 0.35, \varepsilon^2 = 0.0072$ $a = 2.1$ km). Background shows total scattering coefficient g_0 and backscattering coefficient g_π based on coda-excitation measurements for various regions of the world shown in Figure 3.10.

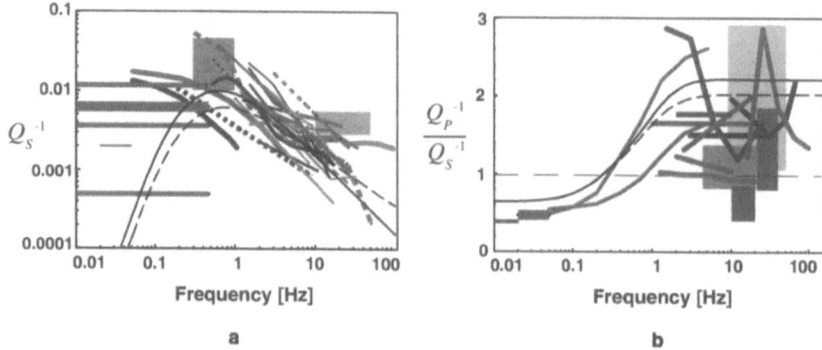

FIGURE 5.11. Values of (a) $^{Tsc}Q_S^{-1}$ and (b) ratio of $^{Tsc}Q_P^{-1}$ to $^{Tsc}Q_S^{-1}$ vs. frequency predicted by the travel-time corrected Born approximation for $\beta_0 = 4$ km/s, $\gamma_0 = \sqrt{3}$, $\nu = 0.8$, and $\upsilon_c = 1/2$: solid curve for the exponential ACF ($\kappa = 0.5$, $\varepsilon^2 = 0.01$ and $a = 2$ km); broken curve for the von Kármán ACF ($\kappa = 0.35$, $\varepsilon^2 = 0.0072$ and $a = 2.1$ km). Background shows regional measurements worldwide as given by Figures 5.1 and 5.3.

comes larger than 1 for higher frequencies. Comparing different scattering modes, we find that SS scattering is dominant in S-wave attenuation [see Sato, 1984a, Fig. 10].

We plot the theoretical S-to-S backscattering coefficient g_π^{SS} (4.57) for the exponential ACF where $\varepsilon^2 = 0.01$ and $a = 2$ km using solid curves along with both backscattering coefficient and total scattering coefficient measured from S-coda excitation of local earthquakes from various regions of the world in Figure 5.10. We plot the theoretical scattering attenuation of S-waves $^{Tsc}Q_S^{-1}$ and ratio $^{Tsc}Q_P^{-1}/^{Tsc}Q_S^{-1}$ along with worldwide observations in Figure 5.11. The theoretical curves provide a good fit to the observed data.

von Kármán ACF

The predicted rate of decrease in Q_S^{-1} with increasing frequency at high frequency given by (5.51) appears to be faster than observations [see Figure 5.11a]. Using data collected in the Kanto area, Japan, Sato [1984b, 1990] and Kinoshita [1994] estimated that $Q_S^{-1} \propto f^{-0.7}$. Introducing the von Kármán ACF, by substituting (2.12) in (5.48), we get frequency dependent scattering attenuation that depends on the order κ for high frequency, $a\omega/\beta_0 \gg 1$ as

$$^{Tsc}Q_P^{-1} \text{ and } ^{Tsc}Q_S^{-1} \propto \varepsilon^2 \left(\frac{\beta_0}{a\omega}\right)^{2\kappa} \tag{5.52}$$

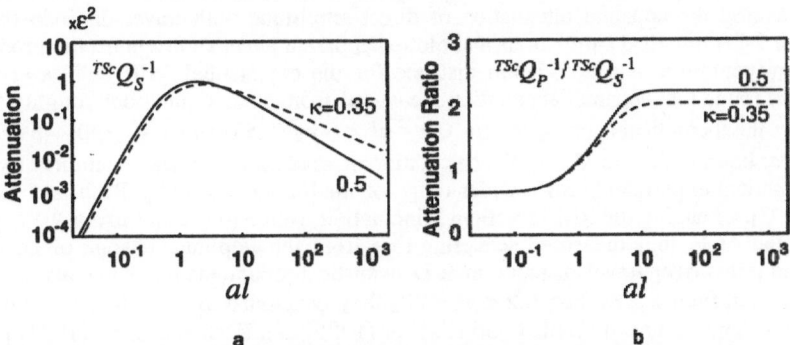

FIGURE 5.12. (a) Frequency-dependence of predicted scattering attenuation for S-wave $^{TSc}Q_S^{-1}$ and (b) ratio $^{TSc}Q_P^{-1}/^{TSc}Q_S^{-1}$ for the von Kármán ACF ($\kappa = 0.35$, broken curve) and the exponential ACF ($\kappa = 0.5$, solid curve) for $\gamma_0 = \sqrt{3}$, $\nu = 0.8$, and $\upsilon_c = 1/2$ ($\psi_c \approx 29°$), where $l = \omega / \beta_0$.

The backscattering coefficient for this case is given by (4.56) and is shown in Figure 4.10 by a broken curve for $\kappa = 0.35$ corresponding to $^{TSc}Q_S^{-1} \propto f^{-0.7}$. Figure 5.12 shows the frequency dependence of S-wave attenuation and the ratio of P- to S-wave attenuation for $\kappa = 0.35$ and $\kappa = 0.5$ (Exponential ACF). For $\kappa = 0.35$ the ratio of P- to S-wave attenuation is 2.03 at high frequencies. The ratio slightly decreases as order κ becomes smaller. In Figures 5.10 and 5.11, broken curves show theoretical predictions for the von Kármán ACF with $\kappa = 0.35$, $\varepsilon^2 = 0.0072$ and $a = 2.1$ km along with data observed throughout the world.

Evaluation of Cutoff Scattering Angle

As shown in Figure 5.5, the choice of the cutoff scattering angle impacts the attenuation predictions made by the single scattering theory. Several investigators have evaluated scattering attenuation using 2-D acoustic finite difference simulation for media having random velocity fluctuation to estimate the cutoff scattering angle, or the lower bound for the integral over the scattering angle [Frankel and Clayton, 1986; Roth and Korn, 1993]. For the 2-D scalar wave equation, the travel-time corrected scattering attenuation is similar to that in 3-D given in (5.20). For the exponential ACF,

$$^{TSc}Q^{-1}(\omega) = \frac{k^2}{\pi} \int_{\psi_c}^{\pi} P\left(2k\sin\frac{\psi}{2}\right) d\psi = \int_{\psi_c}^{\pi} \frac{2\varepsilon^2 a^2 k^2}{\left(1 + 4a^2 k^2 \sin^2\frac{\psi}{2}\right)^{3/2}} d\psi \qquad (5.53)$$

where Ψ_C is the lower bound of the angular integral. Frankel and Clayton [1986] measured the apparent attenuation of direct amplitude with travel distance from their 2-D numerical simulations and plotted apparent attenuation against the product of wavenumber and correlation distance for the exponential ACF with $\varepsilon=10\%$. They found that apparent attenuation measured from their simulations roughly follows the theoretical curve given for $0.2 < ak < 6$ by (5.53) when $\Psi_C=30°-45°$. The lower bound of angle Ψ_C for the calculation of scattering loss was examined using numerical experiments for a wider range of media parameters by Roth and Korn [1993]. Changing the RMS fractional fluctuation from 3 to 9% and using the exponential ACF, they measured scattering loss from the amplitude change of an isolated pulse over travel distance in 2-D acoustic random media. Their results are shown in Figure 5.13. For $0.2 < ak < 20$, they concluded that Ψ_C ranges from 20 to 40°. The studies of Frankel and Clayton [1986] and Roth and Korn [1993] provide evidence supporting the value of 29° proposed for the 3-D case. Fang and Müller [1996] preferred that the lower bound of scattering angle about 20° for scattering attenuation based on measurements of the decay of the envelope maximum and spectral amplitude with travel distance. Only Jannaud et al. [1991b] suggested that the critical scattering angle could be as large as 90°.

For the calculation of scattering attenuation through distributed cracks, the idea of neglecting scattering energy within a cone around the forward direction is useful. From the measurement of amplitude attenuation through an aluminum block containing parallel cylindrical voids, Dubendorff and Menke [1986] found that the apparent attenuation was well fit by the corrected scattering attenuation model when the cutoff scattering angle is 10° for P-waves, 6° for SH-waves, and 15° for SV-waves.

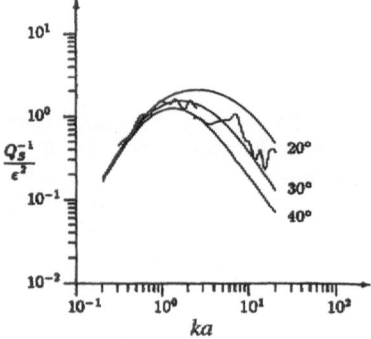

FIGURE 5.13. Scattering attenuation in 2-D acoustic random media. Irregular line shows measurements made from 2-D finite difference simulations. Regular lines show predictions of (5.53) for various values of Ψ_c. Medium is characterized by an exponential ACF with $\varepsilon=9\%$. [From Roth and Korn, 1993, with permission from Blackwell Science, United Kingdom.]

Diffraction Effects

The derivation of scattering attenuation using the Born approximation can be viewed as a kind of differential approach since the correction for travel-time fluctuation is given in differential form in (5.11). Diffraction effects caused by long-wavelength components of the inhomogeneity were neglected. These diffraction effects become increasingly important as travel distance increases.

Shapiro and Kneib [1993] investigated this phenomena in detail for isotropic random acoustic media in 2-D and 3-D. They measured the decay of amplitude of the coherent wavefield $\langle u \rangle$ and that of the mean logarithm of amplitude $\langle \ln A_0 \rangle$ with travel distance in a frequency range dominated by forward scattering. Recall that coherent wavefield is the mean wavefield from many realizations of random media (see Figure 5.6a). Regressions of $\langle \ln A_0 \rangle$ vs. travel distance are common methods for measuring attenuation in the earth. Figure 5.14a shows $\langle \ln A_0 \rangle$ and $\ln \langle u \rangle$ predicted by the parabolic approximation (see Section 8.1.1) at 100 Hz against travel distance for short travel distances in 3-D, where 100 Hz corresponds to a wavelength of 30 m and $ak \approx 4.2$. $\ln \langle u \rangle$ decreases linearly with travel distance. However, $\langle \ln A_0 \rangle$ decreases more slowly and is similar to the curve predicted by Wu's [1982a] approximation for short travel distances that uses $\psi_c = 90°$ and counts scattering only into the back half-space for the estimation of scattering attenuation. This means that attenuation of the coherent wavefield is caused mainly by travel-time fluctuations but that backscattering alone is insufficient to explain the observed amplitude attenuation. Shapiro and Kneib [1993] also measured $\langle \ln A_0 \rangle$

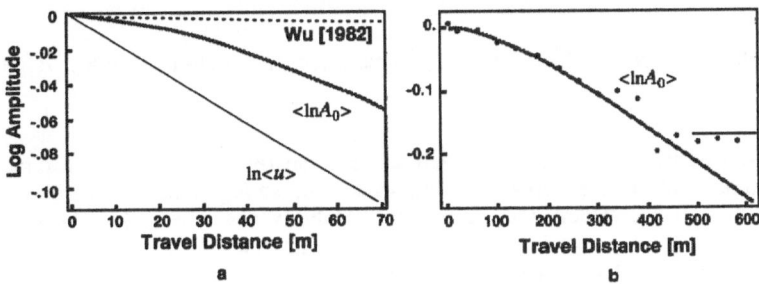

FIGURE 5.14. (a) Plots of logarithm amplitude vs. travel distance of 100 Hz waves in a 3-D acoustic random medium of average velocity V_0=3 km/s having an exponential ACF with a=20 m and ε=3%. The bold convex curve is $\langle \ln A_0 \rangle$ predicted from the parabolic approximation theory. The broken curve is the prediction by Wu [1982a] who used single scattering theory with $\psi_c = 90°$. (b) Plots of logarithm amplitude vs. travel distance of 100 Hz waves in a 2-D acoustic random medium of the same statistical characteristics as in (a). Dots are $\langle \ln A_0 \rangle$ measured from numerical simulations. The bold convex curve is $\langle \ln A_0 \rangle$ predicted from the parabolic approximation theory, and the horizontal bar corresponds to saturation due to strong scattering. [From Shapiro and Kneib, 1993, with permission from Blackwell Science, United Kingdom.]

from 2-D finite difference calculations for the scalar wave equation by taking many point measurements of the wavefield, computing the spectra, and averaging the natural logarithms. Figure 5.14b plots $\langle \ln A_0 \rangle$ vs. distance at long travel distances measured from the numerical simulations. Numerical results, represented by dots, agree well with the bold convex curve, which is predicted to be due to de-focusing of the wavefield by the parabolic approximation for the specific model structure with $ak \gg 1$, where backscattering is neglected. Even though the medium fluctuation is small, at distances larger than 500 m in the simulation, the mean logarithm amplitude stays at the same level irrespective of travel distance because of the dominance of the incoherent wavefield due to diffraction and forward scattering. There is a difference between the global estimate and the local estimate of attenuation. The parabolic approximation does not work for obtaining scattering attenuation over a wide range of frequencies, since it is a high-frequency approximation. However, the results of Shapiro and Kneib [1993] show that logarithms of amplitude spectra do not decay linearly with travel distance over a broad distance range. They raised concern about the careless use of the linear regression to estimate the characteristics of random media.

In Chapter 8 we will discuss the use of the parabolic approximation to model the scattered wavefield for strong forward scattering. Rather than study the first arrival amplitude, we will show that it is more appropriate to model and analyze the shape of the envelope containing the first arriving energy. Strong diffraction causes much slower decay of the peak amplitude of the envelope than exponential. The spectra of random inhomogeneities seems to be very broad in the earth, therefore, we will have to include the contribution of diffraction effects due to long-wavelength inhomogeneities when modeling amplitude attenuation in addition to large angle scattering due to short-wavelength inhomogeneities.

Scattering Attenuation in One-Dimensional Inhomogeneous Elastic Media

As discussed in Section 2.2, well-log data clearly show evidence for random inhomogeneity in the earth. The importance of stratigraphic effects in causing transmission attenuation was raised by O'Doherty and Anstey [1971]. Richards and Menke [1983] studied wave propagation though a 1-D multilayered structure as a model of the heterogeneous crust. Their model is composed of a sequence of layers each having one of two different velocities with thickness distributed as a Poisson process, as illustrated in Figure 5.15a. They investigated waveform characteristics caused by scattering and intrinsic attenuation through stratified media. Figure 5.15b shows the transmission responses for an impulsive source in two cases. The upper trace shows strong excitation of high-frequency coda waves for a medium having no intrinsic attenuation $^{I}Q^{-1} = 0$. The lower trace shows that, for a medium having frequency-independent intrinsic attenuation $^{I}Q^{-1} = 0.01$, high-frequency components in the coda diminish in size. Scattering causes a broadening of the pulse width with increasing travel distance but preserves the high-frequency content of a waveform. Such a modulation effect has been studied as a tool for

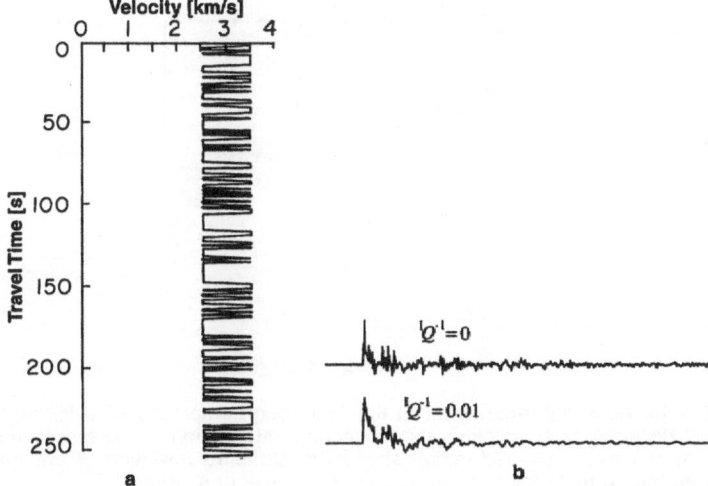

FIGURE 5.15. (a) Acoustic velocity log for a medium composed of layers having one of two velocities; (b) Transmission response for an impulsive signal propagating through the medium illustrated in (a) for two values of intrinsic attenuation. [From Richards and Menke, 1983, copyright by the Seismological Society of America.]

characterizing random media [Lerche and Menke, 1986; Burridge et al., 1988, 1989].

The scattering attenuation for an elastic wave traveling through 1-D random media can be derived using the mean wavefield theory or the Born approximation [Sato, 1982a; Wenzel, 1982; Banik et al., 1985]:

$$^{TSc}Q^{-1}(k) = k P_{\mathrm{Imp.}}(2k) \tag{5.54}$$

where $P_{\mathrm{Imp.}}$ is the PSDF of the fractional fluctuation of impedance and the argument of the PSDF is the exchange wavenumber equal to twice the wavenumber of the incident wave, which corresponds to backward scattering. For 1-D media, using the travel-time correction is equivalent to neglecting forward scattering, which has zero exchange wavenumber. This formula has been used to study scattering loss in a 1-D random structure derived from well-log data [Görich and Müller, 1987]. It has also been derived from a self averaging procedure [Shapiro and Zien, 1993].

Shiomi et al. [1996] numerically examined elastic wave propagation in 1-D random media having PSDF $\propto k^{-1.5}$ for fractional fluctuation of elastic wave velocity with RMS fluctuation of 15% and mass density PSDF $\propto k^{-1.3}$ with RMS fractional fluctuation of 6%. They studied a suite of models, each 1 km long. Each sample medium was divided into 2,000 layers having thickness of 0.5 m. They used an exact solution for the wave equation and numerically simulated wave propagation in time. The amplitude of the incident Ricker wavelets decay with travel distance as coda waves are excited. Apparent attenuation vs. frequency was

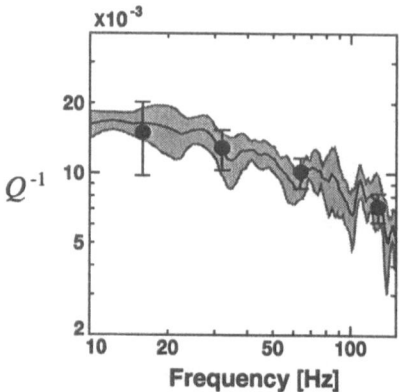

FIGURE 5.16. Apparent attenuation of the first-trough amplitude of a Ricker wavelet with travel distance (solid circles) and scattering attenuation (solid curve) predicted from (5.54), where the shaded region shows the standard deviation of the measurements from the 10 trials for each frequency. [Courtesy of K. Shiomi.]

estimated by measuring the change with propagation distance of the amplitude of the first trough in the first arriving wave packet. Figure 5.16 shows the measured apparent attenuation against frequency where each solid circle is an ensemble average over 10 trials for media having the same statistical characterizations and for Ricker wavelet sources having the corresponding central frequency. The solid curve in Figure 5.16 shows scattering attenuation predicted from (5.54), which is nearly proportional to the –0.5th power of frequency for the given structure. The amplitude decay of the first arriving trough is well explained by the first-order perturbation method; however, they found that the maximum trace amplitude between the first trough and the following peak does not agree with the single scattering theory and they argue that the theory needs to include contributions from multiple reflections to provide a better fit to the later portions of the waveforms.

5.4 SCATTERING ATTENUATION DUE TO DISTRIBUTED CRACKS AND CAVITIES

As discussed in Section 2.1, microscopic cracks are known to be pervasive in crustal rocks. Several models for predicting the influence of cracks and inclusions in rocks on elastic properties have been developed since the pioneering work of Walsh [1965]. Although Walsh's work was based on a static model, other models were based on a dynamic approach, in which the scattering of waves by cracks and inclusions having lengths much smaller than the seismic wavelength was modeled. One such model was due to Budiansky and O'Connell [1976]. This model used a self-consistent approach in which the interaction of cracks was not specifically calculated but the effects of other cracks were incorporated by considering that the media surrounding a given crack have the unknown elastic properties of a rock containing a distribution of cracks. The models of Walsh [1965] and Budiansky and O'Connell [1976] both predict that, to first order, the effect of cracks on the elastic properties of rocks is scaled by a parameter known as the crack density, which is the ratio of the total porosity of all the cracks to the aspect ratio of the cracks. Several models have been developed to predict the effects of empty and fluid-filled cracks on intrinsic seismic attenuation, as discussed in Section 5.2. These models were developed to predict the bulk properties of rocks, so little attention was paid to the character of the scattered wavefield. It is natural to imagine a single crack or a distribution of cracks as the heterogeneity and to investigate the characteristics of the scattered wavefield.

There have been several attempts to solve the boundary value problem for the scattered wavefield caused by plane waves incident on an isolated spherical inclusion [Ying and Truell, 1956; Einspruch et al., 1960; Yamakawa, 1962; Korneev and Johnson, 1993a, b]. Gritto et al. [1995] and Korneev and Johnson [1996] examined conversion scattering characteristics for the incidence of both P- and S-waves on a spherical inclusion. They pointed out the significant amount of P-to-S scattering compared with S-to-P scattering.

Kikuchi [1981] analytically calculated elastic wave attenuation due to distributed cracks of half-length a_c in 2-D space. The crack is geometrically described as the limit of an ellipsoid on which the stress is free. The resultant scattering attenuation Q_P^{-1} for waves arriving normal to the crack plane has a peak at $a_c k \approx 0.64$, and the peak value is a few times larger than the peak value of scattering attenuation Q_S^{-1}. However, it is difficult to imagine open cracks having dimensions large enough to be comparable to regional seismic wavelengths deep within the earth. Kawahara and Yamashita [1992] used an integral equation to examine elastic wave attenuation

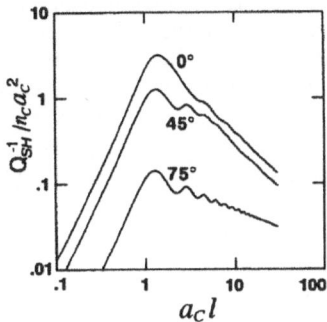

FIGURE 5.17. Scattering loss due to aligned nonopening cracks for SH-waves having various incident angles relative to the plane of the aligned cracks, where n_c is the number density of cracks, and $l=\omega/\beta_0$. [From Kawahara and Yamashita, 1992, with permission of Birkhäuser Verlag AG, Switzerland.]

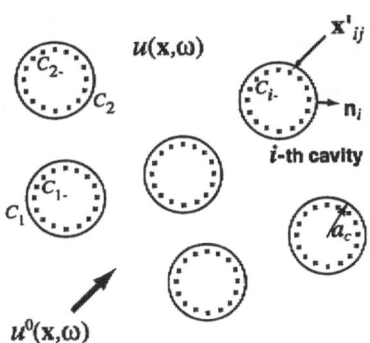

FIGURE 5.18. Schematic illustration of distributed cylindrical cavities and the incident SH-wave $u^0(\mathbf{x},\omega)$, where total wavefield is given by $u(\mathbf{x},\omega)$.

for waves at oblique incidence on a fracture zone containing randomly distributed cracks whose planes are aligned parallel to the fracture plane. Figure 5.17 shows scattering attenuation for SH-waves, where the peak at $a_c l \approx 1.4$ is almost independent of the incidence angle [Kawahara and Yamashita, 1992]. The higher wavenumber asymptote of scattering attenuation is proportional to the reciprocal of the wavenumber. If we fit the scattering attenuation to the predicted peak in observed attenuation at 0.5 Hz discussed in Section 5.1, $a_c \approx 1.8$ km. It is difficult to imagine such large cracks in the earth. Yamashita [1990] used an integral equation to calculate SH-wave scattering attenuation through a medium composed of a distribution of cracks whose half-sizes a_c obey an inverse power of dimension and whose orientations are random. He discussed the change in scattering attenuation in relation to the power of the crack size distribution. Later, using the boundary integral method, Murai et al. [1995] numerically simulated SH-wave propagation through a medium containing 72 distributed parallel plane cracks containing Newtonian viscous fluid. Matsunami [1990] measured attenuation and amplitude fluctuation of acoustic waves propagating through an aluminum plate with many clusters of small open holes. He found a peak in Q^{-1} when the wavelength is about 1.5 times the average diameter of the cluster of holes. For all models investigated, the crack model predicts a peak in Q^{-1} when the wavelength is of the same order as the dimension of the crack.

Scattering attenuation due to distributed cylindrical cavities was analytically studied using a scattering matrix by Varadan et al. [1978]. Numerical synthesis of time domain seismograms for waves incident on a distribution of open cavities was done by Benites et al. [1992]. They used the boundary integral method to deterministically model multiple scattering of SH-waves in 2-D media containing a distribution of randomly spaced cylindrical cavities of radius a_c, as schematically illustrated in Figure 5.18. Since an exact numerical method was employed, they were able to investigate wave scattering in media containing strong velocity contrasts. The boundary integral approach used by Benites et al. [1992] was a frequency-domain implementation. The total SH-wavefield $u(\mathbf{x}, \omega)$ at angular frequency ω obeys the Helmholtz equation:

$$(\Delta + l^2) u(\mathbf{x}, \omega) = 0 \tag{5.55}$$

where wavenumber $l = \omega / \beta_0$. The solution is written in the form of an indirect integral representation where the total wavefield is the sum of the incident wave and scattered waves from sources located along the boundaries of the M cavities:

$$u(\mathbf{x}, \omega) = u^0(\mathbf{x}, \omega) + \sum_{i=1}^{M} \oint_{C_{i-}} G(\mathbf{x} - \mathbf{x}', \omega) \Lambda_i(\mathbf{x}', \omega) dc_i(\mathbf{x}') \tag{5.56}$$

where $\mathbf{x}'$ is the location of a source Λ_i for the scattered wavefield, dc_i is an infinitesimal line element, and a circle C_{i-} is interior to boundary C_i of the ith cavity. Practically they took the radius of C_{i-} as 80% of the radius of C_i to avoid the singularity of the Green function at the location where boundary conditions must be met. The Green function corresponding to outgoing waves at a large distance is written using the Hankel function of the first kind of zeroth order as

$$G(\mathbf{x},\omega) = -\frac{i}{4}H_0^{(1)}(lr) \tag{5.57}$$

where $r \equiv |\mathbf{x}|$. It is the solution of

$$\left(\Delta + l^2\right)G(\mathbf{x},\omega) = \delta(\mathbf{x}) \tag{5.58}$$

It is necessary to describe the boundary condition to solve (5.56), since it is a class of indirect integral representation problems. Benites et al. [1992] discretized the source distribution for the i-th cavity as

$$\Lambda_i(\mathbf{x},\omega) = \sum_{j=1}^{N} A_{ij}(\omega)\delta\left(\left|\mathbf{x} - \mathbf{x}'_{ij}\right|\right) \tag{5.59}$$

where A_{ij} is a complex constant that represents the strength of the source located at the jth point along a circle C_{i-} of the ith cavity $\mathbf{x}_{ij}'$ and N is the number of sources. The minimum number of sources required is $2\pi a_C/(\lambda_s/4)$ for wavelength λ_s. Then, they wrote (5.56) as

$$u(\mathbf{x},\omega) = u^0(\mathbf{x},\omega) + \sum_{i=1}^{M}\sum_{j=1}^{N} A_{ij}(\omega)G\left(\mathbf{x} - \mathbf{x}'_{ij},\omega\right) \tag{5.60}$$

They imposed the Neumann boundary conditions on the cavity surfaces in a least square sense, that is, they minimized the square sum of traction along all boundaries:

$$\sum_{k=1}^{M} \oint_{C_k}\left|\mu\frac{\partial u}{\partial \mathbf{n}_k}\right|^2 dc_k \Rightarrow \text{Min.} \tag{5.61}$$

where $\mathbf{n}_k$ is the outward normal vector to boundary C_k. Substituting (5.60) in (5.61), they got a system of simultaneous linear equations for A_{ij}:

$$\sum_{i=1}^{M}\sum_{j=1}^{N}\left(\sum_{k=1}^{M}\oint_{C_k}\frac{\partial G_{mn}^{*}}{\partial \mathbf{n}_k}\frac{\partial G_{ij}}{\partial \mathbf{n}_k}dc_k\right)A_{ij} = -\sum_{k=1}^{M}\oint_{C_k}\frac{\partial G_{mn}^{*}}{\partial \mathbf{n}_k}\frac{\partial u^0}{\partial \mathbf{n}_k}dc_k \tag{5.62}$$

where asterisk stands for complex conjugate and G_{mn} means $G\left(\mathbf{x} - \mathbf{x}'_{mn},\omega\right)$ for $\mathbf{x}$ on the mth boundary. The right-hand side represents the interaction between the incident wave and nth source of the mth cavity. The left-hand side represents the in-

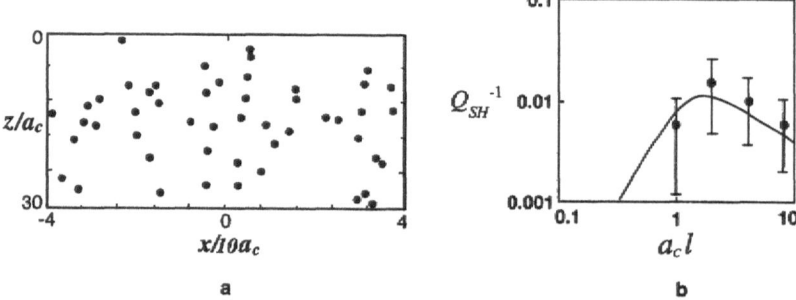

FIGURE 5.19. (a) Configuration of 50 cavities of radius a_C used in numerical simulations, where the SH-wave is incident from below. [From Benites et al., 1992, with permission of Birkhäuser Verlag AG, Switzerland.] (b) Scattering loss of direct SH-waves due to distribution of cavities shown in (a) where $l=\omega/\beta_0$. Vertical bars denote the standard deviation of measurements made from numerical simulations using the boundary integral approach [Benites et al., 1992]. Solid curve shows the scattering loss calculated using the optical theorem. [From Kawahara and Yomogida, 1996, with permission of the Seismological Society of Japan.]

teraction between the nth source of the mth cavity and the jth source of the ith cavity. Solving these simultaneous linear equations in the angular-frequency domain, they get the strength of the sources A_{ij}. By using the inverse Fourier transform of the solutions obtained for many frequencies, they obtained synthetic seismograms in the time domain.

Solving the problem for the incidence of a plane wave or a line source, Benites et al. [1992] synthesized seismograms for the medium containing 50 cavities illustrated in Figure 5.19a. They numerically simulated waves on the earth's surface for a Ricker wavelet incident from below. They found that the S-coda is composed of many wave trains following the direct S-wave arrival. They also measured the amplitude of direct SH-wave against travel distance through an infinite medium that includes a region containing distributed cavities, like that shown in Figure 5.19a, and calculated scattering attenuation. Solid circles in Figure 5.19b show the calculated scattering attenuation for the SH-wave, which has a peak amplitude of the order of 0.01. Their scattering attenuation measurements are well explained by the solid curve which is obtained by using the optical theorem based on the dispersion relationship [Kawahara and Yomogida, 1996].

Another approach for modeling wave propagation in inhomogeneous media is to use the finite difference solution to the wave equation [Alford et al. 1974; Aminzadeh et al., 1994]. This approach is based on a discretization of the medium and the equations of motion describing wave propagation. Although this method is reliable for modeling wave propagation in media having relatively modest spatial variations in elastic properties, it does not work well in media that are strongly heterogeneous since derivatives are calculated as averages over many grid points in the medium, which is equivalent to assuming that the medium varies smoothly rather than having discontinuous variations in medium properties. Finite difference simulations of wave propagation in inhomogeneous media have been conducted by Frankel and Clayton [1986] as described in Section 3.3.2. Figure 2.16 shows a seismogram calculated using a 3-D finite difference simulation of wave propagation in a strongly heterogeneous medium. In this case, the simulation was conducted to provide correct arrival times of signals in the seismogram but little attention was made to calculating the correct amplitudes.

Another approach for doing numerical simulations of wave propagation is the phononic lattice solid method introduced by Mora [1992], in which the medium is discretized and wave motion is transmitted by quasi-particles. The phononic lattice solid method is an extension to wave propagation of the lattice gas [Hardy et al., 1973] and the Boltzmann lattice gas [Holme and Rothman, 1992] approaches that have been used to numerically simulate fluid flow including fluid flow in porous media for geological applications [Rothman, 1988]. In the Boltzmann lattice gas approach to modeling fluid flow, quasi-particles are used to simulate the number density of particles in the fluid. Mora [1992] used quasi-particles traveling between lattice points in a grid describing a medium to simulate wave propagation in heterogeneous media. Huang and Mora [1994] presented an improved version of the phononic lattice solid approach for propagating waves in heterogeneous media that they call the phononic lattice solid by interpolation. Their method is computationally faster and more reliable in strongly heterogeneous media than the phononic lattice solid method of Mora [1992]. Huang and Mora [1994] show how to use a Boltzmann lattice gas approach to model the movement of quasi-particles between lattice points, the scattering of the quasi-particles by medium inhomogeneities, and the interaction of quasi-particles at lattice points. They demonstrate that their approach is equivalent to solving the acoustic wave equation for the inhomogeneous medium in the macroscopic limit. Huang and Mora [1996] used the method to investigate scattering by a medium containing a suite of solid inclusions and one containing empty pores. The inhomogeneities are randomly located and have random sizes ranging between 0.0625 and 0.15 times the wavelength of the incident wave. The solid inclusions have a velocity that is 25% lower than that of the surrounding medium. A point source was located in the center of the medium. They show that the seismograms obtained for the solid inclusion case are dominated by the direct-arriving energy and have little coda whereas the seismograms for the medium with empty pores contain significant coda energy whose amplitudes are similar to those of the direct arrivals.

5.5 POWER-LAW DECAY OF MAXIMUM AMPLITUDE WITH TRAVEL DISTANCE

Models developed in this chapter have been based on the assumption that the first arrival amplitude decays exponentially with increasing travel distance. However, measurements of earthquake magnitude are based on the observation that the maximum amplitude of the first arrival packet decays as a power law with increasing travel distance. Richter's [1935] formula for measuring local magnitudes M_L of earthquakes and subsequent modifications of the formula have often been based on a linear relationship between logarithms of maximum amplitude A_{Max} and hypocentral distance r. This phenomenologically determined relationship applies worldwide. Tsuboi [1954] and later Watanabe [1971] proposed the following relationship for Japan

$$\log A_{Max} = -1.73 \log r + c(M_L) \tag{5.63}$$

Analyzing the maximum amplitude on vertical component seismograms against hypocentral distance for 10–200 km in the Kanto–Tokai area, Japan, Noguchi [1990] reported that the power law

$$\log A_{Max} = -1.96 \log r + c(M_L) \tag{5.64}$$

fits the observed data as shown in Figure 5.20. These observed power-law decay relationships have been considered empirical rules.

Mandelbrot [1983] suggested that, in general, there is a fractal structure underlying the occurrence of power laws. The hypocenter distribution of earthquakes has clustering characteristics [Kagan and Knopoff, 1980]. Figure 5.21 shows the cluster-like epicenter distribution of microearthquakes in Kanto, Japan. The fractal dimension characterizing the clustering is generally less than the Euclidean dimension: 1.6 for the epicenter distribution in Central Asia [Sadovskiy et al., 1984] and 2.2 for hypocenter distribution in Kanto, Japan [Hirata and Imoto, 1991] where the epicenter is the projection onto the earth's surface of the location of the hypocenter. Amplitude attenuation of seismic waves is caused by distributed attenuation bodies

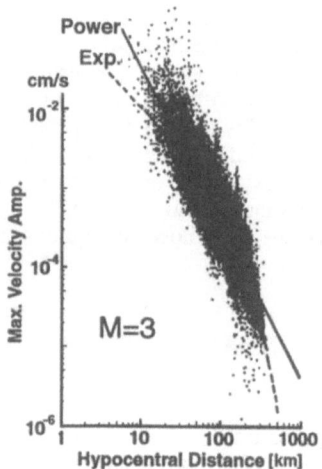

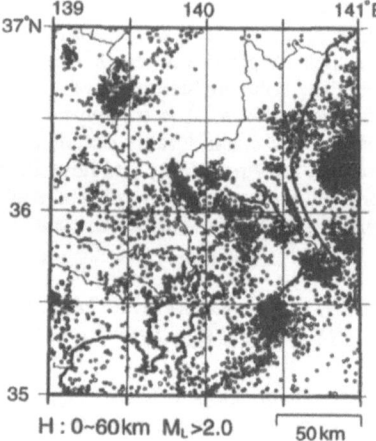

FIGURE 5.20. Maximum amplitude of vertical component seismic waves against travel distance in the Kanto–Tokai region, Japan, where the solid line is (5.64). [From Noguchi, 1990, with permission of NIED, Japan.]

FIGURE 5.21. Epicenter distribution of microearthquakes located in Kanto, Japan by NIED (1982–1988). Note strong clustering of locations. [Data, Courtesy of K. Obara.]

regardless of whether attenuation is dominated by intrinsic mechanisms or scattering. Following the characterization of the hypocenter distribution as a fractal structure, we extend the concept of "homogenous random distribution" to "fractally homogenous random distribution" of attenuation bodies in a statistical sense.

We consider a medium having a fractally homogenous random distribution of attenuation bodies with effective cross section σ_A and number density n_A in 3-D space:

$$n_A(r) = c_A \, r^{D_A - 3} \qquad (5.65)$$

where D_A is the fractal dimension and c_A is a scaling coefficient [Sato, 1988b]. Cross section may be considered that for either scattering or intrinsic absorption. We may take any point as the origin in the fractally homogeneous medium. The number of attenuation bodies within a sphere of radius r is proportional to r^{D_A}.

For seismic energy radiated spherically from the origin, the decrease of energy-flux density $J(r)$ at radius r due to traversing a distance dr is written as

$$\frac{d}{dr}\left[4\pi r^2 J(r)\right] = -\sigma_A n_A(r)\left[4\pi r^2 J(r)\right] \qquad (5.66)$$

Analytically integrating and making the correspondence $J(r) \Rightarrow Amp.^2$, we get

$$J(r) \propto Amp.^2 \propto \begin{cases} \dfrac{1}{4\pi r^2}\exp\left[-c_A\sigma_A r\right] & \text{for } D_A = 3 \\[2mm] \dfrac{1}{4\pi r^2}\exp\left[-\dfrac{c_A\sigma_A}{D_A-2}r^{D_A-2}\right] & \text{for } D_A \neq 2 \\[2mm] \dfrac{1}{4\pi r^{2+c_A\sigma_A}} & \text{for } D_A = 2 \end{cases} \qquad (5.67)$$

For a fractal distribution of attenuation bodies, the amplitude of seismic waves *Amp.* decays exponentially with travel distance for only one value of fractal dimension $D_A = 3$. *Amp.* decays as a power of increasing distance only for $D_A = 2$. Figure 5.22 shows how *Amp.* varies with distance on a logarithmic scale. The power-law decay as given by (5.63) can be interpreted as the case of $D_A = 2$ and $c_A\sigma_A = 1.46$. The curvature of *Amp.* against distance r is very sensitive to the fractal dimension of the distribution.

Applying (5.67) to observed maximum amplitudes of acceleration seismograms from 20 earthquakes in Kanto, Japan, Kishimoto and Kinoshita [1990] estimated fractal dimension D_A for frequencies 3–20 Hz. Fractal dimension D_A was estimated by finding the minimum AIC value for each frequency, where AIC= $-2\times$(maximum likelihood – number of free parameters) [Sakamoto et al., 1986]. The AIC is minimized around $D_A = 2.1$ for frequencies higher than 10 Hz. Their result means that the logarithmic plot of maximum amplitude against travel distance is almost a straight line but a little bit convex. The fractal dimension estimated

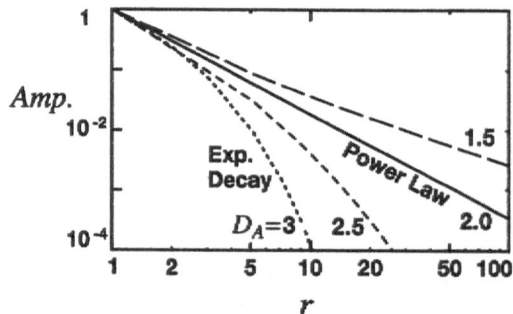

FIGURE 5.22. Plot of *Amp.* against distance r from (5.67), where $c_A\sigma_A = 1.46$. All curves are normalized by their value at $r=1$.

from the dependence of maximum amplitude with travel distance nearly agrees with the fractal dimension of 2.2 for the hypocenter distribution of microearthquakes found for this region by Hirata and Imoto [1991]. Applying (5.67) to the measurement of attenuation in New Mexico, U. S. A. [Carpenter and Sanford, 1985], Godano et al. [1994] estimated that D_A=2.3.

CHAPTER 6

Synthesis of Three-Component Seismogram Envelopes for Earthquakes Using Scattering Amplitudes from the Born Approximation

Figure 2.29 shows that the initial direction of motion of the P-wave, the amplitudes of the direct P- and S-waves, and the high-frequency character of the P- and S-coda envelopes on all three components of motion from a local microearthquake are influenced by the angle between the fault plane and the receiver and the scattering characteristics of the lithosphere. To better understand the effects of scattering on recorded regional seismograms, we desire a method to synthesize three-component seismogram envelopes for realistic earthquake sources in an inhomogeneous medium. The simplest way to synthesize seismograms is to sum up all waves scattered by distributed heterogeneities in the time domain. Craig et al. [1991] synthesized the high-frequency seismograms of a local explosion by summing up all singly scattered waves from distributed spherical obstacles; however, this method needs detailed information about the scatterers and considerable computational work. It is easier to synthesize the MS envelope by adding the power of scattered waves. Assuming the complete incoherence of scattered waves, Malin [1980] synthesized the MS acoustic-wave envelope by summing up scattered wave power. Extending the method of summing up single scattered energy that was developed in Chapter 3, we will introduce a way to synthesize three-component seismogram envelops [Sato, 1984a] resulting from realistic earthquake sources in an infinite inhomogeneous medium. We include the source radiation from a point shear-dislocation and the frequency-dependent nonisotropic scattering resulting from the Born approximation that was discussed in Chapters 4 and 5. Next, we compare the synthesized enve-

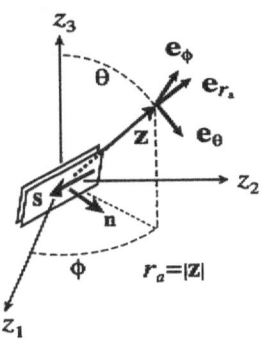

FIGURE 6.1. Geometry of the point shear-dislocation, where **n** is normal to the fault and **s** represents the fault slip direction.

lopes with observed envelopes. In the last section, a simulation method for an inhomogeneous half-space is introduced, and we show a comparison with high-frequency seismogram envelopes of microearthquakes observed by a local network.

6.1 EARTHQUAKE SOURCE

6.1.1 Point Shear-Dislocation

An earthquake is considered to be a rupture along a fault in the earth. We will not concern ourselves with the dynamics of rupture propagation [see e.g. Kostrov and Das, 1988] but instead choose to use a simple model describing nonspherical radiation from the fault region. When the travel distance and the wavelength of radiated waves are much larger than the fault dimension L_F, the earthquake source can be represented as a point shear-dislocation. This source is geometrically characterized by a unit vector n normal to the fault plane and a unit vector s parallel to the direction of fault slip as illustrated in Figure 6.1. Aki and Richards [1980, p. 113] show that the far-field displacement vector in an infinite homogenous elastic medium is related to the seismic moment time function $M(t)$ representing the particle slip along the fault by

$$\mathbf{u}(\mathbf{z},t) = B_r^P \mathbf{e}_{r_a} \frac{2}{\sqrt{15}} \frac{\dot{M}(t - r_a/\alpha_0)}{4\pi\rho_0\alpha_0^3 r_a} + \left[B_\theta^S \mathbf{e}_\theta + B_\phi^S \mathbf{e}_\phi\right]\sqrt{\frac{2}{5}} \frac{\dot{M}(t - r_a/\beta_0)}{4\pi\rho_0\beta_0^3 r_a} \qquad (6.1)$$

where $\mathbf{z} = (r_a, \theta, \phi)$ in spherical coordinates and the overdot means time derivative. The three orthogonal unit vectors of the spherical coordinate system are given by

$$\mathbf{e}_{r_a} \equiv \mathbf{z}/|\mathbf{z}| = \sin\theta\cos\phi\,\mathbf{e}_1 + \sin\theta\sin\phi\,\mathbf{e}_2 + \cos\theta\,\mathbf{e}_3$$

$$\mathbf{e}_\theta = \cos\theta\cos\phi\,\mathbf{e}_1 + \cos\theta\sin\phi\,\mathbf{e}_2 - \sin\theta\,\mathbf{e}_3 \qquad (6.2)$$

$$\mathbf{e}_\phi = -\sin\phi\,\mathbf{e}_1 + \cos\phi\,\mathbf{e}_2$$

The first term in (6.1) is the far-field P-wave and the second term is the far-field S-wave, where B_r^P, B_θ^S and B_ϕ^S are the radiation patterns of P- and S-waves:

$$B_r^P(\theta,\phi;\mathbf{n},\mathbf{s}) = \sqrt{15}(\mathbf{e}_{r_a}\mathbf{n})(\mathbf{e}_{r_a}\mathbf{s})$$

$$B_\theta^S(\theta,\phi;\mathbf{n},\mathbf{s}) = \sqrt{\frac{5}{2}}\left[(\mathbf{e}_\theta\mathbf{n})(\mathbf{e}_{r_a}\mathbf{s}) + (\mathbf{e}_\theta\mathbf{s})(\mathbf{e}_{r_a}\mathbf{n})\right] \qquad (6.3)$$

$$B_\phi^S(\theta,\phi;\mathbf{n},\mathbf{s}) = \sqrt{\frac{5}{2}}\left[(\mathbf{e}_\phi\mathbf{n})(\mathbf{e}_{r_a}\mathbf{s}) + (\mathbf{e}_\phi\mathbf{s})(\mathbf{e}_{r_a}\mathbf{n})\right]$$

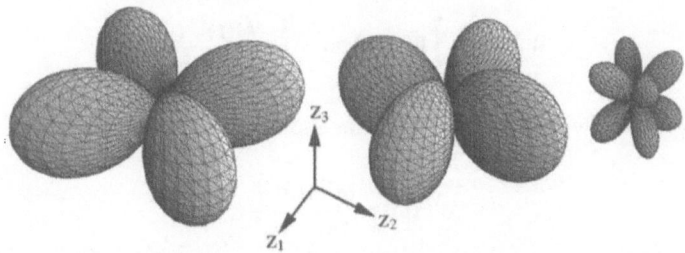

FIGURE 6.2. Radiation patterns of a point shear-dislocation source, where $\mathbf{n}=(1,0,0)$ and $\mathbf{s}=(0,1,0)$. From left to right, $|B_r{}^p(\theta,\phi)|$, $|B_\theta{}^s(\theta,\phi)|$ and $|B_\phi{}^s(\theta,\phi)|$.

The P-wave displacement vector has only a radial component. The S-wave displacement vector appears only on transverse components. Radiation patterns are normalized by

$$\oint |B_r{}^P(\theta,\phi;\mathbf{n},\mathbf{s})|^2 d\Omega(\theta,\phi) = \oint \left[|B_\theta{}^S(\theta,\phi;\mathbf{n},\mathbf{s})|^2 + |B_\phi{}^S(\theta,\phi;\mathbf{n},\mathbf{s})|^2\right] d\Omega(\theta,\phi) = 4\pi \quad (6.4)$$

where $d\Omega(\theta,\phi) = \sin\theta\, d\theta\, d\phi$. For example, for a fault whose plane is parallel to the z_2 - z_3 plane with normal in the direction z_1 and fault slip in the z_2 direction, the fault geometry is given by $\mathbf{n}=(1, 0, 0)$ and $\mathbf{s}=(0, 1, 0)$, and the radiation patterns are given by

$$B_r{}^P(\theta,\phi;\mathbf{n},\mathbf{s}) = \frac{\sqrt{15}}{2}\sin^2\theta\sin 2\phi$$

$$B_\theta{}^S(\theta,\phi;\mathbf{n},\mathbf{s}) = \frac{1}{2}\sqrt{\frac{5}{2}}\sin 2\theta\sin 2\phi \qquad (6.5)$$

$$B_\phi{}^S(\theta,\phi;\mathbf{n},\mathbf{s}) = \sqrt{\frac{5}{2}}\sin\theta\cos 2\phi$$

Figure 6.2 shows 3-D perspective views of the corresponding radiation patterns. $B_r{}^P$ and $B_\phi{}^S$ have four lobes each, and $B_\theta{}^S$ has eight lobes of smaller amplitudes. There is no far-field radiation in the null direction which is given by $\mathbf{n}\times\mathbf{s}$.

6.1.2 Omega-Square Model for the Source Spectrum

Integrating energy-flux density $\alpha_0\rho_0|\dot{u}|^2$ given by (6.1) for P-waves over the surface of a sphere and time, we get the radiated P-wave energy

$$W^P = \frac{1}{2\pi}\int_{-\infty}^{\infty}\widehat{W}^P(\omega)d\omega = \int_0^{T_0}\frac{\ddot{M}(t)^2}{15\pi\rho_0\alpha_0^5}dt \qquad (6.6)$$

where T_0 is the source duration. The P-wave source energy spectral density is given by

$$\widehat{W}^P(\omega) = \frac{\omega^4|M(\omega)|^2}{15\pi\rho_0\alpha_0^5} \qquad (6.7)$$

where $M(\omega)$ is the Fourier transform of the seismic moment time function. Using the same procedure, we get the radiated S-wave energy

$$W^S = \frac{1}{2\pi}\int_{-\infty}^{\infty}\widehat{W}^S(\omega)d\omega = \int_0^{T_0}\frac{\ddot{M}(t)^2}{10\pi\rho_0\beta_0^5}dt \qquad (6.8)$$

where the S-wave source energy spectral density is expressed by

$$\widehat{W}^S(\omega) = \frac{\omega^4|M(\omega)|^2}{10\pi\rho_0\beta_0^5} \qquad (6.9)$$

S-wave energy in a frequency band of width Δf and central frequency f is $\widehat{W}^S(2\pi f)\Delta f$. The ratio of radiated S- to P-wave energy is a function of the ratio of P- to S-wave velocity γ_0:

$$\frac{W^S}{W^P} = \frac{\widehat{W}^S(\omega)}{\widehat{W}^P(\omega)} = \frac{3}{2}\gamma_0^5 \qquad (6.10)$$

The ratio of the energies is about 23.4 for a Poisson solid, $\gamma_0 = \sqrt{3}$.

One well-known source spectrum that has been shown to be consistent with observed radiation from earthquakes is the omega-square model [Aki and Richards, 1980, p. 823]:

$$\omega|M(\omega)| = \frac{M_0}{1+(\omega/2\pi f_c)^2} \qquad (6.11)$$

where f_c is the corner frequency and M_0 is the seismic moment. From (6.1) the spectrum of far-field displacement waveform is flat at low frequencies and rolls off like ω^{-2} at high frequencies. For small earthquakes, there is an empirical rule between M_0 [dyn $\cdot$ cm] and the local magnitude M_L [Thatcher and Hanks, 1973]:

$$\log M_0 = 1.5M_L + 16.0 \qquad (6.12)$$

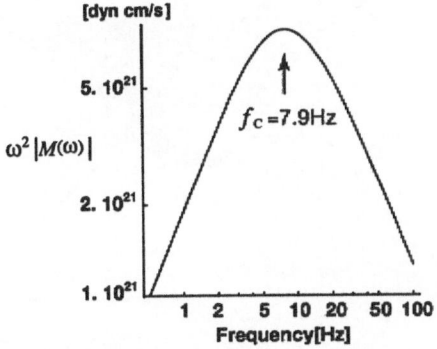

FIGURE 6.3. Velocity source spectrum predicted by the omega-square source model for an M_L=3 earthquake.

There is an empirical rule relating observed corner frequency f_C [Hz] and M_L [Watanabe, 1971]:

$$\log f_C = 1.5 - 0.20 M_L \qquad (6.13)$$

In Figure 6.3, we illustrate the velocity source spectrum $\omega^2 |M(\omega)|$ for an earthquake having M_L=3. The spectrum has a maximum at 7.9 Hz.

6.2 ENVELOPE SYNTHESIS IN AN INFINITE SPACE

6.2.1 Geometry of Source and Receiver

We imagine an ensemble of infinite 3-D inhomogeneous media characterized by their fractional fluctuation of wave velocity $\xi(\mathbf{x})$, where the ensemble average $\langle \xi \rangle = 0$. As discussed in Section 4.2.2, we assume that the fractional fluctuations of P- and S-wave velocities are the same and that of mass density is equal to $\nu \xi$ with ν=0.8. The inhomogeneous elastic media can thus be characterized statistically using one ACF function or its PSDF.

The seismic source is taken as a point shear-dislocation source, as given in Section 6.1. The source is located at the origin and a three-component seismometer is located at distance r along the z_3 axis, so $\mathbf{x} = (0,0,r)$. Particle motions are recorded in directions $\mathbf{e}_1$, $\mathbf{e}_2$, and $\mathbf{e}_3$ as illustrated in Figure 6.4. We divide the inhomogeneous medium into cubes having dimension L with $L > a$ in order to assure the incoherency of waves scattered from different cubes. We consider a ray whose path starts from the source, is singly scattered within a cube having center coordinate $\mathbf{z}$, and reaches the receiver. The distances between the source and scatterer and between the scatterer and receiver are $r_a = |\mathbf{z}|$ and $r_b = |\mathbf{x} - \mathbf{z}|$, respectively. We define

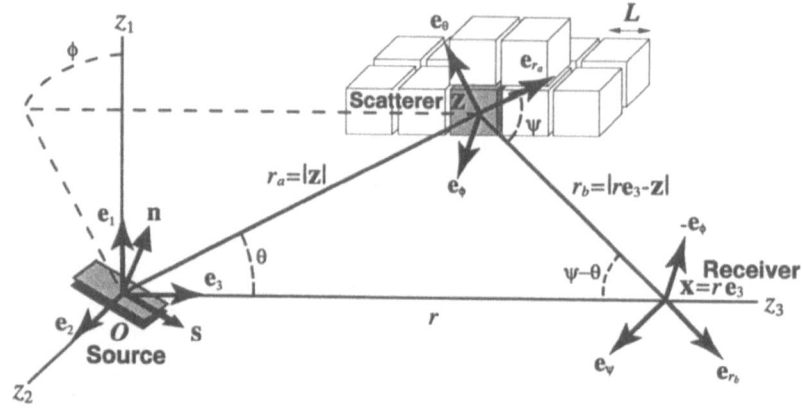

FIGURE 6.4. Geometry of the single scattering process. A point shear-dislocation source is located at the origin. The inhomogeneous medium is broken into cubic elements of length L in which scattering occurs. A three-component seismometer is located at distance r along the z_3 axis.

the unit vector pointing from the scatterer to the receiver by

$$\mathbf{e}_{r_b} \equiv \frac{\mathbf{x} - \mathbf{z}}{|\mathbf{x} - \mathbf{z}|} = \cos\psi \, \mathbf{e}_{r_a} - \sin\psi \, \mathbf{e}_\theta$$

$$= -\sin(\psi - \theta)\cos\phi \, \mathbf{e}_1 - \sin(\psi - \theta)\sin\phi \, \mathbf{e}_2 + \cos(\psi - \theta) \, \mathbf{e}_3$$

(6.14)

The unit vector corresponding to a scattering angle is defined as

$$\mathbf{e}_\psi \equiv -\sin\psi \, \mathbf{e}_{r_a} - \cos\psi \, \mathbf{e}_\theta$$

$$= -\cos(\psi - \theta)\cos\phi \, \mathbf{e}_1 - \cos(\psi - \theta)\sin\phi \, \mathbf{e}_2 - \sin(\psi - \theta) \, \mathbf{e}_3$$

(6.15)

It will be useful in the following to note that $\left(\mathbf{e}_{r_b}, \mathbf{e}_\psi, -\mathbf{e}_\phi\right)$ are orthogonal unit vectors of a spherical coordinate system whose origin is at the scatterer.

6.2.2 Power Spectral Density of Velocity Wavefield at the Receiver

Power Spectral Density

We may define the Fourier transform of the velocity wavefield vector at location $\mathbf{z}$ for a time window having length ΔT beginning at time t as

$$\dot{u}_i(z,t;\omega) = \int_t^{t+\Delta T} \dot{u}_i(z,t+t')e^{i\omega t'}\,dt' \tag{6.16}$$

The corresponding power spectral density of the ith component of the velocity wavefield is a function of both time and angular frequency:

$$S_i(z,t;\omega) = \frac{1}{\Delta T}\left|\dot{u}_i(z,t;\omega)\right|^2 \tag{6.17}$$

We assume that the window length ΔT is longer than the source duration T_0. We expect that the change in S_i is a smooth function of t, that is, we take the time resolution to be of the order of ΔT in the following and assume that the variation of wavefield is quasi-stationary within a window having such a time length.

The time average of the energy density for component i at the receiver is twice the kinetic energy. It is written as

$$E_i(\mathbf{x},t) = \rho_0 \left\langle \left|\dot{u}_i(\mathbf{x},t)\right|^2 \right\rangle_T = \frac{\rho_0}{2\pi} \int_{-\infty}^{\infty} S_i(\mathbf{x},t;\omega)\,d\omega \tag{6.18}$$

where $\langle \cdots \rangle_T$ is a time average over ΔT. From the single scattering model for elastic waves, we can construct the power spectral density at the receiver as the sum of two direct propagation terms and four scattering modes:

$$\begin{aligned} S_i(\mathbf{x},t;\omega) &= S_i^{0P}(\mathbf{x},t;\omega) + S_i^{0S}(\mathbf{x},t;\omega) \\ &+ S_i^{PP}(\mathbf{x},t;\omega) + S_i^{PS}(\mathbf{x},t;\omega) + S_i^{SP}(\mathbf{x},t;\omega) + S_i^{SS}(\mathbf{x},t;\omega) \end{aligned} \tag{6.19}$$

where S_i^{0P} and S_i^{0S} are the power spectral densities of direct waves, S_i^{PS} is a P-wave from the source that is scattered to an S-wave in the medium and recorded as an S-wave at the receiver, etc. To solve for the waveform envelope, we must determine each of the six terms in (6.19) for each component of motion.

Example: Solving for the Contribution from S-to-P Single Scattering

From (6.1), the direct velocity wavefield spectra at angular frequency ω at a scatterer at z are given by

$$\dot{\mathbf{u}}^{0P}(z,t;\omega) = B_r^{\,P}\mathbf{e}_{r_a}\frac{2}{\sqrt{15}}\frac{-\omega^2 M(\omega)}{4\pi\rho_0\alpha_0{}^3 r_a}e^{-Q_P^{-1}(\omega)\frac{\omega r_a}{2\alpha_0}} \quad \text{for } 0 < t - \frac{r_a}{\alpha_0} < \Delta T$$
$$= 0 \qquad\qquad\qquad\qquad\qquad\qquad\qquad\quad \text{otherwise} \tag{6.20a}$$

and

$$\dot{u}^{0S}(z,t;\omega) = \left[B_\theta{}^S e_\theta + B_\phi{}^S e_\phi\right]\sqrt{\frac{2}{5}}\frac{-\omega^2 M(\omega)}{4\pi\rho_0\beta_0{}^3 r_a}e^{-Q_S^{-1}(\omega)\frac{\omega r_a}{2\beta_0}} \quad \text{for } 0 < t - \frac{r_a}{\beta_0} < \Delta T$$

$$= 0 \qquad\qquad\qquad\qquad \text{otherwise} \qquad\qquad (6.20b)$$

where the trailing exponential term accounts for energy loss between the source and scattering location.

We first consider SP single scattering amplitude for each S-wave polarization. Since we consider that the scattering occurs in the far field of the earthquake source, we assume that the spherical wavefront representing the S-waves radiated from the source can be represented as plane waves in a cube of extent L so that $lr_a \gg 1$ and $r_a \gg L$. The plane waves are scattered by the inhomogeneous cube with dimension L. We also assume that the receiver is located in the far field of the scattered P-waves so that $kr_b \gg 1$, and $r_b \gg L$. We also introduce a factor to account for attenuation between the scattering region and the receiver. Then, using (4.43c), the SP scattered wavefield spectra at angular frequency ω at the receiver is given by

$$\dot{u}^{1SP}(x,t;\omega) = \sqrt{\frac{2}{5}}\frac{-\omega^2 M(\omega)}{4\pi\rho_0\beta_0{}^3 r_a\, r_b}\left[B_\theta{}^S(\theta,\phi;n,s)\cdot F_r^{SP}(\psi,\pi;\omega)e_{r_b}\right.$$

$$\left.+ B_\phi{}^S(\theta,\phi;n,s)\cdot F_r^{SP}(\psi,\pi/2;\omega)e_{r_b}\right]e^{-Q_S^{-1}(\omega)\frac{\omega r_a}{2\beta_0}-Q_P^{-1}(\omega)\frac{\omega r_b}{2u_0}}$$

$$\text{for } 0 < t - r_a/\beta_0 - r_a/\alpha_0 < \Delta T \qquad (6.21)$$

$$= 0 \qquad\qquad\qquad \text{otherwise}$$

where scattering angle $\zeta = \pi$ for incident S-waves polarized to e_θ, and $\zeta = \pi/2$ for the polarization to e_ϕ. We discarded the common phase factor e^{ikr_b}. We take time window length $\Delta T \geq L/\beta_0$, since we consider only the power of incoherent scattered waves from scattering cubes of linear dimension L. To evaluate the partition of the scattered energy that contributes to the ith component of motion, we take the inner product of the scattered velocity wavefield vector with the ith unit vector e_i. Using $F_r^{SP}(\psi,\pi/2;\omega) = 0$ as given by (4.46c) and the source energy spectral density given by (6.9), the square of ith velocity wavefield spectra due to S-waves from the source that are singly scattered to P-waves in one scattering cube is given by

$$\left|\dot{u}_i^{1SP}(x,t;\omega)\right|^2 = \frac{1}{\rho_0\beta_0}\frac{\hat{W}^S(\omega)}{4\pi r_a{}^2 r_b{}^2}\left|B_\theta{}^S(\theta,\phi;n,s)\right|^2\left|F_r^{SP}(\psi,\pi;\omega)\right|^2$$

$$\times\left(e_{r_b}\cdot e_i\right)^2 e^{-Q_S^{-1}(\omega)\omega r_a/\beta_0 - Q_P^{-1}(\omega)\omega r_b/\alpha_0}$$

$$\text{for } 0 < t - r_a/\beta_0 - r_a/\alpha_0 < \Delta T \qquad (6.22)$$

$$= 0 \qquad\qquad\qquad \text{otherwise}$$

Dividing (6.22) by window length ΔT and summing up the contribution of all scattering cubes with volume L^3, we have the power spectral density S_i^{SP} at the receiver. Then, we replace the summation over all scattering cubes by an integral over infinite 3-D space and the square of scattering amplitude with its ensemble average. We may replace the time condition with the delta function in time. It is equivalent to assuming that the source energy radiation at angular frequency ω is $\hat{W}^S(\omega)\delta(t)$ for S-waves. Finally, we obtain the power spectral density for single SP-scattering

$$S_i^{SP}(\mathbf{x},t;\omega) = \frac{\hat{W}^S(\omega)}{4\pi\rho_0\beta_0} \int_{-\infty}^{\infty}\int_{-\infty}^{\infty}\int_{-\infty}^{\infty} d\mathbf{z}\ \delta\left(t - \frac{r_a}{\beta_0} - \frac{r_b}{\alpha_0}\right)\frac{1}{r_a^2 r_b^2} B_\theta^S(\theta,\phi;\mathbf{n},\mathbf{s})^2$$

$$\times \left\{\left\langle\left|F_r^{SP}(\psi,\pi;\omega)\right|^2\right\rangle\Big/L^3\right\}\left(\mathbf{e}_{r_b}\cdot\mathbf{e}_i\right)^2 e^{-Q_S^{-1}(\omega)\omega r_a/\beta_0 - Q_P^{-1}(\omega)\omega r_b/\alpha_0} \qquad (6.23a)$$

This final representation is independent of the choice of ΔT. From (4.52c), the square of SP-scattering amplitude per volume L^3 is independent of the choice of dimension L. The integral is practically taken over the isochronal scattering shell corresponding to a given lapse time.

Contributions from Other Scattering Modes

In the same way, we can derive the contribution to the power spectral density of the ith velocity wavefield component at the receiver for the other scattering modes:

$$S_i^{PP}(\mathbf{x},t;\omega) = \frac{\hat{W}^P(\omega)}{4\pi\rho_0\alpha_0} \int_{-\infty}^{\infty}\int_{-\infty}^{\infty}\int_{-\infty}^{\infty} d\mathbf{z}\ \delta\left(t - \frac{r_a}{\alpha_0} - \frac{r_b}{\alpha_0}\right)\frac{1}{r_a^2 r_b^2} B_r^P(\theta,\phi;\mathbf{n},\mathbf{s})^2$$

$$\times \left[\left\langle\left|{}^T F_r^{PP}(\psi,\zeta;\omega)\right|^2\right\rangle\Big/L^3\right]\left(\mathbf{e}_{r_b}\cdot\mathbf{e}_i\right)^2 e^{-\omega Q_P^{-1}(\omega)t} \qquad (6.23b)$$

$$S_i^{PS}(\mathbf{x},t;\omega) = \frac{\hat{W}^P(\omega)}{4\pi\rho_0\alpha_0} \int_{-\infty}^{\infty}\int_{-\infty}^{\infty}\int_{-\infty}^{\infty} d\mathbf{z}\ \delta\left(t - \frac{r_a}{\alpha_0} - \frac{r_b}{\beta_0}\right)\frac{1}{r_a^2 r_b^2} B_r^P(\theta,\phi;\mathbf{n},\mathbf{s})^2$$

$$\times \left[\left\langle\left|F_\psi^{PS}(\psi,\zeta;\omega)\right|^2\right\rangle\Big/L^3\right]\left(\mathbf{e}_\psi\cdot\mathbf{e}_i\right)^2 e^{-\omega Q_P^{-1}(\omega)r_a/\alpha_0 - \omega Q_S^{-1}(\omega)r_b/\beta_0} \qquad (6.23c)$$

$$S_i^{SS}(\mathbf{x},t;\omega) = \frac{\hat{W}^S(\omega)}{4\pi\rho_0\beta_0} \int_{-\infty}^{\infty}\int_{-\infty}^{\infty}\int_{-\infty}^{\infty} d\mathbf{z}\ \delta\left(t - \frac{r_a}{\beta_0} - \frac{r_b}{\beta_0}\right)\frac{1}{r_a^2 r_b^2}$$

$$\times \left[\left\langle\ \left|B_\theta^S(\theta,\phi;\mathbf{n},\mathbf{s})\ {}^T F_\psi^{SS}(\psi,\pi;\omega)(\mathbf{e}_\psi\cdot\mathbf{e}_i)\right.\right.\right. \qquad (6.23d)$$

$$\left.\left.\left.+ B_\phi^S(\theta,\phi;\mathbf{n},\mathbf{s})\ {}^T F_\zeta^{SS}(\psi,\pi/2;\omega)(-\mathbf{e}_\phi\cdot\mathbf{e}_i)\right|^2\right\rangle\Big/L^3\right] e^{-\omega Q_S^{-1}(\omega)t}$$

where we use explicit representation of scattering amplitudes for PP- and SS-scattering as given by (5.44) and (5.45) that were derived by using the travel-time corrected Born approximation since scattering by only the short-wavelength components of random inhomogeneity contributes to the envelope formation. We note that F_ψ^{PS} and $^T F_r^{PP}$ are independent of ζ as shown in (4.46) and (5.37), respectively. Similar to (4.52), we can write the ensemble average of the squared scattering amplitude per unit volume by using the PSDF of the random fluctuation.

Isochronal Scattering Shells in Prolate Spheroidal Coordinates

Since there are delta functions in the volume integrals over 3-D space in (6.23a-d) we can reduce the triple integrals to double integrals by using the prolate spheroidal coordinate system having foci at the source and the receiver as was done in Section 3.1.2. This transformation is necessary to allow us to numerically evaluate the integrals when we want to simulate average seismogram envelopes. Introducing the transformation (3.13), we write the distances between the source and the scatterer and the scatterer to the receiver as in (3.14). Then, delta functions can be written as

$$\delta\left(t - \frac{r_a}{\alpha_0} - \frac{r_b}{\alpha_0}\right) = \frac{\alpha_0}{r}\delta\left(\frac{\alpha_0 t}{r} - v\right) \qquad \text{for PP - scattering}$$

$$\delta\left(t - \frac{r_a}{\alpha_0} - \frac{r_b}{\beta_0}\right) = \frac{2\alpha_0}{(1+\gamma_0)r}\delta\left[\frac{2\alpha_0 t/r + (\gamma_0 - 1)w}{1+\gamma_0} - v\right] \qquad \text{for PS - scattering}$$

$$\delta\left(t - \frac{r_a}{\beta_0} - \frac{r_b}{\alpha_0}\right) = \frac{2\alpha_0}{(1+\gamma_0)r}\delta\left[\frac{2\alpha_0 t/r - (\gamma_0 - 1)w}{1+\gamma_0} - v\right] \qquad \text{for SP - scattering}$$

$$\delta\left(t - \frac{r_a}{\beta_0} - \frac{r_b}{\beta_0}\right) = \frac{1}{\gamma_0}\frac{\alpha_0}{r}\delta\left(\frac{1}{\gamma_0}\frac{\alpha_0 t}{r} - v\right) = \frac{\beta_0}{r}\delta\left(\frac{\beta_0 t}{r} - v\right) \qquad \text{for SS - scattering}$$

(6.24)

where $\alpha_0 t/r$ is the lapse time normalized by the P-wave travel-time. The argument of each delta function is a linear combination of prolate spheroidal coordinates w and v. The isochronal scattering shell, which defines the locus of points where waves arriving at a given lapse time at the receiver must have scattered for a given scattering mode, is found by setting the argument of each delta function equal to zero. The first and last lines of (6.24) are the delta functions in the expressions for PP- and SS-scattering, respectively. The zeros of these two delta functions occur where v is equal to the lapse time divided by the direct wave travel-time, which means that the isochronal scattering shells are spheroidal. For PS- and SP-scattering, the second and third lines of (6.24), respectively, linear combinations of w, v, and the normalized lapse-time determine the isochronal scattering shells, which are axially symmetric around the source-receiver axis and look like egg shells in space. We illustrate cross sections of the isochronal scattering shells for four scattering modes at different normalized lapse-times in Figure 6.5. Their surfaces become larger as lapse time increases. We note that the point-source model, the far-field condition, and the condition for the Fraunhofer zone may not be valid in a strict sense when the lapse time is near the direct P- or S-wave arrival-time since either path length r_a or r_b can take very small values.

Scattering angle ψ and radiation angle θ can be found for each scattering mode using (3.17) with v given as a function of w and normalized lapse time $\alpha_0 t / r$ by setting the argument of (6.24) to zero. Figure 6.6 shows plots of ψ versus θ for different scattering modes at different normalized lapse time $\alpha_0 t / r$.

The infinitesimal volume element dz in (6.23a–d) is given by (3.15) for the prolate spheroidal coordinate system. We rewrite equations (6.23a–d) as follows:

$$S_i^{PP}(\mathbf{x},t;\omega) = H\left(\frac{\alpha_0}{r}t - 1\right)\frac{\widehat{W}^P(\omega)}{2\pi\rho_0 r^2}\int_0^{2\pi}d\phi\int_{-1}^{1}dw\,\frac{1}{v^2 - w^2}B_r^P(\theta,\phi;\mathbf{n},\mathbf{s})^2$$

$$\times\left[\left\langle\left|^T F_r^{PP}(\psi,\zeta;\omega)\right|^2\right\rangle\Big/L^3\right]\left(\mathbf{e}_{r_b}\cdot\mathbf{e}_i\right)^2 e^{-Q_r^{-1}(\omega)\omega t}$$

(6.25a)

where $v = \dfrac{\alpha_0}{r}t$.

$$S_i^{PS}(\mathbf{x},t;\omega) = H\left(\frac{\alpha_0}{r}t - 1\right)\frac{\widehat{W}^P(\omega)}{2\pi\rho_0 r^2}\frac{2}{1+\gamma_0}\int_0^{2\pi}d\phi\int dw\,\frac{1}{v^2 - w^2}B_r^P(\theta,\phi;\mathbf{n},\mathbf{s})^2$$

$$\times\left[\left\langle\left|F_\psi^{PS}(\psi,\zeta;\omega)\right|^2\right\rangle\Big/L^3\right]\left(\mathbf{e}_\psi\cdot\mathbf{e}_i\right)^2 e^{-\omega\left[(v+w)\frac{Q_P^{-1}(\omega)}{\alpha_0}+(v-w)\frac{Q_S^{-1}(\omega)}{\beta_0}\right]\frac{r}{2}}$$

(6.25b)

where $v = \dfrac{2\alpha_0 t/r + (\gamma_0 - 1)w}{1 + \gamma_0}$ and the integral bound for w is

$$\frac{1+\gamma_0 - 2\alpha_0 t/r}{\gamma_0 - 1} \le w \le 1 \quad \text{for } 1 < \frac{\alpha_0}{r}t < \gamma_0 \quad \text{and}$$

$$-1 \le w \le 1 \quad\qquad \text{for } \frac{\alpha_0}{r}t > \gamma_0$$

$$S_i^{SP}(\mathbf{x},t;\omega) = H\left(\frac{\alpha_0}{r}t - 1\right)\frac{\widehat{W}^S(\omega)}{2\pi\rho_0 r^2}\frac{2\gamma_0}{1+\gamma_0}\int_0^{2\pi}d\phi\int dw\,\frac{1}{v^2 - w^2}B_\theta^S(\theta,\phi;\mathbf{n},\mathbf{s})^2$$

$$\times\left[\left\langle\left|F_r^{SP}(\psi,\pi;\omega)\right|^2\right\rangle\Big/L^3\right]\left(\mathbf{e}_{r_b}\cdot\mathbf{e}_i\right)^2 e^{-\omega\left[(v+w)\frac{Q_S^{-1}(\omega)}{\beta_0}+(v-w)\frac{Q_P^{-1}(\omega)}{\alpha_0}\right]\frac{r}{2}}$$

(6.25c)

where $v = \dfrac{2\alpha_0 t/r - (\gamma_0 - 1)w}{1 + \gamma_0}$ and the integral bound for w is

$$-1 \le w \le \frac{2\alpha_0 t/r - (1+\gamma_0)}{\gamma_0 - 1} \quad \text{for } 1 < \frac{\alpha_0}{r}t < \gamma_0 \quad \text{and}$$

$$-1 \le w \le 1 \quad\qquad \text{for } \frac{\alpha_0}{r}t > \gamma_0$$

$$S_i^{SS}(\mathbf{x},t;\omega) = H\left(\frac{\beta_0}{r}t - 1\right)\frac{\widetilde{W}^S(\omega)}{2\pi\rho_0 r^2}\int_0^{2\pi}d\phi\int_{-1}^{1}dw\frac{1}{v^2 - w^2}$$

$$\times\left[\left\langle\left|B_\theta^S(\theta,\phi;\mathbf{n},\mathbf{s})\,{}^TF_\psi^{SS}(\psi,\pi;\omega)(\mathbf{e}_\psi\cdot\mathbf{e}_i)\right.\right.\right. \tag{6.25d}$$

$$\left.\left.\left.+\,B_\phi^S(\theta,\phi;\mathbf{n},\mathbf{s})\,{}^TF_\zeta^{SS}(\psi,\pi/2;\omega)(-\mathbf{e}_\phi\cdot\mathbf{e}_i)\right|^2\right\rangle\middle/L^3\right]e^{-\omega Q_S^{-1}(\omega)t}$$

where $v = \dfrac{\beta_0}{r}t$.

S-Coda Excitation

The predicted S-coda-wave excitation can be compared with the phenomenological predictions made in Chapter 3. For late arriving S-coda waves, $\beta_0 t/r \gg 1$, we may set $\psi \approx \pi$ as shown in Figure 6.6. From (4.49) and (4.52d),

$$\left\langle\left|{}^TF_\psi^{SS}(\pi,\pi;\omega)\right|^2\right\rangle\middle/L^3 = \left\langle\left|{}^TF_\zeta^{SS}(\pi,\pi/2;\omega)\right|^2\right\rangle\middle/L^3$$

$$= -\left\langle{}^TF_\psi^{SS}(\pi,\pi;\omega)^{*}{}^TF_\zeta^{SS}(\pi,\pi/2;\omega)\right\rangle\middle/L^3 = g_\pi^{SS}(\omega)/4\pi \tag{6.26}$$

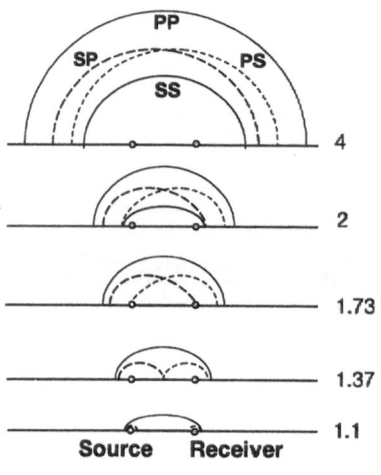

FIGURE 6.5. Isochronal scattering shells projected onto $z_1 - z_3$ plane for different normalized lapse-times $\alpha_0 t/r$. [From Sato, 1984a, copyright by the American Geophysical Union.]

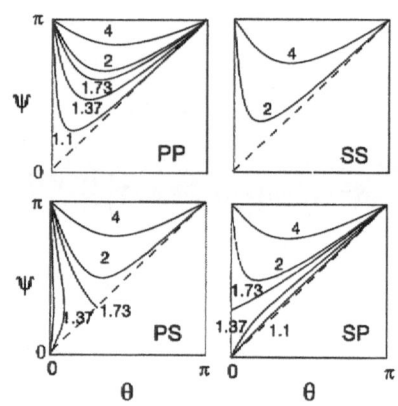

FIGURE 6.6. Scattering angle vs. radiation angle for different normalized lapse times $\alpha_0 t/r$. [From Sato, 1984a, copyright by the American Geophysical Union.]

Setting $e_\psi \approx e_\theta$ in (6.25d), we get the same lapse-time dependence found in the single backscattering model as given by (3.10):

$$S_i(x,t;\omega) \approx S_i^{SS}(x,t;\omega)$$

$$= \frac{\widehat{W}^S(\omega)g_\pi^{SS}(\omega)}{2\pi\rho_0\beta_0^2 t^2} C_i(n,s)e^{-\omega Q_S^{-1}(\omega)t} \quad \text{for } \beta_0 t/r \gg 1 \tag{6.27}$$

However, the partition of the energy into different components is strongly controlled by the source radiation pattern

$$C_i(n,s) \equiv \frac{1}{4\pi}\oint\left[B_\theta{}^S(\theta,\phi;n,s)(e_\theta e_i) + B_\phi{}^S(\theta,\phi;n,s)(e_\phi e_i)\right]^2 d\Omega(\theta,\phi) \tag{6.28}$$

When we take the fault geometry as $n = (0,0,1)$ and $s = (0,1,0)$,

$$C_1 = \frac{2}{21} \quad \text{and} \quad C_2 = C_3 = \frac{19}{42} \tag{6.29}$$

where we note that $C_1 + C_2 + C_3 = 1$. The S-coda amplitude is small in the null direction. If the source radiation is spherical, there is no difference in the coda excitation among components since $C_1 = C_2 = C_3 = 1/3$.

6.2.3 Numerical Simulations

Power Spectral Densities for a Microearthquake

Using (6.20a, b) with (6.3), (6.7) and (6.9), we obtain the power spectral density of the direct velocity wavefield smoothed over the time window ΔT at the receiver on the third axis:

$$S_i^{0P}(x,t;\omega) = \frac{\widehat{W}^P(\omega)}{4\pi\rho_0\alpha_0 r^2\Delta T}15(n_3 s_3)^2 e^{-Q_P^{-1}(\omega)\omega r/\alpha_0}$$

$$\text{for} \quad r/\alpha_0 < t < r/\alpha_0 + \Delta T \quad \text{and } i = 3 \tag{6.30a}$$

$$= 0 \quad \text{otherwise}$$

$$S_i^{0S}(x,t;\omega) = \frac{\widehat{W}^S(\omega)}{4\pi\rho_0\beta_0 r^2\Delta T}\frac{5}{2}(n_i s_3 + n_3 s_i)^2 e^{-Q_S^{-1}(\omega)\omega r/\beta_0}$$

$$\text{for} \quad r/\beta_0 < t < r/\beta_0 + \Delta T \quad \text{and } i = 1, 2 \tag{6.30b}$$

$$= 0 \quad \text{otherwise}$$

The total power spectral density is the sum of power spectral densities due to direct propagation (6.30a-b) and single scattering (6.25a-d) as given by (6.19); therefore, the synthesized time trace is not smooth around the direct wave arrivals.

To illustrate the envelope calculation, we choose an $M_L=3$ earthquake having geometry given by $\mathbf{n}=(0,0,1)$ and $\mathbf{s}=(0,1,0)$. The receiver is located at a distance of 30 km on the z_3 axis. We may consider the first and the second components as transverse and the third component as radial. The unperturbed medium parameters are $\beta_0=4$ km/s, $\gamma_0 = \sqrt{3}$ and $\rho_0 = 3$ g/cm^3 with no intrinsic absorption. The parameters for the exponential ACF describing the medium inhomogeneity are $\varepsilon^2 = 0.01$, $a = 2$km and $\nu = 0.8$. We use theoretical scattering attenuation $^{TSc}Q_S^{-1}$ and $^{TSc}Q_P^{-1}$ based on the travel-time corrected Born approximation as derived in Section 5.3.2. These values agree with the backscattering coefficient of S-waves in the lithosphere that were used to model the S-coda excitation and the frequency-dependent scattering attenuation for S-waves (see solid curves in Figures 5.10 and 5.11).

Dividing the $w \otimes \phi$ space into 50x36 grids, we numerically evaluate the double integrals in (6.25a–d). The power spectral density is calculated at every integer second of lapse time ($\Delta T = 1$ s) from 5 s to 60 s in eight octave-width frequency bands ranging from 0.5 to 64 Hz with central frequency $f_n = 2^{n-2}$ Hz and width $\Delta f_n = \ln 2 f_n$ [Hz] for $n=1–8$. These frequencies cover the frequency range of data recorded by commonly used short period (moving coil type) velocity seismome-

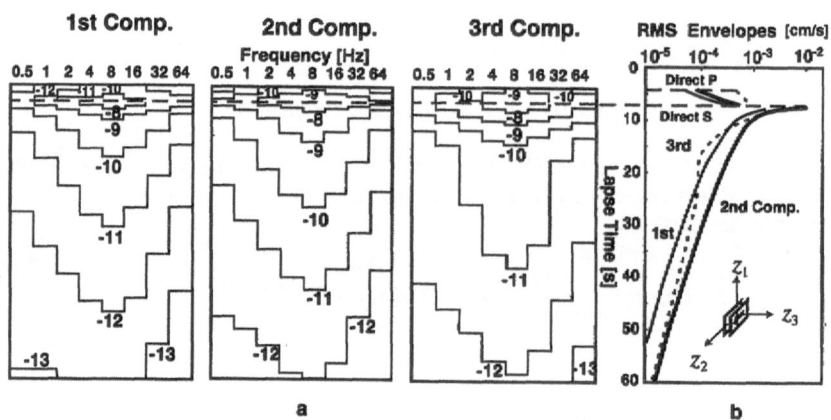

FIGURE 6.7. (a) Contour plot of the logarithm of power spectral densities S_i in CGS units for three components of particle velocity at a distance of 30 km along the third axis from a point shear-dislocation source with $M_L=3$. (b) Temporal traces of RMS velocity amplitude envelopes. Medium perturbation is characterized by an exponential ACF with $\varepsilon^2 = 0.01$, $a = 2$ km, and $\nu = 0.8$. The background elastic medium has $\beta_0 = 4$ km/s, $\gamma_0 = \sqrt{3}$, and $\rho_0 = 3$ g/cm^3. [From Sato, 1984a, copyright by the American Geophysical Union.]

ters. The source duration T_0 for an $M_L=3$ earthquake is smaller than 1 s. Arrival times of direct P- and S-waves at the receiver are 4.3 s and 7.5 s, respectively. The minimum of the scattering angle ψ becomes larger with increasing lapse time for PP- and SS-scattering, as illustrated in Figure 6.6. We used travel-time corrected scattering amplitudes choosing $\psi_C \equiv 2\sin^{-1}(1/4) \approx 29°$. The travel-time correction has impact only when the scattering angle is smaller than the cutoff scattering angle as discussed in Section 5.3.2, the time period 4.3–4.5 s of the P-wave and 7.5–7.8 s of the S-wave. Direct wave contributions dominate the single scattering waves whose travel times are in these time intervals. Using (5.25), the RMS travel-time fluctuation for the direct S-wave at 8 Hz due to the long-wavelength components of velocity fluctuation is about 0.3 s and is about 0.8 s at lapse time 60 s.

Contour plots of the power spectral density for the three components of particle velocity are shown in Figure 6.7a. The second component has the largest amplitude of all three components when the direct S-wave arrives. The S-coda has the maxi-

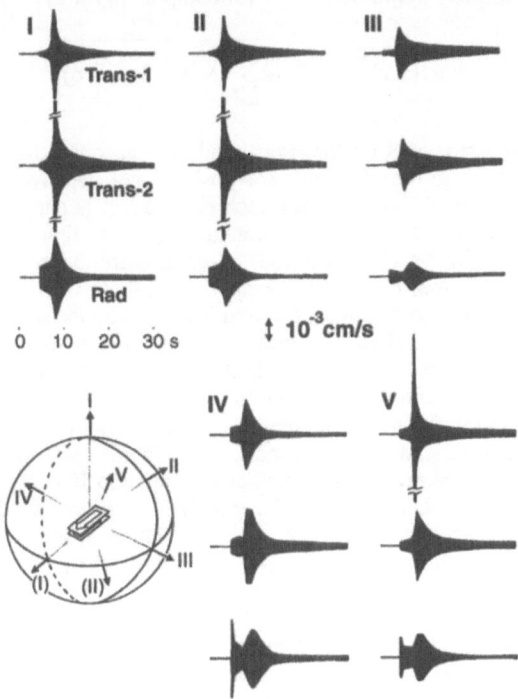

FIGURE 6.8. Three-component velocity seismogram envelopes for the 0.5–64 Hz band at a distance of 30 km in five different directions for a point shear-dislocation earthquake source of $M_L=3$. [From Sato, 1984a, copyright by the American Geophysical Union.]

mum value at the same frequency for every lapse-time since the scattering loss f/Q_s is frequency independent for $f > 1$ Hz. SS-scattering dominates the S-coda on the transverse components because of strong radiation of S-waves from the source. P coda is composed of different kinds of scattering modes. However, SP-scattering dominates the early P-coda on the radial component.

RMS Velocity Amplitude Envelopes

Using the power spectral density of particle velocity, we may estimate the average velocity amplitude for each component on the free surface. Practically, we may consider the first and the second components as horizontal components of motion and the third component as vertical movement on the free surface. Taking the square root of the numerically evaluated values of (6.25a–d) for the eight octave-width frequency bands and multiplying by two to account for surface amplification, we get the RMS velocity amplitude of the ith component for the 0.5–64 Hz band as

$$\sqrt{\left\langle \left| \dot{u}_i(\mathbf{x},t) \right|^2 \right\rangle_T} = 2\sqrt{2\sum_{n=1}^{8} \left\langle S_i(\mathbf{x},t;2\pi f_n) \right\rangle \Delta f_n} \tag{6.31}$$

where the leading factor of 2 on the right-hand side accounts for the free surface effect and factor 2 in the root accounts for positive and negative frequencies. The temporal trace of average amplitude for each component is shown in Figure 6.7b. P-coda amplitudes gradually increase with increasing lapse time. Early S-coda amplitudes are larger than P-coda amplitudes. S-coda amplitudes smoothly decrease with increasing lapse time. Even at long lapse-time, there is a predicted difference among components in S-coda amplitudes that is caused by the radiation pattern.

Figure 6.8 shows a simulation of three-component seismogram envelopes at a distance of 30 km for five different directions from a point shear-dislocation earthquake source of $M_L=3$. Direction I is a nodal direction for direct P-waves and the direction of maximum S-wave radiation. Direction IV is in the direction of maximum P-wave radiation, and a nodal direction of S-wave radiation. Direction III is along the null axis where no energy is radiated. Direction II is at the middle of directions I and III. Directions (I) and (II), which have been added for clarity, are equivalent to directions I and II for the energy radiation, respectively. Direction V is at the middle of directions I and (II). The amplitudes of the P-coda on the transverse components increase with lapse time in directions I, II, and III, and are nearly constant in directions IV and V. The amplitudes of the radial component P-codas increase with lapse time in directions I and II, but decrease with lapse time in directions III, IV, and V. P-coda appears just after the calculated P-arrival in directions I, II, and III even though they are in nodal directions for direct P-wave radiation. For the radial component P-coda, the contribution of SP-conversion scattering near the hypocenter is particularly large. S-coda amplitudes are large on all three components in all directions. Those of the transverse components of motion decrease rapidly with increasing lapse time immediately after the calculated S arrival-time but decrease gradually at large lapse time. S-coda appears on the radial component but shows a more complicated time-dependence than on transverse components. In directions III, IV, and V, radial component S-coda amplitude increases with lapse time and peaks a little later than the calculated S-arrival time. S-coda appears even in the nodal direction IV of S-wave radiation. We note that pseudo P- and S-phases appear due to scattering even in direction III located along the null axis, where the whole seismogram is composed only of scattered waves. For S-coda, SS-scattering contributes significantly to all three components for a wide range of lapse times. Partitions of S-coda energy into different components is controlled by the radiation pattern even in the latter part of S-coda in this simulation.

Comparison with Observed Envelopes

As shown in Figure 2.27, Kuwahara et al. [1991] reported a rapid drop of the semblance coefficient value in the S-coda of a local earthquake indicating that S-coda is composed of scattered waves arriving from many directions. For large lapse times, the radius of the isochronal scattering shells are larger, as shown in Figure 6.5, so scattered waves can arrive simultaneously from many directions in the S-coda. In radial-component P-coda, the contribution of SP-conversion scattering is dominant and scattered waves come from around the hypocenter, as illustrated by the scattering isochrons shown for small lapse times in Figure 6.5. The predicted importance of SP-conversion scattering agrees with the semblance analysis of P-coda as shown in Figure 2.27. Kuwahara et al. [1991] found that P-coda in the vertical component has almost the same propagation direction as that of the direct P-wave arriving from the epicenter of a local earthquake which means that the P-coda is also composed of P-waves propagating from the hypocentral region [see also Wagner and Owens, 1993]. Kuwahara et al. [1997] analyzed and discussed the large contribution to P-coda, especially in the vertical component, of S-to-P conversion scattering occurring at lateral impedance contrasts in the crust. Using f-k analysis, Dainty and Schultz [1995] also showed the importance of lateral heterogeneities for the excitation of P-coda.

Figure 2.29 shows an example of observed three-component seismograms recorded from a shallow strike-slip earthquake that occurred east of the Izu Peninsula, Japan. These seismograms were recorded at stations with epicentral distances between 10 and 60 km. The fault plane solution of the earthquake is well determined by observed initial P-wave motions shown in the figure. Station NRY is near the P-wave nodal line corresponding to directions I–II in Figure 6.8. The P-phase is unclear in the vertical component and the S-phase is large in the horizontal components at this station. The P-coda amplitudes gradually increase with time for all three components. Stations YMK and SMD correspond to direction IV, and JIZ corresponds to directions IV–V. The direct P-wave in the vertical component of JIZ is one pulse with the same period of 0.2 s as the S-pulse in the NS component of NRY. Vertical-component P-coda envelopes first decrease in amplitude, then increase between the P- and S-phases for these three stations. Scattering has increased the apparent duration of the S-phases in the horizontal components on YMK, SMD, and JIZ. The S-wave peak in the vertical component occurs later than those in the horizontal components at these three stations. Station HTS corresponds to directions II–V, where P- and S-phases appear in all three components. Good agreement is due to the fact that SH-waves dominate the horizontal-component S-waves for shallow strike-slip earthquakes. The agreement between model and theory is not so good for reverse fault earthquakes since the ray take-off angle and hence the mix of SH- and SV-motion depend strongly of the crustal velocity model. The good qualitative agreement between the observed seismograms in Figure 2.29 and the predicted envelopes in Figure 6.8 shows the validity of the envelope synthesis based on the addition of the power of single scattered waves. However, more quantitative analysis is necessary to examine how S-coda excitation levels of different components depend on the radiation pattern.

We have used scattering amplitudes for random elastic inhomogeneity based on the travel-time corrected Born approximation; however, note that the formulation developed here accepts scattering amplitudes from any kind of obstacles; for example cracks, cavities, and/or low-velocity anomalies.

6.3 ENVELOPE SYNTHESIS IN A HALF-SPACE

For comparison with observed three-component seismogram envelopes of regional earthquakes, it is necessary to reliably account for the contribution of the free surface since it acts as a reflector for seismic body waves. To account for the effect of the free surface on the amplitudes of seismogram envelopes, we multiplied the infinite space amplitudes by two in (6.31). The factor two is valid only for S-waves polarized parallel to the free surface or for normally incident waves of any type. Here, we will briefly introduce a more realistic approach for modeling the contribution of the free surface that was developed by Yoshimoto [1995] and Yoshimoto et al. [1997a, b].

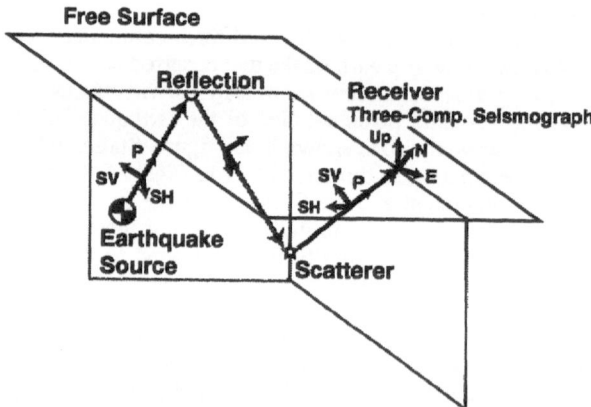

FIGURE 6.9. Geometry of scattered ray undergoing reflection at the free surface. Ray is singly scattered within the medium.

6.3.1 Effects of the Free Surface

We imagine a point shear-dislocation source in a randomly inhomogeneous elastic half-space and a three-component seismograph installed on the free surface, as illustrated in Figure 6.9. The description of the scattering process of elastic vector waves is the same as described in Section 6.2. The reflection and mode conversion at the free surface are calculated from the incident angle and the polarization using Snell's law [Aki and Richards, 1980, p. 133], where the incident wave is assumed to be a plane wave near the reflection point. It is convenient to decompose the S-waves into two components with orthogonal polarization directions. The SH-wave is polarized parallel to the flat free surface and the SV-wave is polarized perpendicular to the SH-wave. For a flat surface, an incident SH-wave generates only a reflected SH-wave. However, incident P- or SV-waves generate a combination of reflected P- and SV-waves. The geometrical spreading factors and the conversion reflection effects can be calculated from the incidence angle, the reflection angle, the background medium velocities, and the distances between the reflection point to the source and the scattering point and between the scattering point and the receiver [Cerveny and Ravindra, 1971]. We will account for only the total reflection and neglect the excitation of inhomogeneous waves.

In addition to the four fundamental modes of scattering, PP, PS, SP, and SS, as discussed for an infinite space, we have to consider eight additional modes corresponding to the single scattering of waves reflected by the free surface: PPP, PPS, PSP, PSS, SPP, SPS, SSP and SSS, where the first phase name indicates the mode of the direct wave from the source to the free surface. For example, SPS means that the direct S-wave radiated from the source is converted to a P-wave upon reflection by the free surface and then scattered to an S-wave by conversion scattering within the inhomogeneous medium. For the four fundamental scattering

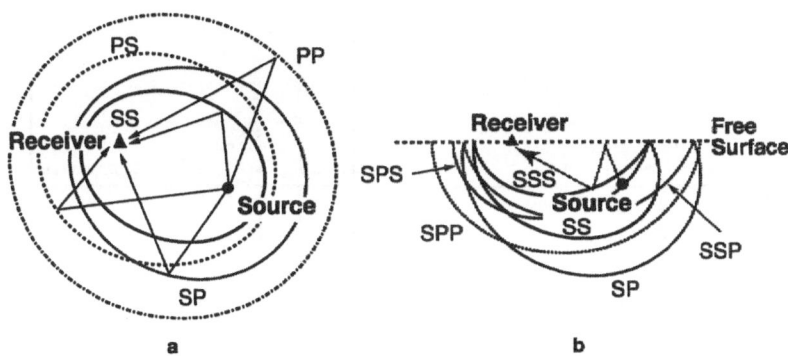

FIGURE 6.10. (a) Slice through a plane containing the source, scattering points, and the receiver showing isochronal scattering shells for the four fundamental scattering modes PP, SS, PS, and SP in infinite space. (b) Vertical slice through a plane containing the source, reflection points, scattering points, and the receiver showing isochronal scattering shells for scattering modes with reflection at the free surface. Shells for S-wave radiation from the source are shown. [Courtesy of K. Yoshimoto.]

modes, the source, scattering point, and receiver are located in a plane, as in Section 6.2 and the isochronal scattering shells are axially symmetric around the source–receiver line (see Figure 6.10a). The geometry of the eight isochronal scattering shells that include free surface reflections are complex; the source, free surface reflection point, and scattering point are located on one plane, but free surface reflection point, scattering point and the receiver are generally located on a different plane, as illustrated in Figure 6.9. Figure 6.10b illustrates the isochronal scattering shells for a special case when one plane contains the source, reflection points, scattering points, and the receiver for S-energy source radiation, as an example.

6.3.2 Numerical Simulations

Yoshimoto [1995] and Yoshimoto et al. [1997a] numerically synthesized seismogram envelopes for an M_L=2 earthquake with a focal depth of 3.5 km in an elastic inhomogeneous half-space. The fault geometry and the configuration of source and receiver are shown in Figure 6.11. The source velocity spectrum has a peak around 13 Hz. The medium inhomogeneity is characterized by an exponential ACF with ε=4% and a rather short correlation distance a=100 m compared with the case studied in the last section. The background parameters for the elastic medium are α_0=6 km/s, γ_0=1.71 (β_0=3.5 km/s), ρ_0=2.7 g/cm³ and ν=0.8. The inhomogeneous half-space was divided into cubes of 200 m on a side. The power spectral density of each component of particle velocity on the free surface was calculated as the sum of the power spectral densities of scattered waves arriving from discretized cubes located on isochronal scattering shells for all scattering modes for a given lapse time in each frequency band as a numerical extension of (6.23a–d).

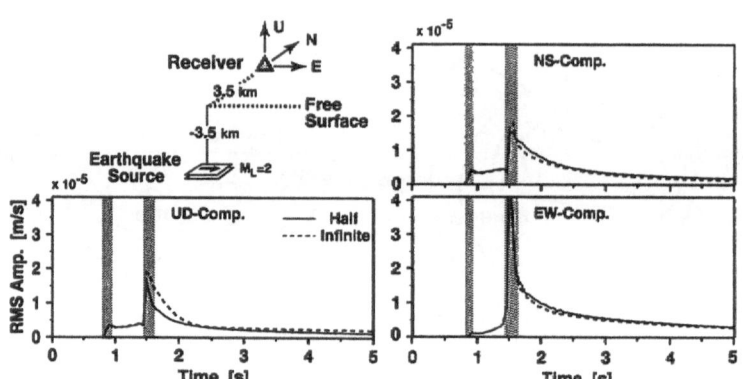

FIGURE 6.11. Synthesized RMS velocity amplitude envelopes on the free surface of an elastic inhomogeneous half-space (solid curve) and twice those in an infinite medium (broken curve) in the 2–64 Hz band. Traces are single scattered wave amplitudes only, and direct waves are not shown. Portions of envelopes dominated by direct waves are shaded. [Courtesy of K. Yoshimoto.]

The RMS velocity amplitude envelope for each component was obtained by taking the square root of the sum of the power spectral densities over the 2–64 Hz band. Figure 6.11 shows RMS velocity envelopes on the free surface (solid curves) and twice the velocity envelopes for an infinite medium obtained in Section 6.2.3 (broken curves). There are few differences in the P-coda excitation between the full-space and half-space results. However, a clear difference is found in the early S-coda in the vertical component of motion, where SS scattered waves dominate, because the multiplication by two overestimates the amplitude for large incidence angles to the free surface.

6.3.3 Crustal Inhomogeneity in the Nikko Area, Northern Kanto, Japan

A large-scale array observation experiment in the Nikko area, northern Kanto, Japan in 1993 provided the opportunity to analyze three-component seismogram envelopes of regional microearthquakes with well-determined focal plane solutions. Yoshimoto [1995] analyzed some of the data collected using the forward modeling procedure discussed in Section 6.3.1 to determine statistical properties of the medium required to fit seismogram envelopes. There are many volcanoes in this area resulting in high seismicity in the shallow crust. Most of earthquakes took place at depths shallower than 15 km. There are reports of a midcrustal seismic reflector that might be related to the presence of partially molten bodies near some active

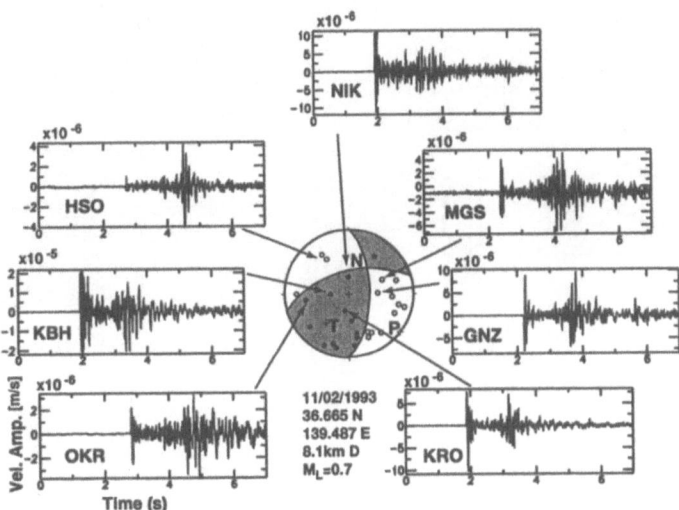

FIGURE 6.12. Vertical-component velocity seismograms (2–16 Hz) of a shallow M_L=0.7 microearthquake observed in the Nikko area, northern Kanto, Japan. [Courtesy of K. Yoshimoto.]

volcanoes [Matsumoto and Hasegawa, 1996]. Waveforms from 38 three-component seismometers were digitized at 200 Hz. Hypocenter locations and focal plane solutions were determined using a uniform half-space model with $\alpha_0 = 6$ km/s and $\alpha_0 / \beta_0 = 1.73$. Yoshimoto [1995] studied 10 earthquakes that had high-quality records and well-determined focal mechanisms.

Most of the earthquakes have similar reverse fault type focal mechanisms in harmony with a regional compressional stress oriented in the NW-SE direction in northern Kanto. The three-component velocity seismograms show variations in envelopes. Examples of observed UD (vertical) component traces are shown in Figure 6.12. Yoshimoto [1995] analyzed the RMS velocity amplitude envelopes in the 2–16 Hz band. RMS envelopes of the traces shown in Figure 6.12 are plotted using solid curves in Figure 6.13, where the ordinate is a logarithmic scale.

For analysis, the source spectra of events were estimated from the spectra of direct P- and S-waves using data from stations having hypocentral distances smaller than 15 km. Site amplification factors at each station in this frequency band were found by the coda-normalization technique for S-waves discussed in Section 3.4.1 and by comparison of P-wave spectra of teleseisms. Station GNZ was chosen as the reference station. Q_P^{-1} and Q_S^{-1} were chosen to have the same value of 0.01 for 2-8 Hz and were taken as 0.005 and 0.003 for 8-16 Hz, respectively. The velocity fluctuation of the crust is characterized by an exponential ACF with two parameters ε^2 and a.

Using the above parameters for characterizing the source and the random medium, Yoshimoto [1995] numerically synthesized three-component RMS velocity

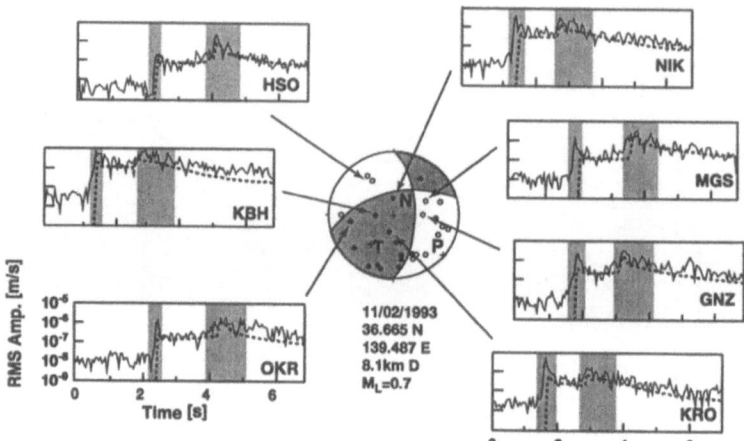

FIGURE 6.13. Observed vertical-component RMS velocity amplitude envelopes (solid curves) for the microearthquake shown in Figure 6.12 and the best fit theoretical curves (dotted curves) for the 2–16 Hz band. Shaded portions around the direct P- and S-phases were not included in calculation of residuals to find the best fitting model. The half-space inhomogeneous medium is characterized by an exponential ACF with $\varepsilon = 5.7\%$ and $a = 400$ m. [Courtesy of K. Yoshimoto.]

amplitude envelopes by discretizing the medium into cubes of dimension 0.5 km on each side, where the small cube size was chosen to make the synthesized envelopes smooth. Values of ε^2 and a were varied to minimize the sum of the square difference between the synthesized and observed three-component RMS seismogram envelopes for a time window of 7 s long starting from the P-wave onset. The misfits at each time were weighted by the square of lapse time giving more weight to the later portions of the seismograms. The residual was calculated using only the portions of the envelopes dominated by scattered waves, neglecting the direct wave windows indicated by shading in Figure 6.13. A total of 149 traces from the 10 earthquakes were analyzed. Inhomogeneity having $\varepsilon > 8\%$ produces scattering attenuation larger than the values assumed for total attenuation, and inhomogeneity having $\varepsilon < 4\%$ is not strong enough to excite sufficient S-coda amplitude to match observed data. The residual was minimized when $\varepsilon^2/a \approx 8 \times 10^{-3}$ km^{-1}. Yoshimoto [1995] concluded that the inhomogeneity in the shallow crust of the study area is characterized by the exponential ACF with $\varepsilon = 5-8\%$ and $a = 300-800$ m.

Figure 6.13 shows the predicted logarithm of RMS velocity amplitude envelopes in the 2–16 Hz band for the UD component (broken curves) using the best fitting model $\varepsilon = 5.7\%$ and $a = 400$ m along with the smoothed observed envelope (solid curves). Figure 6.14 shows logarithmic plots of the three-component envelopes at station OKR. Although the later portion of the S-coda was weighted more heavily in fitting the data, there is a good agreement between observed and synthesized P-coda amplitude. To investigate differences in the characteristics of the inhomogeneity with the frequency band of seismic waves, Yoshimoto et al. [1997b] examined the Nikko data using only the frequency band 8–16 Hz. They obtained the same result for ε and a that Yoshimoto [1995] obtained from the analysis of data in the 2–16 Hz band. The correlation distance estimated here is shorter than estimated from the analysis discussed in Chapter 5 since the region under investigation here is shallower and more complex because of the presence of active volcanoes.

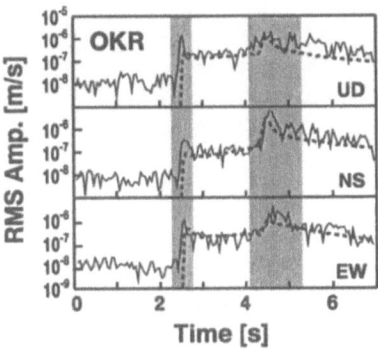

FIGURE 6.14. Three-component RMS velocity amplitude envelopes observed (solid curves) for the microearthquake shown in Figure 6.12 and the theoretically predicted envelopes (dotted curves) for the 2–16 Hz band at station OKR. [Courtesy of K. Yoshimoto.]

CHAPTER 7

Envelope Synthesis Based on the Radiative Transfer Theory: Multiple Scattering Models

In Chapter 3, we found that the squared sum of incoherent S-waves that are singly scattered by distributed random heterogeneities can provide an adequate first-order model of the MS envelope of S-wave seismograms. As lapse time increases, however, we expect that double, triple, and higher-order multiple scattering contribute more than the single scattering process, so we need a model that includes the effects of multiple scattering. Kopnichev [1977] and Gao et al. [1983] modeled multiple scattering effects by summing up higher order scattering terms as a simple extension of the single scattering model; however, their models do not conserve total energy.

A systematic approach for modeling the multiple scattering process is to use the radiative transfer theory (energy transport theory) for the energy density. The basic equation of this approach is called the equation of transfer and is equivalent to Boltzmann's equation used in the kinetic theory of gases and neutron scattering. This theory has been successfully employed for modeling atmospheric and underwater visibility and the propagation of light in the atmospheres of planets [Chandrasekhar, 1960]. It deals directly with the transport of energy through a medium containing scatterers. It is assumed that the addition of power holds rather than the addition of wavefields. The development of the radiative transfer theory is rather heuristic, and it lacks a rigorous basis on the wave equation. There is a way to include some information about the correlation of wavefields into the radiative transfer theory [Foldy, 1945; Frisch, 1968; Ishimaru, 1978, Chapter 14]; however, we restrict ourselves to a rather phenomenological treatment that deals only with energy transport.

Initial seismological models using the radiative transfer theory were borrowed from other fields and did not include mode conversion between P- and S-waves upon scattering. Wu [1985] and Wu and Aki [1988a] first explored the use of the radiative transfer theory as a model for high-frequency seismogram envelopes of local earthquakes. They applied the stationary-state solution for media having isotropic scattering [Lin and Ishimaru, 1974] to the estimation of seismic albedo, which Wu [1985] defined as the ratio of scattering attenuation to total attenuation. Shang and Gao [1988] first formulated the multiple isotropic scattering process in

2-D space as an integral equation for the nonstationary state appropriate for the case of impulsive radiation. Later, Zeng et al. [1991] extended the nonstationary case to 3-D space, and Sato [1993] and Wu [1993] analytically solved the problem in 1-D space. The solutions obtained with the radiative transfer theory conserve total energy for media having no intrinsic absorption, isotropic scattering, and spherical radiation from the source. Sato et al. [1997] used the radiative transfer theory to investigate the multiple isotropic scattering process for nonspherical source radiation. Their results showed that the azimuthal dependence of coda excitation diminishes with increasing lapse time even for nonspherical radiation from a point shear-dislocation. Using the concept of the specific intensity [see Ishimaru, 1978, p. 148], Sato [1994b, 1995a] investigated the multiple nonisotropic scattering process in the framework of the radiative transfer theory. In the case of large forward scattering, this model predicts a concentration of scattered energy just after the direct wave arrival and smaller excitation of coda energy compared with the isotropic scattering case. Note that a uniform distribution of coda energy density for large lapse time is predicted for the multiple isotropic scattering process in a medium having a fractal distribution of scatterers [Sato, 1995b]. There have been a few studies using the radiative transfer theory to investigate whole seismogram envelopes including P- and S-phases. Supposing isotropic scattering for PP-, PS-, SP- and SS-scattering, Zeng [1993] and Sato [1994a] synthesized time traces of the energy density for the case of spherical source radiation.

In parallel with the analytical studies, several investigations using Monte Carlo simulation of the multiple scattering process for envelope synthesis were conducted [Gusev and Abubakirov, 1987; Abubakirov and Gusev, 1990; Hoshiba, 1991]. Parallel to the theoretical development of the radiative transfer theory, practical analysis of S-wave seismogram envelopes was undertaken and used to investigate the limitations of the theory. The radiative transfer approach provided a framework for measuring seismic albedo, scattering attenuation, and intrinsic attenuation which has led to measurements throughout the world [Fehler et al., 1992].

In this chapter, we introduce the application of the radiative transfer theory to model the multiple scattering process and several extensions of the theory that are appropriate for modeling lithospheric seismic propagation processes. We will also discuss the multiple lapse-time window analysis for the estimation of seismic albedo, intrinsic attenuation, and scattering attenuation of S-waves.

7.1 MULTIPLE ISOTROPIC SCATTERING PROCESS FOR SPHERICAL SOURCE RADIATION

As discussed in Chapter 3, we assume that point-like isotropic scatterers of cross section σ_0 are distributed randomly and homogeneously in infinite space with number density n, where $g_0 \equiv n\sigma_0$ is the total scattering coefficient characterizing the scattering power per unit volume. The background medium propagation velocity V_0 is constant.

7.1.1 Three-Dimensional Case

Formulation of the temporal evolution of the multiple scattering process in 3-D space was done by Zeng et al. [1991]. We assume an impulsive spherical source radiating energy at time zero from the hypocenter located at the origin. Source energy is described by $W\delta(t)$. We schematically show the configuration of the source, the receiver, and the last scattering point in Figure 7.1. The energy density due to the propagation of coherent waves from the source is $We^{-(V_0g_0+b)t} \times$ $\delta(t-|\mathbf{x}|/V_0)/4\pi V_0|\mathbf{x}|^2$, where we include the exponential decay term V_0g_0 to account for scattering attenuation and b for intrinsic attenuation per time in addition to the geometrical spreading term $1/4\pi|\mathbf{x}|^2$. Generation of scattered energy per unit time from a unit volume at scattering point $(\mathbf{x}',t')$ is a product of g_0, V_0, and energy density $E(\mathbf{x}',t')$. Including geometrical spreading and the time lag due to propagation between the scatterer and the receiver at $(\mathbf{x},t)$, we get the energy-flux density at the receiver due to scattered waves from a unit volume as $\int dt'\, V_0\, E(\mathbf{x}',t')g_0\,\delta(t-t'-|\mathbf{x}-\mathbf{x}'|/V_0)e^{-(V_0g_0+b)(t-t')}/4\pi|\mathbf{x}-\mathbf{x}'|^2$. Integrating $\mathbf{x}'$ over the entire space and dividing by V_0, we get the total contribution of scattered energy density from infinite space. Adding the energy density of direct propagation of coherent waves from the source to that of the scattered waves, we arrive at the following integral equation:

$$E(\mathbf{x},t) = W\,G(\mathbf{x},t) + V_0\,g_0 \int\limits_{-\infty}^{\infty}\int\limits_{-\infty}^{\infty}\int\limits_{-\infty}^{\infty}\int\limits_{-\infty}^{\infty} G(\mathbf{x}-\mathbf{x}',t-t')E(\mathbf{x}',t')dt'd\mathbf{x}' \qquad (7.1)$$

The Green function for the coherent wave energy is given by

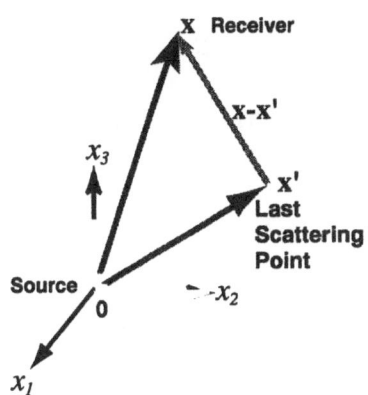

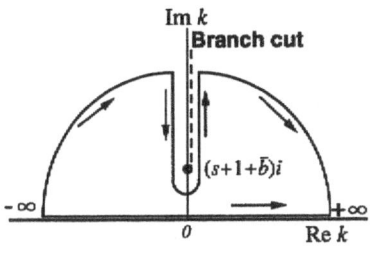

FIGURE 7.1. Configuration of the source, the receiver, and the last scattering point in a scattering medium.

FIGURE 7.2. Integral contour in the complex k-plane to evaluate the single scattering term.

$$G(\mathbf{x},t) = \frac{1}{4\pi V_0 r^2} H(t)\delta(t - r/V_0)e^{-(V_0 g_0 + b)t} \tag{7.2}$$

where $r = |\mathbf{x}|$. We may explicitly write the argument of the Green function using scalar r. This function rigorously corresponds to the propagator for the coherent intensity [Ishimaru, 1978, p. 261]. The scattering system is assumed to be exactly described by (7.1) and (7.2), not only in the far field but also in near field since scatterers are assumed to be point-like. Function (7.2) is not the solution to a differential equation with a boundary condition; however, we dare to call that function "Green function" for simplicity. The solution of (7.1) for unit energy radiation practically gives "the effective Green function" for an impulsive source radiation in the scattering medium. We can evaluate the energy density for any type of radiation time function as a convolution integral by using "the effective Green function".

Substituting the explicit representation (7.2) in (7.1) and integrating over time, we can get the following integral form

$$E(\mathbf{x},t) = W\frac{e^{-(g_0 V_0 + b)t}}{4\pi V_0 r^2}\delta(t - r/V_0)$$
$$+ \frac{g_0}{4\pi}\int_0^{V_0 t} dr''' \oint d\Omega(\mathbf{q}''')e^{-(g_0 + b/V_0)r'''} E(\mathbf{x} + r'''\mathbf{q}''', t - r'''/V_0) \tag{7.3}$$

where $r''' \equiv |\mathbf{x}' - \mathbf{x}|$. The integral over the solid angle in the second line is taken with respect to direction $\mathbf{q}''' \equiv (\mathbf{x}' - \mathbf{x})/r'''$, which is the unit vector pointing to the last scatterer from the receiver. This is an extension of the radiative transfer equation for the multiple isotropic scattering process for the stationary state as given in Ishimaru [1978, p. 161].

For simplicity, we scale quantities using the following

$$\bar{t} = V_0 g_0 t, \quad \bar{\mathbf{x}} = g_0 \mathbf{x}, \quad \bar{b} = \frac{b}{V_0 g_0}, \quad \bar{G} = \frac{G}{g_0^3}, \text{ and } \bar{E} = \frac{E}{W g_0^3} \tag{7.4}$$

where the overbar means the nondimensional quantity. Then, (7.1) and (7.2) are written as

$$\bar{E}(\bar{\mathbf{x}},\bar{t}) = \bar{G}(\bar{\mathbf{x}},\bar{t}) + \int\int\int\int_{-\infty-\infty-\infty-\infty}^{\infty\ \infty\ \infty\ \infty} \bar{G}(\bar{\mathbf{x}} - \bar{\mathbf{x}}', \bar{t} - \bar{t}')\bar{E}(\bar{\mathbf{x}}', \bar{t}')d\bar{t}'d\bar{\mathbf{x}}' \tag{7.5}$$

and

$$\bar{G}(\bar{\mathbf{x}},\bar{t}) = \frac{1}{4\pi\bar{r}^2} H(\bar{t})\delta(\bar{t} - \bar{r})e^{-(1+\bar{b})\bar{t}} \tag{7.6}$$

where $\bar{r} = |\bar{\mathbf{x}}|$. To solve these equations, first we take the Fourier transform in space and the Laplace transform in time of (7.6):

$$\hat{\tilde{G}}(k,s) = \int_{-\infty}^{\infty}\int_{-\infty}^{\infty}\int_{-\infty}^{\infty} d\bar{\mathbf{x}}\, e^{-ik\bar{\mathbf{x}}} \int_{0}^{\infty} d\bar{t}\, e^{-s\bar{t}} G(\bar{\mathbf{x}},\bar{t}) = \frac{1}{2}\int_{0}^{\pi}\sin\theta\, d\theta \int_{0}^{\infty} dr\, e^{-(ik\cos\theta+s+1+\bar{b})\bar{r}}$$

$$= \frac{1}{k}\left(\frac{1}{2i}\ln\frac{1+i\dfrac{k}{s+1+\bar{b}}}{1-i\dfrac{k}{s+1+\bar{b}}}\right) = \frac{1}{k}\tan^{-1}\frac{k}{s+1+\bar{b}} \tag{7.7}$$

where the argument is scalar $k=|\mathbf{k}|$ because there is no specific orientation. The tilde and the caret mean the Fourier transform in 3-D space and the Laplace transform in time, respectively. We note that $\tan^{-1}z \equiv (1/2i)\ln[(1+iz)/(1-iz)]$ [Abramowitz and Stegun, 1990, p. 80]. Taking the Fourier–Laplace transform of the convolution integral (7.5),

$$\hat{\tilde{E}}(k,s) = \frac{\hat{\tilde{G}}(k,s)}{1-\hat{\tilde{G}}(k,s)} \tag{7.8}$$

The energy density in space-time written as the inverse Fourier–Laplace transform of (7.8) is given by

$$\bar{E}(\bar{\mathbf{x}},\bar{t}) = \frac{1}{2\pi i}\int_{-i\infty}^{i\infty} ds\, e^{s\bar{t}}\, \frac{1}{(2\pi)^3}\int_{-\infty}^{\infty}\int_{-\infty}^{\infty}\int_{-\infty}^{\infty} d\mathbf{k}\, e^{ik\bar{\mathbf{x}}}\hat{\tilde{E}}(k,s)$$

$$= \frac{1}{2\pi i}\int_{-i\infty}^{i\infty} ds\, e^{s\bar{t}}\, \frac{1}{(2\pi)^2}\int_{0}^{\infty} k^2 dk\, \hat{\tilde{E}}(k,s)\int_{0}^{\pi} e^{ik\bar{r}\cos\theta_k}\sin\theta_k d\theta_k$$

$$= \frac{1}{2\pi i}\int_{-i\infty}^{i\infty} ds\, e^{s\bar{t}}\, \frac{1}{(2\pi)^2 i\bar{r}}\int_{0}^{\infty} k dk\, \hat{\tilde{E}}(k,s)\left(e^{ik\bar{r}} - e^{-ik\bar{r}}\right) \tag{7.9}$$

$$= \frac{1}{2\pi i}\int_{-i\infty}^{i\infty} ds\, e^{s\bar{t}}\left[\frac{1}{(2\pi)^2 i\bar{r}}\int_{-\infty}^{\infty} dk\, k\, e^{ik\bar{r}}\hat{\tilde{E}}(k,s)\right]$$

$$= \frac{1}{\bar{r}}\frac{1}{(2\pi)^2}\int_{-\infty}^{\infty}\int_{-\infty}^{\infty} d\omega\, dk\, e^{-i\omega\bar{t}-ik\bar{r}}\left[\frac{ik}{2\pi}\hat{\tilde{E}}(-k,s\to-i\omega)\right]$$

where we used $\hat{\tilde{E}}(k,s)=\hat{\tilde{E}}(-k,s)$ since $\hat{\tilde{G}}(k,s)=\hat{\tilde{G}}(-k,s)$. There is no specific orientation, so $\bar{E}$ depends on $\bar{r}=|\bar{\mathbf{x}}|$. Therefore, we may write the argument of $\bar{E}$ as scalar $\bar{r}$. The path of integration in the complex s-plane runs parallel to the imaginary axis in positive real space so that all the singularities of the integrand are on the left-hand side in the complex s-plane. We have substituted the inverse Fourier transform for the inverse Laplace transform in the last step of (7.9) for convenience of numerical evaluation with a 2-D FFT [Zeng et al., 1991].

Conservation of Total Energy

We can derive the temporal dependence of total energy defined as the space integral of energy density as a function of time, which is given by the Fourier transform of (7.8) at wavenumber $k = 0$

$$\hat{\bar{E}}(0,s) = \frac{\hat{\bar{G}}(0,s)}{1-\hat{\bar{G}}(0,s)} = \frac{1}{s+\bar{b}} \tag{7.10}$$

This means that the total energy decreases exponentially with increasing lapse time only due to intrinsic absorption:

$$\int\limits_{-\infty}^{\infty}\int\limits_{-\infty}^{\infty}\int\limits_{-\infty}^{\infty} d\bar{x}\, \bar{E}(\bar{x},\bar{t}) = e^{-\bar{b}\bar{t}} \tag{7.11}$$

The conservation of total energy with no intrinsic absorption ($b=0$) confirms the mathematical consistency of the formulation.

Analytical Representation of the Single Scattering Term

By using the inverse Fourier–Laplace transformation (7.9), we have formally derived the space-time distribution of energy density. But the convergence of the numerical integration is slow since the integral kernel oscillates rapidly for large wavenumbers. Therefore, first we formally decompose the energy density into three terms corresponding to direct, single scattering, and multiple scattering of order greater than or equal to two:

$$\bar{E}(\bar{x},\bar{t}) = \bar{G}(\bar{x},\bar{t}) + \bar{E}^1(\bar{x},\bar{t}) + \bar{E}^M(\bar{x},\bar{t}) \tag{7.12}$$

In parallel, we rewrite (7.8) as

$$\hat{\bar{E}}(k,s) = \hat{\bar{G}}(k,s) + \hat{\bar{E}}^1(k,s) + \hat{\bar{E}}^M(k,s)$$

$$= \hat{\bar{G}}(k,s) + \hat{\bar{G}}(k,s)^2 + \frac{\hat{\bar{G}}(k,s)^3}{1-\hat{\bar{G}}(k,s)} \tag{7.13}$$

Following Zeng et al. [1991], we analytically evaluate the single scattering term after by writing it as an integral on the real k-axis:

$$\hat{\bar{E}}^1(\bar{x},s) = \frac{1}{(2\pi)^3} \int_{-\infty}^{\infty}\int_{-\infty}^{\infty}\int_{-\infty}^{\infty} \hat{\bar{\bar{G}}}(k,s)^2 e^{ik\bar{x}} d\mathbf{k} = \frac{1}{(2\pi)^2 i\bar{r}} \int_{-\infty}^{\infty} \hat{\bar{\bar{G}}}(k,s)^2 e^{ik\bar{r}} k dk$$

$$= \frac{1}{(2\pi)^2 i\bar{r}} \left(\int_{\substack{\text{Half}\\\text{Circle}}} \hat{\bar{\bar{G}}}(k,s)^2 e^{ik\bar{r}} k dk + \int_{\substack{\text{Along}\\\text{Branch Cut}}} \hat{\bar{\bar{G}}}(k,s)^2 e^{ik\bar{r}} k dk \right) \qquad (7.14)$$

Closing the contour of integration at infinity in the upper complex k-plane, we adopt the technique of residue and branch cut integration. We take a branch cut from the branch point $k = (s+1+\bar{b})i$ to infinity on the imaginary k axis. The integral contour is taken as a half circle (two quarter circles) and along the branch cut on the imaginary axis, as schematically illustrated in Figure 7.2. As $|k| \to \infty$, the integral along a half circle vanishes since $\lim_{|z|\to\infty} \tan^{-1} z = \pm\pi/2$:

$$\lim_{\substack{|k|\to\infty\\\text{Half}\\\text{Circle}}} \int \hat{\bar{\bar{G}}}(k,s)^2 e^{ik\bar{r}} k dk \approx \lim_{|k|\to\infty} \int_{\pi}^{0} \left(\pm\frac{\pi}{2k}\right)^2 e^{ik\bar{r}} k dk \le \frac{\pi^2}{4} \lim_{|k|\to\infty} \left| \int_{\pi}^{0} \frac{e^{ik\bar{r}}}{k} dk \right| = 0 \quad (7.15)$$

Therefore, the integral is given only by the integral along the branch cut, which is written as a Laplace transform with respect to time:

$$\hat{\bar{E}}^1(x,s) = \frac{1}{(2\pi)^2 i\bar{r}} \int_{\substack{\text{Along}\\\text{Branch Cut}}} \hat{\bar{\bar{G}}}(k,s)^2 e^{ik\bar{r}} k dk$$

$$= \frac{1}{(2\pi)^2 i\bar{r}} \left[\int_{(1+s+\bar{b})\,i+0}^{\infty i+0} \hat{\bar{\bar{G}}}(k,s)^2 e^{ik\bar{r}} k dk - \int_{(1+s+\bar{b})\,i-0}^{\infty i-0} \hat{\bar{\bar{G}}}(k,s)^2 e^{ik\bar{r}} k dk \right]$$

$$= \frac{1}{(2\pi)^2 i\bar{r}} \int_{1}^{\infty} e^{-(s+1+\bar{b})\bar{r}u} \frac{1}{u} \left[\left(i\tanh^{-1}\frac{1}{u} + \frac{\pi}{2} \right)^2 - \left(i\tanh^{-1}\frac{1}{u} - \frac{\pi}{2} \right)^2 \right] du \quad (7.16)$$

$$= \frac{1}{4\pi\bar{r}} \int_{0}^{\infty} \left(\frac{2}{u}\tanh^{-1}\frac{1}{u} \right) H(u-1) e^{-(s+1+\bar{b})\bar{r}u} du$$

$$= \int_{0}^{\infty} \left[\frac{1}{4\pi\bar{r}^2} K\left(\frac{\bar{t}}{\bar{r}}\right) H(\bar{t}-\bar{r}) e^{-(1+\bar{b})\bar{t}} \right] e^{-s\bar{t}} d\bar{t}$$

where we used $\tan^{-1} iz = i \tanh^{-1}(1/z) \pm \pi/2$ on the right/left of the branch cut, and K is given by (3.19a). The integral kernel of (7.16) gives the solution in time:

$$\bar{E}^1(\bar{x},\bar{t}) = \frac{1}{4\pi\bar{r}^2} K\left(\frac{\bar{t}}{\bar{r}}\right) H(\bar{t}-\bar{r}) e^{-(1+\bar{b})\bar{t}} \qquad (7.17)$$

This is the same as (3.22) and makes a connection to the single scattering theory developed in Chapter 3. The exponential scattering attenuation term is necessary even when there is no intrinsic absorption.

Spatiotemporal Variation of the Energy Density

The multiple scattering term $\bar{E}^M(\bar{x},\bar{t})$ can be numerically evaluated using (7.9):

$$\bar{E}^M(\bar{x},\bar{t}) = \frac{1}{\bar{r}} \frac{1}{(2\pi)^2} \int_{-\infty}^{\infty}\int_{-\infty}^{\infty} d\omega\, dk\, e^{-i\omega\bar{t}-ik\bar{r}} \left[\frac{ik}{2\pi} \frac{\hat{\bar{G}}(-k,-i\omega)^3}{1-\hat{\bar{G}}(-k,-i\omega)} \right] \qquad (7.18)$$

where the integrand is the last term of (7.13). Adding the direct, the single scattering, and the multiple scattering terms, we obtain the energy density.

Figure 7.3 shows results of numerical simulations illustrating the spatiotemporal dependence of the normalized energy density for the case of no intrinsic absorption ($b = \bar{b} = 0$). The traces in Figure 7.3a show temporal variations of normalized energy density at various distances from the source. At small distances from the source the energy density decreases rapidly after the direct arrival, as predicted by the single scattering approximation. However, as distance increases, $\bar{r} = 3.2$, energy density of early coda becomes stationary because of multiple scattering. Figure 7.3b shows spatial variations in normalized energy density at various normalized lapse times. There is a concentration of energy density around the source. As lapse time increases, the energy density distribution asymptotically approaches a Gaussian curve corresponding to the diffusion solutions given by (3.30) and Figure 3.7b. Zeng [1991] showed that the time dependence of the energy density for the multiple scattering model asymptotically converges to the diffusion solution: the temporal decay is proportional to the -1.5 th power of lapse time.

Figure 7.4 shows the temporal change in the spatial integrals of normalized energy density for direct, single scattering, and multiple scattering. The contribution of multiple scattering dominates over the single scattering for lapse times larger than the mean free time, $\bar{t} \gg 1$. The sum of all three terms equals 1 as given by (7.11) since energy is conserved when there is no intrinsic attenuation.

Interpretation of Q_C^{-1}

The radiative transfer theory for the case of spherical source radiation, a uniform distribution of isotropic scatterers, and no intrinsic attenuation predicts that the MS coda envelope decays according to the -1.5th power of lapse time. Thus, with intrinsic attenuation, the shape of the MS coda envelope is given by $t^{-1.5}\exp(-{}^I Q_S^{-1}\omega t)$. If we use (3.36) with $n=2$ to find coda attenuation Q_C^{-1}, we find that it is smaller than ${}^I Q_S^{-1}$ and may even be less than zero if intrinsic attenuation is small. When intrinsic attenuation is large, Q_c^{-1} is dominated by intrinsic attenuation. This relationship between Q_c^{-1} and $'Q_s^{-1}$ found from the radiative transfer theory was pointed out by Hoshiba [1991] and Wennerberg [1993]. Based on the observation that $Q_c^{-1} > 0$ in field data, Wennerberg [1993] argued that the close agreement between Q_s^{-1} and Q_c^{-1} reported by Aki [1980a] implies that S-wave attenuation is dominated by intrinsic mechanisms. Hoshiba [1991] argued that the relation is imprecise due to the restriction of the model that scatterers are uniformly distributed.

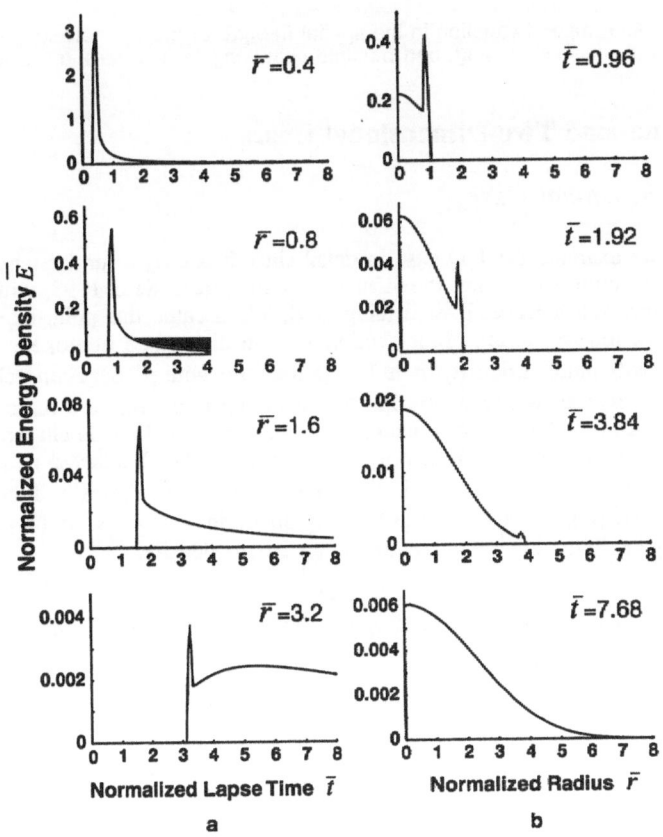

FIGURE 7.3. Spatiotemporal change in the normalized energy density for spherically symmetric source radiation and no intrinsic absorption. **(a)** Time traces at different distances. **(b)** Spatial distribution at different lapse times. For the calculation, the 2-D FFT in (7.18) was done over 200x200 points for ((0–16), (0–16)) in the normalized space-time $(\bar{r}, \bar{t})$. The source duration is taken as 0.16 and is two samples long.

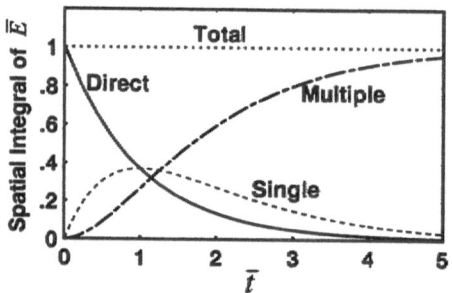

FIGURE 7.4. Temporal variation in the spatial integral of the normalized energy density for direct, single scattering, and multiple scattering for the case of no intrinsic attenuation.

7.1.2 One- and Two-Dimensional Cases

One-Dimensional Case

Here we examine the 1-D case in detail since it is easy to understand how the analytical solution approaches the diffusion solution [after Sato, 1993, with permission of Blackwell Science, United Kingdom]. We assume that point-like isotropic scatterers of total scattering cross section σ_0 are distributed homogeneously and randomly with number density n in a 1-D medium with propagation velocity V_0. In 1-D, isotropic scattering means that the scattering in the forward direction is the same as that in the backward direction. The scattering medium is characterized by constant total scattering coefficient $g_0 = n\sigma_0$. Energy $W/2$ radiated in one direction from the source at the origin is the same as that in the opposite direction. The multiple scattering process for the case of no intrinsic absorption $(b = 0)$ is described by the following integral equation for energy density:

$$E(x,t) = W\,G(x,t) + V_0\,g_0 \int\!\!\int_{-\infty}^{\infty} G(x - x',t - t')E(x',t')\,dt'\,dx' \qquad (7.19)$$

where the Green function for the coherent part is given by

$$G(x,t) = \frac{1}{2V_0} H(t)\delta\!\left(t - |x|/V_0\right)e^{-V_0 g_0 t} \qquad (7.20)$$

Let us keep the dimensional quantities here. First we take the Fourier transform in space and the Laplace transform in time of the Green function (7.20) as

$$\hat{\hat{G}}(k,s) = \int_{-\infty}^{\infty} dx e^{-ikx} \int_{0}^{\infty} dt e^{-st} G(x,t)$$

$$= \frac{1}{2V_0} \int_{-\infty}^{\infty} dx e^{-(s+g_0V_0)|x|/V_0 - ikx} = \frac{s+g_0V_0}{(s+g_0V_0)^2 + V_0^2 k^2}$$

(7.21)

Solving the integral equation (7.19) in the Fourier–Laplace space, we get the energy density as

$$\hat{\hat{E}}(k,s) = \frac{W\hat{\hat{G}}(k,s)}{1 - V_0 g_0 \hat{\hat{G}}(k,s)}$$

(7.22)

Note that this form is different from (7.8) because now we are working with dimensional quantities. Next, we take the inverse Fourier transform of (7.22) with respect to space coordinate x [see Gradshteyn and Ryzhik, 1994, p. 445]:

$$\hat{E}(x,s) = \frac{W}{2\pi} \int_{-\infty}^{\infty} \frac{(s+g_0V_0)}{(s+g_0V_0)^2 + V_0^2 k^2 - (s+g_0V_0)g_0V_0} e^{ikx} dk$$

$$= \frac{W(s+g_0V_0)}{2V_0} \frac{1}{\sqrt{s(s+g_0V_0)}} e^{-\sqrt{s(s+g_0V_0)}|x|/V_0}$$

(7.23)

By using a Laplace transform formula [Abramowitz and Stegun, 1970, p. 1027],

$$\frac{e^{-\sqrt{s(s+g_0V_0)}|x|/V_0}}{\sqrt{s(s+g_0V_0)}} = \int_{0}^{\infty} dt e^{-st} e^{-g_0V_0t/2} I_0\left(\frac{g_0V_0}{2}\sqrt{t^2 - (x/V_0)^2}\right) H(t - |x|/V_0)$$

(7.24)

Substituting this in (7.23) and using partial integration, we obtain

$$E(x,t) = \frac{W}{2V_0}\delta(t - |x|/V_0)e^{-\left(\frac{g_0V_0}{2}\right)t} + \frac{W}{4}g_0 e^{-\left(\frac{g_0V_0}{2}\right)t} H(t - |x|/V_0)$$

$$\times \left\{ I_0\left(\frac{g_0V_0}{2}\sqrt{t^2 - \left(\frac{x}{V_0}\right)^2}\right) + \frac{t}{\sqrt{t^2 - (x/V_0)^2}} I_1\left(\frac{g_0V_0}{2}\sqrt{t^2 - \left(\frac{x}{V_0}\right)^2}\right) \right\}$$

(7.25)

where I_0 and I_1 are modified Bessel functions and $I_0'(z) = I_1(z)$ [Abramowitz and Stegun, 1970, p. 376]. The first term on the right-hand side of (7.25) is composed of the direct and forward-scattered energy. The exponential factor for the direct energy density in (7.25) is half that in (7.20) since forward-scattered energy arrives at the same time as the direct energy in 1-D. In Figure 7.5, we show temporal traces

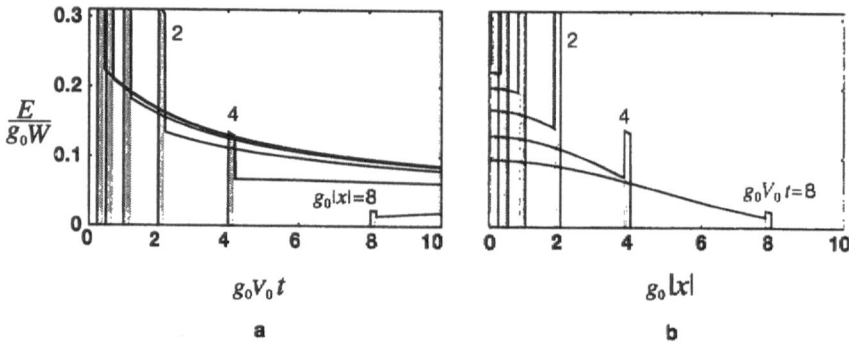

FIGURE 7.5. Normalized energy density for 1-D multiple scattering against (a) normalized lapse time for selected normalized distances and (b) normalized distance for selected normalized times, where the normalized source duration is $g_0 V_0 T_0 = 0.2$. [From Sato, 1993, with permission of Blackwell Science, United Kingdom.]

of the normalized energy density at different source-receiver distances and spatial sections at different lapse times.

Knowing the asymptotic behavior of the modified Bessel functions [Abramowitz and Stegun, 1970, p. 377]

$$I_0(z) \approx I_1(z) \approx \frac{e^z}{\sqrt{2\pi z}} \qquad \text{for } z \gg 1 \qquad (7.26)$$

we get the asymptotic expansion for energy density at large lapse times

$$E(x,t) \approx \frac{W}{2V_0} \delta(t - |x|/V_0) e^{-\left(g_0 V_0/2\right)t} + \frac{W}{\sqrt{4\pi D_c t}} e^{-\frac{|x|^2}{4 D_c t}} H(t - |x|/V_0)$$

$$\text{for } g_0 V_0 \sqrt{t^2 - (x/V_0)^2} \gg 1 \text{ and } t \gg |x|/V_0 \qquad (7.27)$$

where diffusivity $D_c = V_0/g_0$ for the 1-D case. The second term exactly coincides with the so called 1-D diffusion solution for $t > |x|/V_0$. The smooth distribution of energy density due to multiple scattering is described by the diffusion equation for a continuous limit of a discretized random walk. We note that the solution for the multiple scattering process (7.25) satisfies causality. A similar expression for the energy density in 1-D was obtained by Wu [1993]. Wu and Xie [1994] found an agreement between the energy density propagation in scattering media based on the radiative transfer theory and numerical simulations using a full-wave theory in random media.

Two-Dimensional Case

The multiple isotropic scattering process in 2-D space was first formulated by Shang and Gao [1988] in advance of the formulation in 3-D space. With dimensions included, the master equation for the multiple isotropic scattering process with no intrinsic absorption is

$$E(\mathbf{x},t) = WG(\mathbf{x},t) + V_0 g_0 \int_{-\infty}^{\infty} \int_{-\infty}^{\infty} \int_{-\infty}^{\infty} G(\mathbf{x}-\mathbf{x}',t-t')E(\mathbf{x}',t')dt'd\mathbf{x}' \qquad (7.28)$$

and

$$G(\mathbf{x},t) = \frac{1}{2\pi V_0 r} H(t)\delta\left(t - \frac{r}{V_0}\right)e^{-V_0 g_0 t} \qquad (7.29)$$

where $r = |\mathbf{x}|$. Shang and Gao [1988] heuristically obtained the analytical solution in the space-time domain, and Sato [1993] solved (7.28) and (7.29) using the same procedure taken in the 1-D case to obtain

$$E(\mathbf{x},t) = \frac{W}{2\pi V_0 r}\delta\left(t - \frac{r}{V_0}\right)e^{-g_0 V_0 t} + \frac{W g_0}{2\pi\sqrt{V_0^2 t^2 - r^2}}e^{g_0\left(\sqrt{V_0^2 t^2 - r^2} - V_0 t\right)}H\left(t - \frac{r}{V_0}\right) \quad (7.30)$$

The second term is asymptotically close to the diffusion solution for $V_0 t \gg r$. We find that the first-order perturbation of the second term coincides with (3.26), which is the single scattering term derived by Kopnichev [1975] for 2-D space.

7.1.3 Nonuniform Distribution of Scatterers

Observed spatial variations in velocity structure caused by geological processes as discussed in Section 2.1 and the nonuniform distribution of microearthquakes discussed in Section 5.5 lead to the idea that the distribution of heterogeneities in the lithosphere may not be uniform. Possible models for the distribution of scatterers include that it is depth-dependent or that it is a fractal distribution. The former conceptually agrees with the conventional image of the lithosphere in which the velocity inhomogeneity decreases with depth. The latter is appropriate for introducing into the framework of the radiative transfer theory.

Depth-Dependent Distribution of Scatterers

Hoshiba [1994] used a Monte Carlo method to numerically simulate coda power envelopes for a structure where the distribution of scatterers decreases with depth. He assumed spherical radiation from the source, isotropic scattering, and only one wave mode. The structural model he used has a constant velocity and no

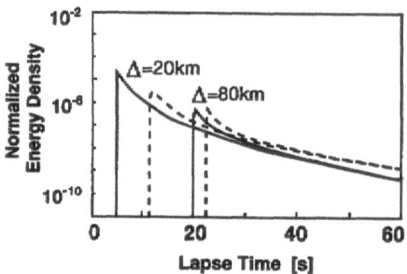

FIGURE 7.6. Temporal evolution of normalized energy density in a medium with depth-dependent scattering coefficient. Results for two sources with focal depths on the surface (solid) and at 40 km (broken) are shown. Medium velocity is β_o=4 km/s, intrinsic attenuation is zero throughout the medium, and scattering coefficient g_0=0.02 km^{-1} for 0-2 km depth, 0.01 km^{-1} for 2-6 km, and 10^{-5} km^{-1} at depths greater than 6 km. [From Hoshiba, 1994, copyright by the American Geophysical Union.]

intrinsic absorption. Figure 7.6 shows temporal traces of the normalized energy density at epicentral distances of 20 km and 80 km for two sources: one located on the surface (solid line) and the other located at a depth of 40 km (broken line). The computed normalized energy density asymptotically approaches a common curve irrespective of epicentral distance as the lapse time increases for each focal depth. However, the asymptotic energy levels of coda are not always the same at long lapse time when the focal depths are different; the difference is as much as 10 dB at 60 s lapse time. He cautioned about the use of the conventional coda-normalization method for estimating the source energy. His simulation shows that the synthesized envelopes give Q_c^{-1} decreasing with increasing lapse time as predicted by the single scattering model of Gusev [1995a] and consistent with many field observations. Distributing more than a hundred cavities in a 2-D half-space composed of two layers having different amounts of intrinsic absorption, Yomogida et al. [1996] used the boundary integral method to synthesize time traces of SH-waves for a point source radiation. They reported that Q_c^{-1} depends strongly on the intrinsic attenuation in the lower layer when the scattering strength of the lower layer is above a certain value.

Fractal Distribution of Scatterers

As shown in Section 5.5, a medium containing a distribution of attenuation bodies with fractal dimension two has a power-law decay of amplitudes with travel distance in agreement with observations that the maximum amplitude on regional seismograms obeys a power-law decay with travel distance. We can mathematically introduce the concept of fractal distribution for scatterers [Sato, 1988a, 1995b] or fractal inhomogeneity [Wu and Aki, 1985a; Main et al., 1990; Shapiro, 1992] into the framework of the radiative transfer theory and/or elastic wave theory. Scattering due to fractally distributed reflectors has been studied in atmospheric radar experiments in meteorology [Rastogi and Scheucher, 1990]. We will show that a medium having a fractal distribution of scatterers with a small fractal dimension has a smaller excitation of coda compared to a medium having a uniform distribution of scatterers.

Consider a fractally homogeneous random distribution of isotropic scatterers of cross section σ_0 with fractal dimension D_s in 3-D space. In the spherical coordinate system, we choose the number density of scatterers at radius r as

$$n(r) = n_0 \left[1 + \left(g_0 r/h\right)^2 \right]^{\frac{D_s-3}{2}}$$

$$\approx \begin{cases} n_0 \left(g_0 r/h\right)^{D_s-3} & \text{for} \quad g_0 r \gg h \\ n_0 & \text{for} \quad g_0 r \ll h \end{cases} \tag{7.31}$$

where $g_0 = \sigma_0 n_0$ and h is a nondimensional cutoff parameter in length. This is one of the simplest functions that is constant for small radius and decays with the power of distance for large distance. The total number of scatterers within a sphere is proportional to the D_s power of radius for $g_0 r \gg h$. This is an extension of (5.65), where the power of radius means the fractal dimension. The most familiar example is the Levi dust [Mandelbrot, 1983]. We suppose that the distribution (7.31) is valid for any origin but can be realized only statistically.

As a natural extension of the radiative transfer equation (7.1), the energy density is written as a convolution integral as follows:

$$E(\mathbf{x},t) = W G(\mathbf{x},t)$$

$$+ \int\int\int\int_{-\infty-\infty-\infty-\infty}^{\infty\ \infty\ \infty\ \infty} G(\mathbf{x}-\mathbf{x}',t-t')V_0\,\sigma_0\,n(|\mathbf{x}-\mathbf{x}'|)E(\mathbf{x}',t')d\mathbf{x}'\,dt' \tag{7.32}$$

where $n(|\mathbf{x}-\mathbf{x}'|)$ means the number density of scatterers at distance $|\mathbf{x}-\mathbf{x}'|$ from the receiver. We choose to ignore intrinsic absorption. The Green function for the coherent part becomes

$$G(\mathbf{x},t) = \frac{1}{4\pi V_0 r^2} H(t)\delta\left(t - \frac{r}{V_0}\right)e^{-B(r)} \tag{7.33}$$

where $r = |\mathbf{x}|$. The scattering attenuation term, given by a line integral, is written using the Gauss hypergeometric function $_2F_1$ [Gradshteyn and Ryzhik, 1994, p. 1066] as

$$B(r) = \int_0^r \sigma_0 n(r')dr' = g_0 r\ _2F_1\left(\frac{3-D_s}{2},\frac{1}{2};\frac{3}{2};-\left(\frac{g_0 r}{h}\right)^2\right) \tag{7.34}$$

where $e^{-B} = e^{-g_0 r}$ for $D_s = 3$ and $e^{-B} \approx \left(h/2g_0 r\right)^h$ for $g_0 r \gg h$ and $D_s = 2$. Taking the Fourier transform in space and the Laplace transform in time, we can solve the integral equation (7.32) with (7.33) and confirm the conservation of total energy.

Figure 7.7 shows spatiotemporal plots of the normalized energy density for $D_s=2$ (solid curve) and $D_s=3$ (broken curve) [Sato, 1995b]. Consistent with Figure 5.22, the direct energy density when $D_s=2$ decays more slowly with increasing distance than when $D_s=3$. Recall from Section 5.5 that the larger fractal dimension gave an exponential decay of amplitude whereas the lower fractal dimension led to a power-law dependence on travel distance. These simulations are qualitative. However, we may say that coda excitation for $D_s=2$ is smaller than that for $D_s=3$ and is spatially uniform at long lapse time agreeing with the observed uniform distribution of coda energy that forms the basic assumption for the coda-normalization method discussed in Section 3.4.

When intrinsic attenuation is nonzero and the distribution of intrinsic attenuation bodies is uniform, we modify the energy density by simply multiplying the temporal dependence of the energy density by an exponential factor for the intrinsic absorption. However, if the distribution of intrinsic attenuation bodies is also fractal, we have to add the decay term for the intrinsic absorption in (7.34).

The uniform distribution of coda energy at long lapse time and the power-law decay of maximum amplitude with travel distance are observations that might be explained by a fractal distribution of scatterers and/or intrinsic absorbers. It is left for us to examine quantitatively the evidence for such a fractal heterogeneity in the real earth.

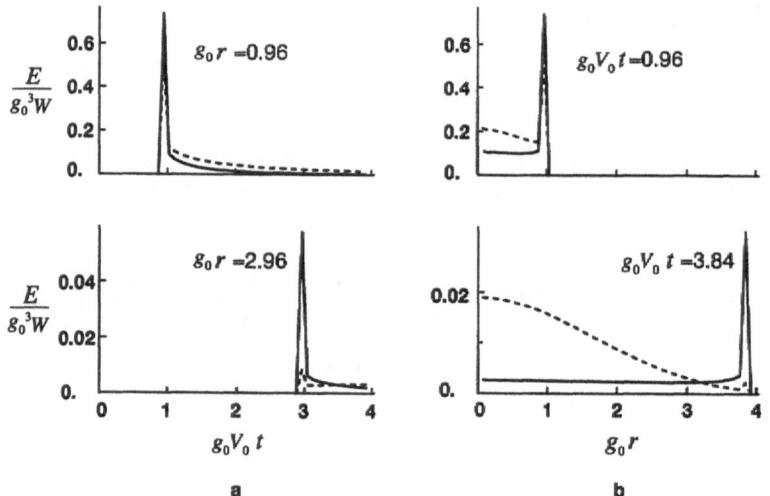

FIGURE 7.7. Solid curves show the (a) normalized energy density against normalized lapse time and (b) normalized energy density against normalized distance for $D_s=2$ and cutoff parameter $h=0.25$ in the case of no intrinsic absorption. Normalized source duration is $g_0 V_0 T_0=0.08$. The broken curve corresponds to $D_s=3$.

7.2 SEPARATION OF SCATTERING AND INTRINSIC ATTENUATION OF S-WAVES

We will now discuss the application to data of the radiative transfer theory that was developed in Section 7.1. The theory has been applied to data in a number of ways. Focusing on a procedure known as the multiple lapse-time window (MLTW) analysis method of Fehler et al. [1992], we will summarize the method and the results of its application to data.

7.2.1 Seismic Albedo

As discussed in Sections 5.1 and 5.2, total S-wave attenuation Q_S^{-1} is a combination of intrinsic or anelastic attenuation ${}^I Q_S^{-1}$ and scattering attenuation ${}^{Sc} Q_S^{-1}$ as given by (5.3). The estimation of parameters characterizing the random heterogeneity of the lithosphere, discussed in Section 5.3.2, that uses the frequency-domain comparison between theoretical scattering attenuation curves and observed attenuation data is an extreme limit that neglects intrinsic absorption. The multiple isotropic scattering model gives us a way to estimate quantitatively the amount of intrinsic and scattering attenuation from regional seismic data. Scattering attenuation is written using the total scattering coefficient for S-waves at angular frequency ω as ${}^{Sc} Q_S^{-1} \equiv g_0 \beta_0 / \omega$ (see (3.3)). Wu [1985] introduced the concept of seismic albedo, which is the ratio of scattering attenuation to total attenuation:

$$B_0 \equiv \frac{{}^{Sc} Q_S^{-1}}{{}^{Sc} Q_S^{-1} + {}^I Q_S^{-1}} = \frac{g_0 \beta_0}{g_0 \beta_0 + {}^I Q_S^{-1} \omega} \tag{7.35}$$

The term seismic albedo is due to the origin of the radiative transfer theory. Seismic albedo ranges from 0 to 1; media with strong heterogeneity and no-intrinsic absorption have high albedo, and homogeneous media have zero seismic albedo.

7.2.2 Multiple Lapse-Time Window Analysis

Several methods have been proposed to determine the amount of total attenuation caused by scattering and intrinsic mechanisms [Aki, 1980a; Taylor et al., 1986; Jacobson, 1987; Frankel and Wennerberg, 1987]. The MLTW method of Fehler et al. [1992] is based on two observations: the early portion of an S-wave seismogram is dominated by the direct S-wave whose amplitude is controlled by the total attenuation of the media; and S-coda is composed entirely of scattered S-waves whose amplitudes are controlled by the total scattering coefficient.

Method Based on the Stationary Solution of the Radiative Transfer Theory

The initial application of the radiative transfer theory to seismology was done by Wu [1985], who calculated a suite of curves showing the variation of the integrated energy density with source–receiver distance for various values of media parameters for the case of isotropic scattering. Since Wu's [1985] work was based on a stationary solution [see Ishimaru, 1978, Chapter 12], the resulting curves can be compared with data only if energy in observed seismograms can be integrated over infinite time. Wu and Aki [1988a], however, using the curves derived from the stationary solution, estimated attenuation parameters from the energy density integrated over a finite time in seismograms. After using the coda-normalization method discussed in Section 3.4 to correct for source and site effects, they arrived at mean curves for the observed shape of the integrated seismic energy density vs. source–receiver distance. The shapes of the observed curves were compared with the shapes predicted from the theory for various values of seismic albedo. Wu and Aki [1988a] concluded that scattering is the dominant cause of attenuation for frequencies lower than 2 Hz and intrinsic absorption dominates for frequencies above 2 Hz in the Hindu–Kush region in Pakistan. Since energy was integrated over a limited time in each seismogram (about 30 s), there may be an inconsistency between their result that scattering is a dominant mechanism of attenuation for lower frequencies and their comparison with the stationary theory. If scattering is dominant, seismograms will have long codas, and the effect of ignoring the energy in the seismogram arriving later than the window over which energy was integrated may be substantial.

Multiple Station Approach Using Time-Domain Solution

To remove the uncertainty from comparing finite seismograms with a theory for infinite lapse-time, Fehler et al. [1992] proposed the MLTW method. This

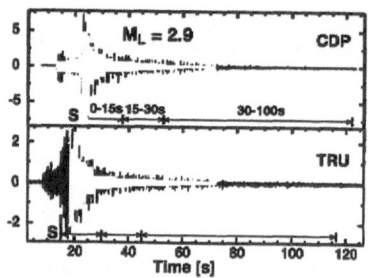

FIGURE 7.8. Velocity seismograms of a regional earthquake recorded by two stations in the Kanto-Tokai region, Japan and three time windows used for the MLTW analysis. [From Fehler et al., 1992, with permission from Blackwell Science, United Kingdom.]

method was made possible by the development of time-domain solutions to the radiative transfer theory, as described in Section 7.1, in which scattering is assumed to be isotropic and radiation is spherically symmetric. Fehler et al. [1992] used the Monte Carlo calculations of the energy density for the multiple isotropic scattering process corresponding to (7.1–7.2) by Hoshiba et al. [1991], which evaluates the integrated energy density for finite time windows as a function of source-receiver distance and medium parameters. The Monte Carlo calculations give results identical to those obtained in Section 7.1.1. The MLTW method is based on the observation that the integrated energy density in various time windows is influenced by the relative amounts of scattering and intrinsic attenuation. Further, the variation of integrated energy density vs. distance is controlled by the absolute amount of scattering and intrinsic attenuation. Fehler et al. [1992] defined three time windows, 0–15 s, 15–30 s, and 30–100 s over which they integrated energy density, where the lapse time is measured from the S-wave onset. Figure 7.8 shows the three time windows on velocity seismograms of a regional earthquake recorded in the Kanto-Tokai region, Japan. For three time windows, the integral $EI_{1,2,3}(f)_{kj}$ for event k at site j was calculated using

$$EI_1(f)_{kj} = \rho_0 \int_0^{15s} \left| \dot{u}_{kj}^S(t; f) \right|^2 dt, \quad EI_2(f)_{kj} = \rho_0 \int_{15s}^{30s} \left| \dot{u}_{kj}^S(t; f) \right|^2 dt, \text{ and}$$

$$EI_3(f)_{kj} = \rho_0 \int_{15s}^{100s} \left| \dot{u}_{kj}^S(t; f) \right|^2 dt \tag{7.36}$$

where $\dot{u}_{kj}^S(t; f)$ is the S-wave velocity seismogram in a frequency band centered at f Hz at lapse time t. Figure 7.9 shows the integrated energy density vs. source–receiver distance for a medium with $\beta_0 = 4$ km/s and $Q_S^{-1}\omega/\beta_0 = g_0 + {}'Q_S^{-1}\omega/\beta_0 = 0.03 \text{km}^{-1}$ calculated by the Monte Carlo simulations. Figure 7.9a shows the energy density integrated over the first 15 s after the S-wave arrival at each distance, and Figure 7.9b shows that over the time period 30–100 s after the S-arrival. Energy density integrated in each plot has been corrected for geometric spreading by multiplying by distance squared. In each plot, results for various values of seismic albedo B_0 are shown. When seismic albedo is small, indicating little scattering compared to intrinsic attenuation, there is a nearly linear dependence of log integrated energy density for the first time-window on distance indicating no scattering of energy to and from the primary wavefield. For the later time window, the curves for small albedo have low amplitude compared to the early time window. Small seismic albedo means little scattering so there is little energy in the coda.

Fehler et al. [1992] analyzed three-component velocity seismograms of 20 local earthquakes recorded at 66 stations of the NIED seismic network, which covers an area of about $350 \times 200 \text{ km}^2$ in the Kanto–Tokai region, Japan. The local magnitude of earthquakes used for the analysis ranges from 2 to 6.1 and their focal depths are less than 50 km.

For practical analysis, Fehler et al. [1992] corrected the integrated energy density for source and site amplification factors using

$$_N EI_1(f)_{kj} = \frac{EI_1(f)_{kj}}{W_k^S(f)\left|N_j^S(f)\right|^2}, \quad _N EI_2(f)_{kj} = \frac{EI_2(f)_{kj}}{W_k^S(f)\left|N_j^S(f)\right|^2}, \text{ and}$$

$$_N EI_3(f)_{kj} = \frac{EI_3(f)_{kj}}{W_k^S(f)\left|N_j^S(f)\right|^2} \tag{7.37}$$

where $W_k^S(f)$ is the S-wave source radiation energy and $N_j^S(f)$ is the S-wave site amplification factor. Relative source factors were determined using the coda-normalization method as described in Section 3.4 by finding the mean coda amplitude in 10 s windows beginning at 50 s lapse-time and averaging over a number of stations that recorded a given event. The average obtained for each event was normalized by the average obtained for a reference event to scale all source factors to this single reference event. Similarly, relative site amplifications were obtained by the coda-normalization method by calculating the mean coda amplitude in 10 s windows centered at the same lapse time at each station for a given earthquake. This mean was normalized by the amplitude recorded by reference station TRU in the middle of the network, at the same lapse time for the same event. The average of the normalized amplitudes obtained using many 10 s lapse-time windows for many events gave a stable estimate of the site amplification relative to the reference site.

Integrated energy density from individual seismograms is plotted vs. source–receiver distance. Fehler et al. [1992] took a running mean over 15 km windows to

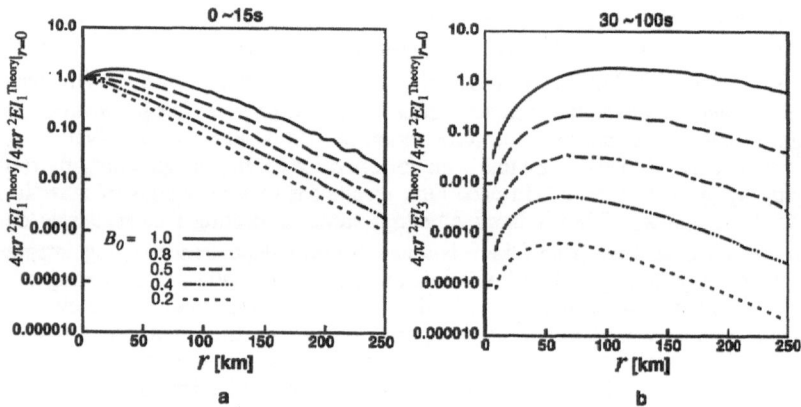

FIGURE 7.9. Integrated energy density corrected for geometrical spreading against source–receiver distance for two time windows calculated from Monte Carlo simulations of the multiple scattering process, (a) 0–15 s and (b) 30–100 s where $g_0 + \omega' Q_S^{-1}/\beta_0 = 0.03$ km^{-1} and $\beta_0 = 4$ km/s. Each trace is normalized by the integral for the first time window at $r=0$. [From Fehler et al., 1992, with permission from Blackwell Science, United Kingdom.]

find curves representing $_NEI_1$, $_NEI_2$, $_NEI_3$ vs. source-receiver distance. These means will be denoted as $\langle _NEI_1 \rangle_D$, $\langle _NEI_2 \rangle_D$, and $\langle _NEI_3 \rangle_D$. The shapes and relative differences of the curves for each time window of integration are compared with the theory to find the values of scattering and intrinsic attenuation. The comparison between data and theory can be done in many ways. The most direct way is to overlay $\langle _NEI_1 \rangle_D$, $\langle _NEI_2 \rangle_D$, and $\langle _NEI_3 \rangle_D$ with curves generated from the theory until acceptable fits are found to both shape and relative amplitude of the curves. Fehler et al. [1992] observed that the difference between $\langle _NEI_1 \rangle_D$ and $\langle _NEI_3 \rangle_D$ is dominated by the amount of scattering. They also observed that the slope of $\langle _NEI_1 \rangle_D$ is dominated by the amount of total attenuation. They defined a measure of the difference

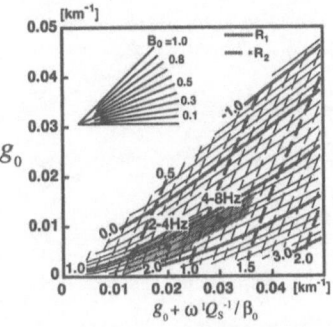

FIGURE 7.10. Total scattering coefficient against total attenuation per distance, where solid and broken curves correspond to constant R_1 and R_2 defined in (7.38a) and (7.38b), respectively. Dark and light shaded areas are the estimates for two frequency bands in the Kanto–Tokai region, Japan. [From Fehler et al., 1992, with permission from Blackwell Science, United Kingdom.]

between $\langle _NEI_1 \rangle_D$ and $\langle _NEI_3 \rangle_D$ at the same distance as

$$R_1 = \log \left[\frac{\langle _NEI_1 \rangle_D \big|_{r=150 \text{ km}}}{\langle _NEI_3 \rangle_D \big|_{r=150 \text{ km}}} \right] \qquad (7.38a)$$

They also defined a measure of the slope of $\langle _NEI_1 \rangle_D$ as

$$R_2 = \log \left[\frac{4\pi r^2 \langle _NEI_1 \rangle_D \big|_{r=50 \text{ km}}}{4\pi r^2 \langle _NEI_1 \rangle_D \big|_{r=150 \text{ km}}} \right] \qquad (7.38b)$$

They used Monte Carlo simulations [Hoshiba et al., 1991] of the multiple isotropic scattering process to make theoretical characteristic curves for the scattering attenuation vs. total attenuation per distance for various values of R_1 and R_2. Examples of such curves are shown in Figure 7.10. By making measurements of R_1 and R_2 from data curves and comparing with values derived from theory, the medium parameters can be obtained. They estimated that g_0=0.004 km^{-1} and B_0=0.45 for 1–2 Hz, g_0=0.0065 km^{-1} and B_0=0.34 at 2–4 Hz, g_0=0.01 km^{-1} and B_0=0.33 for 4–8 Hz. Figure 7.11 shows the log of integrated energy density corrected for geometri-

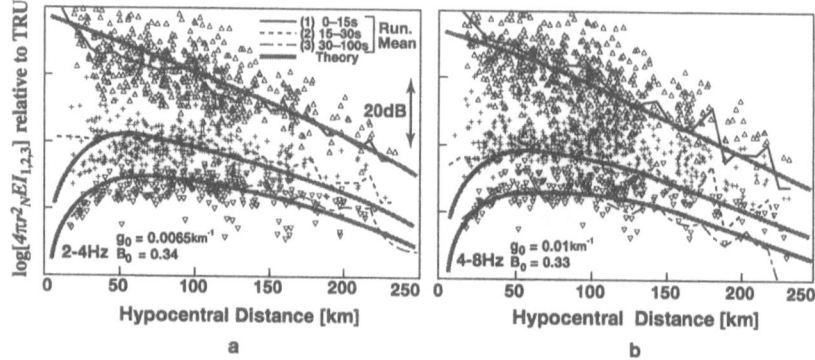

FIGURE 7.11. Plots of normalized integrated energy density with geometrical spreading correction vs. hypocentral distance in the Kanto–Tokai region, Japan, relative to the value at a borehole hard rock site TRU for vertical component data, where running means over a 15 km window and best fit theoretical curves are shown by fine lines and bold curves, respectively : **(a)** for 2–4 Hz band; **(b)** for 4–8 Hz band. [From Fehler et al., 1992, with permission from Blackwell Science, United Kingdom.]

cal spreading plotted vs. hypocentral distance. Only the shapes and relative differences in amplitude among three curves are important. The running means of the data over 15 km vs. distance $\langle _N EI_1 \rangle_D$, $\langle _N EI_2 \rangle_D$, and $\langle _N EI_3 \rangle_D$ are plotted in fine lines, and the bold lines show the fit to the data from the theory obtained using Figure 7.10 and measurements of R_1 and R_2 from the data.

Single Station Approach Using Time-Domain Solution

Hoshiba [1993] proposed a method to use data from only one station to develop curves $\langle _N EI_1 \rangle_D$, $\langle _N EI_2 \rangle_D$, and $\langle _N EI_3 \rangle_D$ similar to those shown in Figure 7.11. When data from a single station are analyzed, the site amplification term $N_j^S(f)$ is not necessary since it will be the same for all events. By developing curves for individual stations, the spatial variability of results can be investigated. Hoshiba [1993] integrated energy density $EI_{1,2,3}(f)_{kj}$ as in (7.36), where he defined three time-windows having equal lengths of 15 s. He corrected the integrated energy density for the source radiation factor using the coda energy density at 60 s lapse-time measured from the earthquake origin time $E^{SCoda}(f)_{kj}$. The observed integrated energy density was compared with the integrated energy density $EI_{1,2,3}^{Theory}(g_0, B_0, f)_{kj}$ calculated from Monte Carlo simulations for the three time windows for various values of g_0 and B_0, and the following residual was minimized for each station j:

Residual$[g_0, B_0, f]_j =$

$$\sum_{n=1}^{3}\sum_{k=1}^{N}\left\{\log\left[\frac{4\pi r_{kj}^2 EI_n(f)_{kj}}{E^{\text{S Coda}}(f)_{kj}}\right] - \log\left[\frac{4\pi r_{kj}^2 EI_n^{\text{Theory}}(g_0, B_0, f)_{kj}}{C_j}\right]\right\}^2 \quad (7.39)$$

where r_{kj} is the hypocentral distance of the kth earthquake and N the total number of earthquakes. Offset C_j, which is related to the unknown site factor $N_j^s(f)$ and an adjustment for dimension, was introduced to minimize the residual since only the relative differences of curves are important. Analyzing data at individual stations in Japan, Hoshiba [1993] concluded that the total attenuation is smaller in regions away from active volcanic zones.

Results of Studies Based on the MLTW Analysis

Mayeda et al. [1992] analyzed data from many stations in Hawaii, central California, and Long Valley, California. Jin et al. [1994] analyzed data from five stations in southern California by fitting data for various time windows of integration using curves calculated from the analytic solution of Zeng et al. [1991], as given in Section 7.1. Jin et al. [1994] concluded that for the five southern California stations that span a region having about 200 km radius, the seismic albedo and scattering attenuation show a spatial dependence for frequencies below about 6 Hz but are similar above 6 Hz. They found that scattering attenuation below 6 Hz is larger near fault zones. They argued that the waves above 6 Hz preferentially travel in the more homogeneous lower portion of the lithosphere resulting in less spatial variation in estimated parameters. They concluded that there is little difference in intrinsic attenuation among the stations. Akinci et al. [1995] reported the predominance of scattering attenuation for frequencies less than 4 Hz but the predominance of intrinsic attenuation for frequencies larger than 8 Hz in southern Spain. Most of these studies are based on the single station method. Figure 7.12 summarizes the results of these measurements. The results indicate a wide variety of relations between scattering and intrinsic attenuation although the general trend is that scattering and intrinsic attenuation decrease with increasing frequency over the range of 1–20 Hz. On average g_0 is estimated to be on the order of 10^{-2} km^{-1}.

The MLTW method was applied to high-frequency (400-1600 Hz) data from a mine in Canada [Feustel et al., 1996]. The investigators inferred that the observed increase in scattering up to 1000 Hz is caused by the 4–6 m characteristic length of the mapped fractures.

Examination of the data in Figure 7.11 reveals that $_N EI_1$, has significantly more scatter than $_N EI_3$ or $_N EI_2$. Since the first time window contains the direct S-wave arrival, the increased scatter is probably caused by nonspherical radiation from the source. We have assumed spherically symmetric radiation from the source in our analysis even though we know that radiated energy is nonspherical as predicted from a fault source model. Nonspherical source radiation will be studied further in the following section.

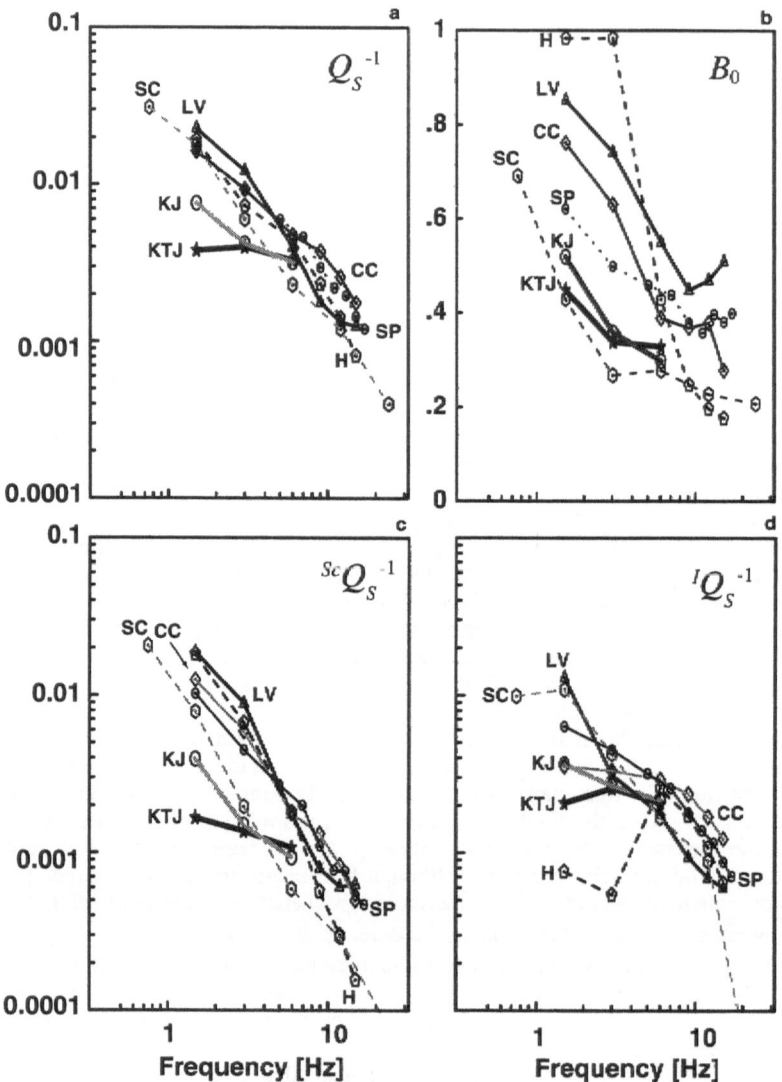

FIGURE 7.12. Summary of the application of the MLTW analysis method to various regions vs. frequency: **(a)** total attenuation and **(b)** seismic albedo of S-waves; **(c)** scattering attenuation and **(d)** intrinsic attenuation of S-waves. Results are from: KTJ, Kanto–Tokai, Japan [Fehler et al., 1992]; KJ, station Kakio2 in Kanto, Japan [Hoshiba, 1993]; LV, Long Valley, CC, central California, H, Hawaii [Mayeda et al., 1992]; SC, station SVD in southern California [Jin et al., 1994]; SP, 0–170 km range data in southern Spain [Akinci et al., 1995].

7.3 MULTIPLE ISOTROPIC SCATTERING PROCESS FOR NONSPHERICAL SOURCE RADIATION

As discussed in Section 6.1.1, radiation of seismic energy from an earthquake source is not spherically symmetric. Figure 7.13 shows MS seismogram envelopes, recorded at a hard rock site, of regional earthquakes having nearly equal hypocentral distances. Amplitudes have been normalized by the average coda amplitude at 60 s lapse-time indicated by an arrow. We find a large variation in amplitudes near the direct S-arrival. However, the variation reduces as lapse time increases and shows no dependence on lapse time after about 30 s. The variation in amplitude near the direct S-arrival is caused by the nonspherical radiation from the earthquake source and is well documented from studies using different types of seismic data. Here, we show how the energy density corresponding to MS seismogram

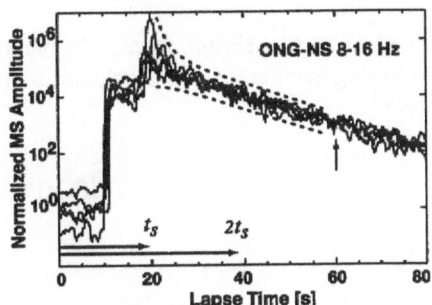

FIGURE 7.13. Horizontal-component seismogram envelopes in the 8–16 Hz band recorded at a hard rock site, Onagawa, near Sendai, Japan of regional earthquakes having nearly the same hypocentral distance. Amplitudes are normalized by the coda amplitudes at 60s lapse-time, which is indicated by an arrow. [From Sato et al., 1997, with permission from Elsevier Science - NL, Sara Burgerhartsraat 25, 1055 KV, Amsterdam, The Netherlands.]

envelope loses its memory of the nonspherical radiation pattern from the source with increasing lapse time [After Sato et al., 1997, with permission from Elsevier Science - NL, Sara Burgerhartsraat 25, 1055 KV, Amsterdam, The Netherlands].

7.3.1 Formulation

We assume nonspherical radiation of energy $\Psi(\theta,\phi)$ from a point source located at the origin in a scattering medium, where the distribution of point-like isotropic scatterers is randomly homogenous, as mentioned in Section 7.1. We use the point shear-dislocation model for an earthquake source, as discussed in Section 6.1. Figure 7.14 shows the angular dependence of the far-field radiation of S-wave energy from the source derived from the square of amplitude radiation patterns given by (6.3) and the geometry of the receiver and the last scattering point. Substituting this radiation pattern in the direct propagation term of (7.5), we may describe the energy propagation through the scattering medium by

$$\overline{E}(\overline{x},\overline{t}) = \Psi(\theta,\phi)\overline{G}(\overline{x},\overline{t}) + \int_{-\infty}^{\infty}\int_{-\infty}^{\infty}\int_{-\infty}^{\infty}\int_{-\infty}^{\infty} \overline{G}(\overline{x}-\overline{x}',\overline{t}-\overline{t}')\overline{E}(\overline{x}',\overline{t}')d\overline{t}'d\overline{x}' \qquad (7.40)$$

The radiation pattern is normalized as

$$\oint d\Omega(\theta,\phi)\,\Psi(\theta,\phi) = 4\pi \qquad (7.41)$$

where $d\Omega(\theta,\phi) = \sin\theta\,d\theta\,d\phi$ and the normalized Green function is given by (7.6). Following Section 7.1.1, the solution of (7.40) in the Fourier–Laplace domain is given by

$$\hat{\bar{E}}(\mathbf{k},s) = \frac{\hat{\bar{\bar{G}}}_{\Psi}(\mathbf{k},s)}{1 - \hat{\bar{G}}(k,s)} \qquad (7.42)$$

where

$$\hat{\bar{\bar{G}}}_{\Psi}(\mathbf{k},s) \equiv \int\limits_{-\infty}^{\infty}\int\limits_{-\infty}^{\infty}\int\limits_{-\infty}^{\infty} d\bar{\mathbf{x}}\,e^{-i\mathbf{k}\bar{\mathbf{x}}}\Psi(\theta,\phi)\int\limits_{0}^{\infty} d\bar{t}\,e^{-s\bar{t}}\overline{G}(\bar{\mathbf{x}},\bar{t}) \qquad (7.43)$$

We note that the argument is not scalar k but vector $\mathbf{k}$ because there is a specific orientation of the source radiation.

Measuring angle θ from the third axis and angle ϕ from the first axis, we introduce spherical harmonic functions $Y_{lm}(\theta,\phi)$ (see the Appendix) for the decomposition of the radiation pattern as

$$\Psi(\theta,\phi) = \sum_{l=0}^{\infty}\sum_{m=-l}^{l}\Psi_{lm}\,Y_{lm}(\theta,\phi) \qquad (7.44)$$

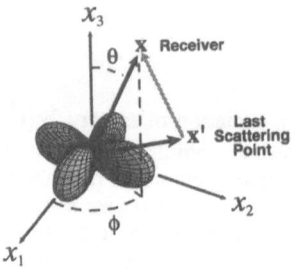

FIGURE 7.14. Geometry of the multiple scattering process for nonspherical radiation of S-wave energy from a point shear-dislocation source located at the origin. The unit normal vector to the fault plane (see Figure 6.1) is $\mathbf{n} = (1,0,0)$ and the unit slip vector is $\mathbf{s} = (0,1,0)$. The third axis is the null axis. [From Sato et al., 1997, with permission from Elsevier Science - NL, Sara Burgerhartsraat 25, 1055 KV, Amsterdam, The Netherlands.]

where $\Psi_{00} = \sqrt{4\pi}$ from the normalization condition (7.41). We take the phase factor of the spherical harmonic functions according to Landau and Lifshitz [1989]. The integral (7.43) can be written as a spherical harmonic expansion:

$$\hat{\bar{\bar{G}}}_\Psi(\mathbf{k},s) = \sum_{l=0}^{\infty} \sum_{m=-l}^{l} \Psi_{lm} \int_0^\infty \bar{r}^2 d\bar{r} \oint d\Omega(\theta,\phi)\, e^{-i k\bar{r}} Y_{lm}(\theta,\phi) \frac{e^{-(s+1+\bar{b})\bar{r}}}{4\pi\bar{r}^2}$$

$$= \sum_{l=0}^{\infty} \sum_{m=-l}^{l} \Psi_{lm} \int_0^\infty d\bar{r}\, e^{-(s+1+\bar{b})\bar{r}} \sum_{l'=0}^{\infty} \sum_{m'=-l'}^{l'} i^{l'} j_{l'}(-k\bar{r})\, Y_{l'm'}(\theta_k,\phi_k)$$

$$\times \oint d\Omega(\theta,\phi)\, Y_{l'm'}{}^*(\theta,\phi) Y_{lm}(\theta,\phi) \tag{7.45}$$

$$= \sum_{l=0}^{\infty} \sum_{m=-l}^{l} (-i)^l \int_0^\infty d\bar{r}\, e^{-(s+1+\bar{b})\bar{r}} j_l(k\bar{r})\, \Psi_{lm} Y_{lm}(\theta_k,\phi_k)$$

$$= \sum_{l=0}^{\infty} \sum_{m=-l}^{l} (-i)^l \bar{\bar{G}}_l(k,s) \Psi_{lm} Y_{lm}(\theta_k,\phi_k)$$

where $\mathbf{k} = (k,\theta_k,\phi_k)$ in spherical coordinates in the wavenumber vector space and we used (A.7). Function $\bar{\bar{G}}_l(k,s)$, which is defined as a Laplace transform of the spherical Bessel function, is explicitly written by using the Gauss hypergeometric function [Gradshteyn and Ryzhik, 1994, p. 732]:

$$\bar{\bar{G}}_l(k,s) \equiv \int_0^\infty j_l(k\bar{r})\, e^{-(s+1+\bar{b})\bar{r}}\, d\bar{r}$$

$$= \frac{1}{k} \left[\frac{k}{2(s+1+\bar{b})} \right]^{l+1} \frac{\sqrt{\pi}\,\Gamma(l+1)}{\Gamma(l+3/2)}\, {}_2F_1 \left[\frac{l+1}{2}, \frac{l+2}{2}, l+\frac{3}{2}; -\left(\frac{k}{s+1+\bar{b}} \right)^2 \right] \tag{7.46}$$

where the argument is scalar k and we note that

$$\hat{\bar{\bar{G}}}(k,s) = \bar{\bar{G}}_0(k,s) = \frac{1}{k} \tan^{-1} \frac{k}{s+1+\bar{b}} \tag{7.47}$$

and

$$\bar{\bar{G}}_1(k,s) = \frac{1}{k} \left(1 - \frac{s+1+\bar{b}}{k} \tan^{-1} \frac{k}{s+1+\bar{b}} \right) \tag{7.48}$$

By using the recurrence formula for the spherical Bessel function [Abramowitz and Stegun, 1970, p. 439]

$$(2l+1) \frac{d}{dz} j_l(z) = l j_{l-1}(z) - (l+1) j_{l+1}(z) \tag{7.49}$$

we get the following recurrence relationship for $\overline{\overline{G}}_l(k,s)$:

$$\overline{\overline{G}}_l(k,s) = -\frac{(2l-1)}{l}\frac{\left(s+1+\overline{b}\right)}{k}\overline{\overline{G}}_{l-1}(k,s) + \frac{(l-1)}{l}\overline{\overline{G}}_{l-2}(k,s) \quad \text{for} \quad l \geq 2 \quad (7.50)$$

We can get every higher order term by using this recurrence relationship. Substituting (7.45) and (7.47) in (7.42) and taking the inverse Fourier transform,

$$\hat{E}(\overline{x},s) = \frac{1}{(2\pi)^3}\int_0^\infty k^2 dk \oint d\Omega(\theta_k,\phi_k) \sum_{l=0}^\infty \sum_{m=-l}^l (-i)^l \Psi_{lm} Y_{lm}(\theta_k,\phi_k) \frac{\overline{\overline{G}}_l(k,s)}{1-\overline{\overline{G}}_0(k,s)}$$

$$\times 4\pi \sum_{l'=0}^\infty \sum_{m'=-l'}^{l'} i^{l'} j_{l'}(k\overline{r}) Y_{l'm'}(\theta,\phi) Y_{l'm'}^*(\theta_k,\phi_k)$$

$$= \sum_{l=0}^\infty \sum_{m=-l}^l \Psi_{lm} Y_{lm}(\theta,\phi) \frac{1}{2\pi^2} \int_0^\infty k^2 dk \frac{\overline{\overline{G}}_l(k,s)}{1-\overline{\overline{G}}_0(k,s)} j_l(k\overline{r}) \qquad (7.51)$$

$$= \sum_{l=0}^\infty \sum_{m=-l}^l \Psi_{lm} Y_{lm}(\theta,\phi) \hat{E}_l(\overline{r},s)$$

where $d\Omega(\theta_k,\phi_k) = \sin\theta_k \, d\theta_k \, d\phi_k$ in the wavenumber space. We can write the spherical Bessel function [Arfken and Weber, 1995, p. 681] in the following form:

$$j_l(z) = \frac{1}{2iz}\left[e^{iz}w_l(z) - (-1)^l e^{-iz}w_l(-z)\right] \quad \text{where} \quad w_l(z) \equiv \sum_{r=0}^l \frac{i^{r-l}(l+r)!}{r!(l-r)!(2z)^r} \quad (7.52)$$

By using this, we have the lth order term in (7.51) as

$$\hat{E}_l(\overline{r},s) \equiv \frac{1}{2\pi^2} \int_0^\infty k^2 dk \frac{\overline{\overline{G}}_l(k,s) j_l(k\overline{r})}{1-\overline{\overline{G}}_0(k,s)} = \frac{1}{\overline{r}}\frac{1}{2\pi}\int_{-\infty}^\infty dk\, e^{ik\overline{r}}\frac{k}{2\pi i}\frac{\overline{\overline{G}}_l(k,s)w_l(k\overline{r})}{1-\overline{\overline{G}}_0(k,s)} \quad (7.53)$$

where we used $\overline{\overline{G}}_l(-k,s) = (-1)^l \overline{\overline{G}}_l(k,s)$. Taking the inverse Laplace transform of (7.51), we get the normalized energy density in space and time:

$$\overline{E}(\overline{x},\overline{t}) = \sum_{l=0}^\infty \sum_{m=-l}^l \Psi_{lm} Y_{lm}(\theta,\phi) \overline{E}_l(\overline{r},\overline{t}) \qquad (7.54)$$

The lth order term can be formally written using a 2-D Fourier transform:

$$\overline{E}_l(\overline{r},\overline{t}) = \frac{1}{\overline{r}}\frac{1}{(2\pi)^2 i}\int_{-i\infty}^{i\infty} ds\, e^{s\overline{t}} \int_{-\infty}^\infty dk\, e^{ik\overline{r}}\frac{k}{2\pi i}\frac{\overline{\overline{G}}_l(k,s)w_l(k\overline{r})}{1-\overline{\overline{G}}_0(k,s)}$$

$$= \frac{1}{\overline{r}}\frac{1}{(2\pi)^2}\int_{-\infty}^\infty\int_{-\infty}^\infty d\omega dk\, e^{-i\omega\overline{t}-ik\overline{r}}\frac{ik}{2\pi}\frac{\overline{\overline{G}}_l(-k,-i\omega)w_l(-k\overline{r})}{1-\overline{\overline{G}}_0(-k,-i\omega)} \qquad (7.55)$$

Using $w_0(z) = 1$ from (7.52), we find that the integral kernel of (7.55) for $l=0$ reduces to (7.8) which was obtained for the case of spherical radiation from the source. Taking the limit $\mathbf{k} \to 0$, we can easily confirm the conservation of total energy in the case of no intrinsic absorption ($b=0$), when the contribution due to source radiation comes only from the lowest order term, $l=0$.

To practically evaluate $\overline{E}_l(\bar{r},\bar{t})$, we follow Section 7.1.1 and formally decompose it into three terms: the direct energy density, the single scattered energy density, and the energy density of multiple scattering of order greater than or equal to two:

$$\overline{E}_l(\bar{r},\bar{t}) = \overline{G}(\bar{r},\bar{t}) + \overline{E}_l^1(\bar{r},\bar{t}) + \overline{E}_l^M(\bar{r},\bar{t}) \tag{7.56}$$

Solving the first-order perturbation of (7.40) in the space-time domain, we get the single scattering term

$$
\begin{aligned}
\overline{E}^1(\overline{\mathbf{x}},\bar{t}) &= \int\!\!\int\!\!\int\!\!\int_{-\infty}^{\infty} \overline{G}(\overline{\mathbf{x}} - \overline{\mathbf{x}}', \bar{t} - \bar{t}')\Psi(\theta',\phi')\overline{G}(\overline{\mathbf{x}}',\bar{t}')d\bar{t}'d\overline{\mathbf{x}}' \\
&= e^{-(1+\bar{b})\bar{t}}\int\!\!\int\!\!\int_{-\infty}^{\infty} \frac{\delta(\bar{t} - \bar{r}' - \bar{r}'')}{(4\pi)^2 \bar{r}'^2 \bar{r}''^2}\Psi(\theta',\phi')d\overline{\mathbf{x}}' \\
&= \frac{e^{-(1+\bar{b})\bar{t}}}{(4\pi)^2}\oint d\Omega(\theta',\phi')\Psi(\theta',\phi')\int_0^{\infty}\bar{r}'^2\, d\bar{r}' \frac{\delta(\bar{t} - \bar{r}' - \bar{r}'')}{\bar{r}'^2\bar{r}''^2} \\
&= \frac{e^{-(1+\bar{b})\bar{t}}}{4\pi\bar{r}^2}H(\bar{t} - \bar{r})\frac{1}{2\pi}\oint d\Omega(\theta',\phi')\frac{\Psi(\theta',\phi')}{(\bar{t}/\bar{r})^2 + 1 - 2(\bar{t}/\bar{r})\cos\theta_0}
\end{aligned}
\tag{7.57}
$$

where $\bar{r}' = |\overline{\mathbf{x}}'|$, $\bar{r}'' = |\overline{\mathbf{x}} - \overline{\mathbf{x}}'|$ and $\cos\theta_0 = \cos\theta\cos\theta' + \sin\theta\sin\theta'\cos(\phi - \phi')$. We also used

$$
\begin{aligned}
\delta(\bar{t} - \bar{r}' - \bar{r}'') &= \delta\left(\bar{t} - \bar{r}' - \sqrt{\bar{r}^2 + \bar{r}'^2 - 2\bar{r}\bar{r}'\cos\theta_0}\right) \\
&= \frac{\bar{t}^2 - 2\bar{t}\bar{r}\cos\theta_0 + \bar{r}^2}{2(\bar{t} - \bar{r}\cos\theta_0)^2}\delta\left[\bar{r}' - \frac{\bar{t}^2 - \bar{r}^2}{2(\bar{t} - \bar{r}\cos\theta_0)}\right]
\end{aligned}
\tag{7.58}
$$

By using

$$\frac{1}{x - y} = \sum_{n=0}^{\infty}(2n+1)P_n(y)Q_n(x) \quad \text{for} \quad |x| > 1 > |y| \tag{7.59}$$

where function $Q_n(x)$ is the Legendre polynomial of the second kind [Abramowitz and Stegun, 1970, p.334], and the addition theorem for the Legendre polynomials (A.6),

$$\bar{E}^1(\bar{\mathbf{x}},\bar{t}) = \frac{e^{-(1+\bar{b})\bar{t}}}{4\pi\bar{r}^2} H(\bar{t}-\bar{r}) \sum_{l=0}^{\infty} \sum_{m=-l}^{l} \Psi_{lm} \frac{\bar{r}}{\bar{t}} \frac{1}{4\pi} \oint d\Omega(\theta',\phi') \frac{Y_{lm}(\theta',\phi')}{\dfrac{(\bar{t}/\bar{r})^2+1}{2(\bar{t}/\bar{r})} - \cos\theta_0}$$

$$= \frac{e^{-(1+\bar{b})\bar{t}}}{4\pi\bar{r}^2} H(\bar{t}-\bar{r}) \sum_{l=0}^{\infty} \sum_{m=-l}^{l} \Psi_{lm} \sum_{r=0}^{\infty} \frac{2l'+1}{4\pi} \frac{\bar{r}}{\bar{t}} Q_{l'} \left[\frac{(\bar{t}/\bar{r})^2+1}{2(\bar{t}/\bar{r})} \right]$$

$$\times \oint d\Omega(\theta',\phi') Y_{lm}(\theta',\phi') P_{l'}(\cos\theta_0)$$

$$= \sum_{l=0}^{\infty} \frac{e^{-(1+\bar{b})\bar{t}}}{4\pi\bar{r}^2} H(\bar{t}-\bar{r}) \frac{\bar{r}}{\bar{t}} Q_l \left[\frac{(\bar{t}/\bar{r})^2+1}{2(\bar{t}/\bar{r})} \right] \sum_{m=-l}^{l} \Psi_{lm} Y_{lm}(\theta,\phi)$$

$$= \sum_{l=0}^{\infty} \bar{E}_l^1(\bar{r},\bar{t}) \sum_{m=-l}^{l} \Psi_{lm} Y_{lm}(\theta,\phi)$$

$$(7.60)$$

where we used the orthogonality condition for spherical harmonic functions, (A.5). Finally we obtain

$$\bar{E}_l^1(\bar{r},\bar{t}) = \frac{e^{-(1+\bar{b})\bar{t}}}{4\pi\bar{r}^2} H(\bar{t}-\bar{r}) \frac{1}{\left(\dfrac{\bar{t}}{\bar{r}}\right)} Q_l \left[\frac{\left(\dfrac{\bar{t}}{\bar{r}}\right)^2+1}{2\left(\dfrac{\bar{t}}{\bar{r}}\right)} \right] \qquad (7.61)$$

Eq. (7.60) with (7.61) is the energy density derived for single isotropic scattering

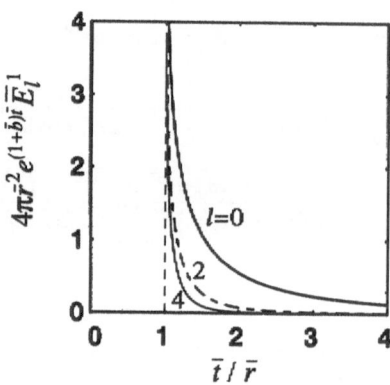

FIGURE 7.15. Temporal change in the single scattering energy density for nonspherical radiation of different orders. As source radiation departs from spherically symmetric, radiation having order l increases in importance but the effect on the energy density decreases more rapidly with increasing lapse time than for order 0. [From Sato et al., 1997, with permission from Elsevier Science - NL, Sara Burgerhartsraat 25, 1055 KV, Amsterdam, The Netherlands.]

from a point shear-dislocation source radiation [Sato, 1982c]. The lowest order term having $l=0$ corresponds to spherical source radiation. Then, substituting the explicit form for Q_0, we obtain

$$\bar{E}_0^1(\bar{r},\bar{t}) = \frac{1}{4\pi\bar{r}^2} K\left(\frac{\bar{t}}{\bar{r}}\right) H(\bar{t}-\bar{r}) e^{-(1+\bar{b})\bar{t}} \tag{7.62}$$

which is the same as (7.17). Figure 7.15 shows time traces of $4\pi\bar{r}^2 e^{(1+\bar{b})\bar{t}} \bar{E}_l^1(\bar{r},\bar{t})$ for selected even orders l. We find that terms of order $l > 0$ decrease more rapidly with increasing lapse time than the lowest order term, that is, the spherical radiation term corresponding to $l=0$ dominates at long lapse times.

Using (7.55), we obtain the multiple scattering energy density for the lth order as

$$\bar{E}_l^M(\bar{r},\bar{t}) = \frac{1}{\bar{r}} \frac{1}{(2\pi)^2} \int\limits_{-\infty}^{\infty}\int\limits_{-\infty}^{\infty} d\omega dk\, e^{-i\omega\bar{t}-ik\bar{r}} \frac{ik}{2\pi} \frac{\overline{\overline{G}}_l(-k,-i\omega)\overline{\overline{G}}_0(-k,-i\omega)^2}{1-\overline{\overline{G}}_0(-k,-i\omega)} w_l(-k\bar{r}) \tag{7.63}$$

We can evaluate (7.63) by using a 2-D FFT. Substituting (7.61) and (7.63) in (7.56), and putting the sum into (7.54), we get the energy density.

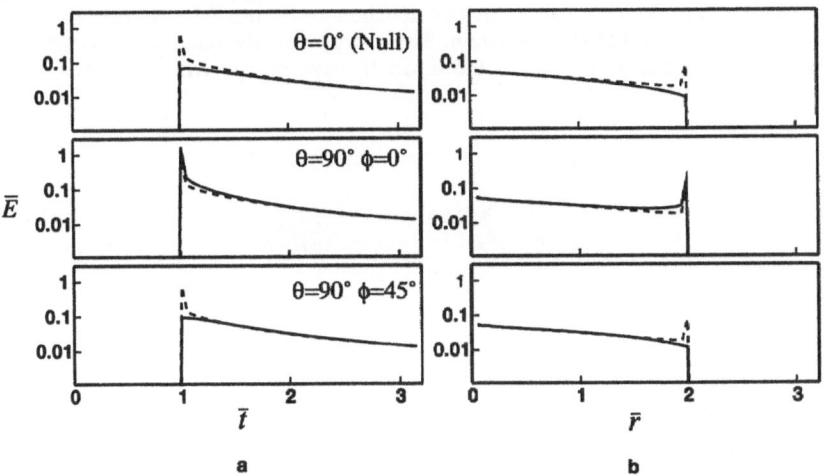

FIGURE 7.16. Spatiotemporal changes in energy density at different azimuths defined in Figure 7.14 for nonspherical radiation from a point shear-dislocation source, where the broken curve corresponds to results for spherical source radiation: (a) time traces at $\bar{r}=1$; (b) spatial sections at $\bar{t}=2$. [From Sato et al., 1997, with permission from Elsevier Science - NL, Sara Burgerhartsraat 25, 1055 KV, Amsterdam, The Netherlands.]

7.3.2 Simulation for a Point Shear-Dislocation Source

For a point shear-dislocation source having unit normal vector $\mathbf{n} = (1,0,0)$ and unit slip vector $\mathbf{s} = (0,1,0)$, we can decompose the radiation of S-wave energy that is schematically illustrated in Figure 7.14 as follows

$$
\begin{aligned}
\Psi(\theta,\phi) &= \left|B_\theta^S(\theta,\phi;\mathbf{n},\mathbf{s})\right|^2 + \left|B_\phi^S(\theta,\phi;\mathbf{n},\mathbf{s})\right|^2 \\
&= \frac{5}{2}\left(\sin^2\theta\cos^2\theta\sin^2 2\phi + \sin^2\theta\cos^2 2\phi\right) \\
&= \sqrt{4\pi}Y_{0,0}(\theta,\phi) + \frac{5}{7}\sqrt{\frac{4\pi}{5}}Y_{2,0}(\theta,\phi) - \frac{2}{7}\sqrt{\frac{4\pi}{9}}Y_{4,0}(\theta,\phi) \\
&\quad + \frac{\sqrt{280\pi}}{21}\left[Y_{4,4}(\theta,\phi) + Y_{4,-4}(\theta,\phi)\right]
\end{aligned}
\tag{7.64}
$$

where the null vector is taken in the direction of the third axis and the radiation patterns for S-waves B_θ^S and B_ϕ^S are given by (6.5).

For the numerical simulation, we choose $\bar{b} = 0$. Figure 7.16 shows the temporal evolution of the normalized energy density at $\bar{r} = 1$ and spatial distributions of energy density at $\bar{t} = 2$ in three different directions from the source. The broken curves show results for spherical source radiation as a reference. Figure 7.17 shows the angular distribution of energy density at $\bar{r} = 1$ on the equatorial plane $\theta = \pi/2$ at various normalized lapse times. Energy densities in Figure 7.17 are normalized by that for the spherical source radiation. The energy density for $\bar{t} = 1$ represents the nonspherical radiation pattern of the direct S-wave energy from the point shear-

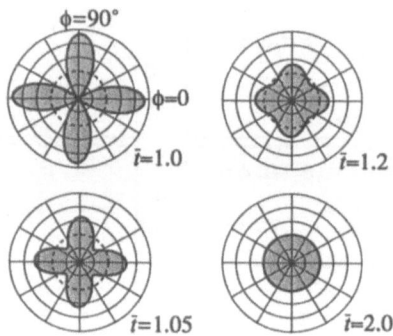

FIGURE 7.17. Energy density on the equatorial plane ($\theta = \pi/2$ in Figure 7.14) at $\bar{r} = 1$ for nonspherical radiation from a point shear-dislocation at various normalized lapse times. The energy density is normalized by that for the case of spherical source radiation, which is shown by a dashed curve in each plot. [From Sato et al., 1997, with permission from Elsevier Science - NL, Sara Burgerhartsraat 25, 1055 KV, Amsterdam, The Netherlands.]

dislocation. However, the azimuth dependence diminishes as lapse time increases and the energy density asymptotically converges to that obtained for spherical source radiation in Section 7.1.1, which corresponds to the lowest mode, $l=0$. The angular variation in energy density becomes less than 3% when the lapse time exceeds twice the direct wave travel-time at a distance of $\bar{r}=1$. This simulation qualitatively agrees with the observed radiation pattern independence of coda amplitudes at long lapse time as shown in Figure 7.13 and the empirical observations that form the basis for the coda-normalization method described in Section 3.4.

7.3.3 Using the Radiative Transfer Theory to Invert for the High-Frequency Radiation from an Earthquake

We have modeled the earthquake source using a point shear-dislocation, which is a reasonable representation for a small earthquake with a simple geometry. For a larger earthquake, it is necessary to consider variations in radiation from different portions of the fault plane and the timing of the radiation from fault segments due to the finite rupture propagation velocity. It has been shown that the amount of slip and the timing of slip along a fault plane can be determined from low-frequency

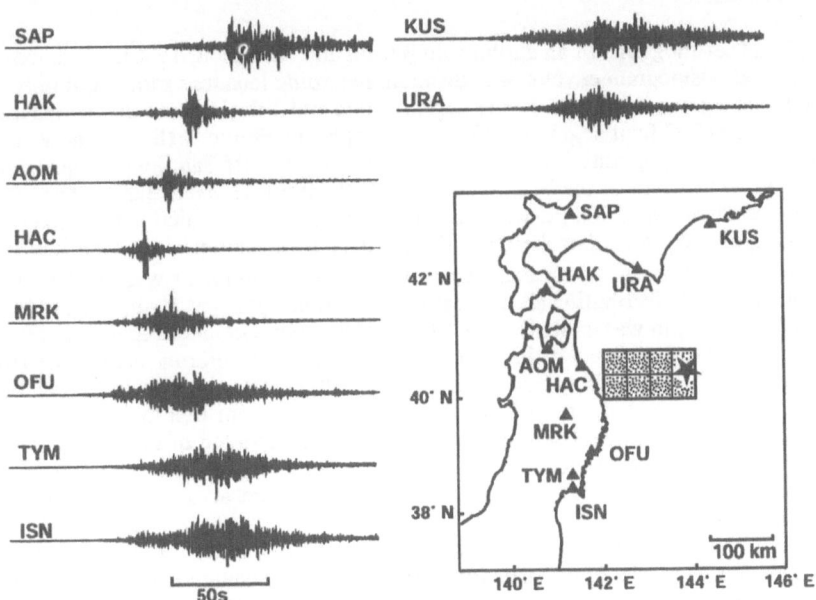

FIGURE 7.18. EW component velocity seismograms ($f >1$ Hz) of the 1994 far east off Sanriku earthquake, Japan, where each trace is normalized by its maximum value. The location of the initial break is indicated by a star at the east end of the main fault (bold rectangle), where small boxes are eight subfaults used in the inversion study. [Courtesy of H. Nakahara.]

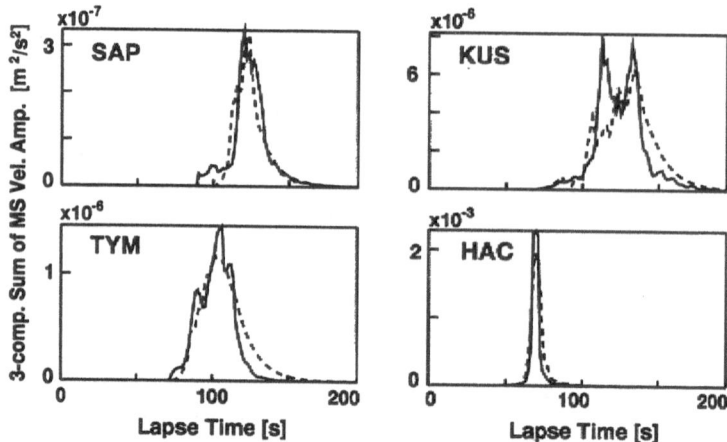

FIGURE 7.19. Sum of the observed three-component MS velocity envelopes at four stations in the 4–8 Hz band (solid) for the 1994 far east off Sanriku earthquake, Japan and the synthesized envelopes (broken) obtained using the effective Green function from the radiative transfer theory for a shear-dislocation source. [Courtesy of H. Nakahara.]

regional seismograms of an earthquake [Olson and Anderson, 1988]. Variations in observed seismogram envelopes at different recording locations reflect not only the fault plane geometry but the rupture propagation and differing amounts of radiation from individual fault segments. For an example, in Figure 7.18, we show high-frequency seismograms for the 1994 M_W=7.7 far east off Sanriku earthquake that took place offshore of the Pacific coast of northeastern Honshu, Japan. The maximum acceleration of 604 gal was recorded at station HAC, which is the nearest recording site to the earthquake fault. From analysis of teleseismic waves, the fault plane solution was a reverse fault type and the seismic moment was 4×10^{20} Nm. The aftershock distribution shows that the fault dimension was about 160 km by 80 km and the depth was as shallow as 13 km at the east end and deepened to 50 km on the west end. Results of inversion of long-period waveforms using data from station TYM located about 300 km from the earthquake, as shown in Figure 7.18, were consistent with the model that the fault ruptured from east to west having an average slip of 0.4 m [Nishimura et al., 1996]. The duration of the seismic signal was very short at HAC located in front of the rupture propagation; however, the signal has longer duration at OFU and URK which are at large angles from the rupture direction.

For assessing hazards due to future earthquakes, it is useful to determine the temporal dependence of the radiation of high-frequency energy from the fault since the temporal dependence controls constructive and destructive interference at various sites and hence the amplitude and duration of high-frequency shaking. Seismic waves with frequencies higher than 1 Hz are rather incoherent, particularly at the long source–receiver distances of observations of this earthquake, and the short-wavelength crustal inhomogeneities are complex, as shown in the previous chapters. Therefore, the conventional inversion method for estimating the rupture process was not successful when applied to high-frequency waves. This suggests that an appropriate analytical procedure may be to disregard the phase information and focus instead on seismogram envelopes.

Zeng et al. [1993] mapped the high-frequency radiation from the fault plane for the 1989 Loma Prieta earthquake from analysis of seismogram envelopes. They used a Green function derived from geometrical ray theory. Gusev and Pavlov [1991] proposed a method to invert for the radiation history from a fault where MS seismogram envelopes of small aftershocks are used as empirical Green functions. Kakehi and Irikura

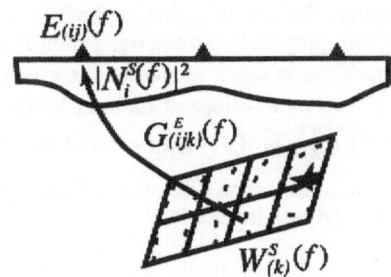

FIGURE 7.20. Geometry of the inversion scheme for the high-frequency energy radiation for the kth subfault showing station i, time j in the f Hz band. The star indicates the initial break and broken curves are isochrons of rupture propagation along the main fault.

[1996] estimated the high-frequency wave radiation from the fault plane of the 1993 Kushiro–Oki earthquake (M_w=7.6) by using RMS acceleration seismogram envelopes of small aftershocks as empirical Green functions.

Here, we introduce a method proposed by Nakahara [1996] in which the energy density for a point shear-dislocation source obtained in Sections 7.3.1 and 7.3.2 is used as the effective Green function for the high-frequency MS seismogram envelope. By using knowledge of the size of the fault plane, its orientation, its location, and the location where rupture initiated, Nakahara [1996] sought to find the rupture velocity, the average duration and the amount of high-frequency energy radiation on each portion of the fault. His approach is based on the assumption that the whole seismogram starting from the S-wave onset is composed only of direct and scattered S-waves and conversions between S- and surface waves are ignored; however, the advantage of this method is that a small number of parameters describe the entire model. Nakahara [1996] considered only S-waves in octave-width frequency bands having central frequency f Hz. Solid curves in Figure 7.19 show the sum of the three components of observed MS velocity envelopes for the 4–8 Hz band. Multiplying the mass density by the sum of the three components of observed MS velocity envelopes in each octave-width frequency band, he obtained a smoothed time trace of observed energy density.

For modeling the energy density, Nakahara [1996] used the configuration of the earthquake fault plane and seismic stations that is schematically illustrated in Figure 7.20. The rupture propagates from the initial break, whose location is indicated by a star on the fault. Positions of the rupture front along the fault are indicated by broken curves. He let $W_{(k)}^{S}(f)$ be the S-wave energy radiated from the subfault k, and $E_{(ij)}(f)$ the energy density recorded at station i at time j measured from the earthquake origin time. $G_{(ijk)}^{E}(f)$ is the effective Green function for unit energy radiation from subfault k as given by (7.54) at station i and time j for the known fault geometry. The effective Green function is delayed for each subfault according to the rupture propagation velocity, which is assumed constant along the fault. The site amplification factor at station i is given by $N_i^S(f)$. Then the predicted S-wave energy density envelope at station i for frequency f and time j is written as the product $|N_i^S(f)|^2 \sum_k G_{(ijk)}^E(f) W_{(k)}^S(f)$. To fit the data, Nakahara [1996] sought to find $W_{(k)}^S(f)$ that minimizes the difference of the square of the residual of the synthesized and observed envelopes:

$$\sum_{i,j} \frac{1}{\sigma_i^2} \left| E_{(ij)}(f) - |N_i^S(f)|^2 \sum_k G_{(ijk)}^E(f) W_{(k)}^S(f) \right|^2 \Rightarrow \text{Min} \qquad (7.65)$$

where σ_i^{-2} is a weight for station i.

Nakahara [1996] divided the main-fault into eight subfaults of size 44x44 km^2, and chose the site amplification factor as 1 at reference station TYM, whose seismometer is located on hard rock. The fault geometry was taken from the Harvard Centroid Moment Tensor (CMT) solution determined from long-period waves. He used values of total scattering coefficient and intrinsic absorption estimated for this area [Hoshiba, 1993] using the MLTW analysis (see Section 7.2). The contribution of each station was equalized by choosing the weight σ_i^{-2} equal to the reciprocal of the peak value of the observed energy density at the station. The solution was found by an iterative method to fit the observed energy density envelope data in four frequency bands (1–2, 2–4, 4–8, and 8–16 Hz) at ten stations. The minimum residual was obtained by choosing rupture velocity as 2.7 km/s and the duration of energy radiation from each subfault as 6 s. The broken curves in Figure 7.19 show the synthesized envelopes at four stations in the 4–8 Hz band. The coincidence between observation and model is very good for station HAC with a short duration packet and for stations with long duration packets, such as SAP, TYM and KUS. Figure 7.21a shows histograms of energy radiated from different subfaults for the 4–8 Hz band. He concluded that more than 90% of high-frequency energy was radiated

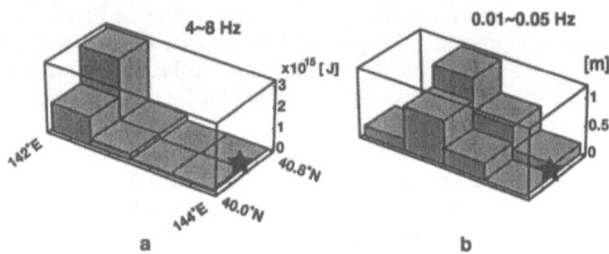

FIGURE 7.21. (a) Histogram of energy radiated in the 4–8 Hz band from the eight subfaults used to model the 1994 far east off Sanriku earthquake, Japan. (b) Histogram of slip based on an inversion study using long-period records at station TYM by Nishimura et al. [1996]. Star indicates location of initial rupture. [Courtesy of H. Nakahara and T. Nishimura.]

from the western half of the fault, where a large amount of slip took place. A total of 50% of the high-frequency energy radiation was from the subfault at the NW end of the fault, at the end of rupture. Madariaga [1977] used a dynamic model of faulting to investigate the high-frequency radiation from a fault with irregular rupture velocity and concluded that any abrupt change in the rupture velocity causes large high-frequency radiation. Thus a region where fault strength varies may dominate the radiation of high-frequency from the fault. The maximum fault displacement occurred on a subfault neighboring the center of the fault as shown in Figure 7.21b. We note that the subfault on the NW end coincides with the location where the largest amount of slip and the largest high-frequency radiation occurred during the 1968 Ms=7.9 off Tokachi earthquake [Mori and Shimazaki, 1985]. Dividing the main-fault into 32 subfaults, Nakahara et al. [1997] made a more precise analysis of the radiation from the earthquake.

7.4 MULTIPLE NONISOTROPIC SCATTERING PROCESS FOR SPHERICAL SOURCE RADIATION

In Sections 7.1 and 7.3, we examined the multiple isotropic scattering process including the effects of nonspherical source radiation; however, as shown in Figure 4.4b, forward-scattering strength increases with increasing frequency. Scattering patterns of elastic vector waves are nonisotropic in general as illustrated in Figure 4.8. How does nonisotropic scattering affect the seismogram envelope? What is the effect of multiple nonisotropic scattering at long lapse time? We will now examine these issues. The first analysis using the radiative transfer theory for nonisotropic scattering was done for the 2-D case of surface wave scattering; it was later studied for the 3-D case [Sato, 1994b, 1995a]. We introduce the formulation for the 3-D case in the following, which corrects an error in Sato [1995a].

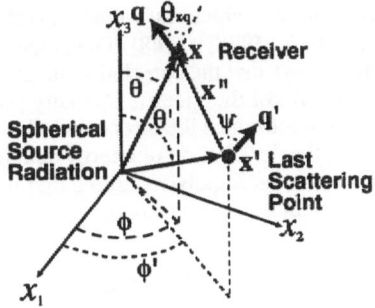

FIGURE 7.22. Geometry used for the study of multiple nonisotropic scattering.

7.4.1 Formulation

Similar to Section 7.1, we assume a homogenous and random distribution of point-like scatterers in 3-D space; however, we assume nonisotropic scattering. We describe the axial symmetric nonisotropic scattering power per volume by scattering coefficient $g(\psi)$ for scattering angle ψ. The total scattering coefficient is given by the average over solid angle (see (3.3)):

$$g_0 \equiv \frac{1}{4\pi}\oint g(\psi)d\Omega(\psi) = \frac{1}{2}\int_0^\pi g(\psi)\sin\psi d\psi \qquad (7.66)$$

We formulate the multiple nonisotropic scattering process in 3-D space after Sato [1995a, with permission from Blackwell Science, United Kingdom]. The background medium is characterized by wave velocity V_0, and the source radiation is spherically symmetric. We define $f(\mathbf{x},t;\mathbf{q})$ as the directional distribution of energy density at $\mathbf{x}$ in the direction given by unit vector $\mathbf{q}$ at time t, where $\mathbf{q} = (1,\theta_q,\phi_q)$ in spherical coordinates. Integrating f over the solid angle with respect to $\mathbf{q}$, the energy density

$$E(\mathbf{x},t) = \oint f(\mathbf{x},t;\mathbf{q})d\Omega(\theta_q,\phi_q) \qquad (7.67)$$

where $d\Omega(\theta_q,\phi_q) = \sin\theta_q d\theta_q d\phi_q$. $f(\mathbf{x},t;\mathbf{q})V_0\mathbf{q}$ gives the energy-flux density in direction $\mathbf{q}$, which is called specific intensity in the radiative transfer theory [Ishimaru, 1978, p. 148]. The stochastic study of the wave equation in random media shows that the correlation function of the wavefield corresponds to the Fourier transform of the specific intensity [Ishimaru, 1978, p. 275].

We introduce the directional Green function for coherent wave energy. It represents the distribution of energy density in the direction $\mathbf{q}$ at location $\mathbf{x}$ due to radiation from an impulsive source that is spherically symmetric, has unit energy, and is located at the origin at $t=0$:

$$G(\mathbf{x},t;\mathbf{q}) = \frac{e^{-(g_0 V_0 + b)t}}{4\pi V_0 r^2}\delta\left(t - \frac{r}{V_0}\right)H(t)\delta_\Omega(\mathbf{x};\mathbf{q}) \qquad (7.68)$$

where $r = |\mathbf{x}|$, b is for intrinsic absorption, and δ_Ω is the delta function for the solid angle as given by (A.9) and (A.10). The delta-function term means that energy-flux density exists only in the radial direction. The total scattering coefficient g_0 appearing in the exponent accounts for the scattering attenuation.

We derive the spatiotemporal change in f when the total energy W is spherically radiated instantaneously from the origin at $t=0$. As an extension of (7.1), the integral equation describing the multiple nonisotropic scattering process is

$$f(\mathbf{x},t;\mathbf{q}) = W\,G(\mathbf{x},t;\mathbf{q})$$
$$+ \int_{-\infty}^{\infty}\int_{-\infty}^{\infty}\int_{-\infty}^{\infty}\int_{-\infty}^{\infty}\oint G(\mathbf{x}'',t-t';\mathbf{q})V_0\,g(\psi)f(\mathbf{x}',t';\mathbf{q}')dx'\,dt'\,d\Omega(\theta_q,\phi_q) \qquad (7.69)$$

where $\mathbf{x}'' = \mathbf{x} - \mathbf{x}'$ is the vector from the last scatterer at $\mathbf{x}'$ to the receiver at $\mathbf{x}$, as illustrated in Figure 7.22. The second term in (7.69) is a convolution integral which means that the energy flux in the direction q' ($|q'| = 1$) hits the last scatterer generating scattered energy flux proportional to $g(\psi)$ in the direction x'', where ψ is the angle between directions q' and x''. The directional Green function appears in the convolution integral to account for propagation from the last scatterer to the receiver. When scattering is isotropic, $g = g_0$ integrating (7.69) over the solid angle, we get (7.1). Equations (7.66)–(7.69) can be considered a natural extension of the radiative transfer theory for isotropic scattering to the case of nonisotropic scattering.

We normalize time, length, and related quantities by V_0 and g_0 as done in (7.4):

$$\bar{t} = V_0 g_0 t, \quad \bar{x} = g_0 x, \quad \bar{b} = \frac{b}{V_0 g_0}, \quad \bar{g} = \frac{g}{4\pi g_0},$$

$$\bar{f} = \frac{f}{W g_0^3}, \quad \bar{G} = \frac{G}{g_0^3} \quad \text{and} \quad \bar{E} = \frac{E}{W g_0^3}$$

(7.70)

The normalization of (7.66) becomes $\oint \bar{g} \, d\Omega = 1$. Then, the integral equation (7.69) is rewritten as

$$\bar{f}(\bar{x}, \bar{t}; q) = \bar{G}(\bar{x}, \bar{t}; q)$$

$$+ 4\pi \int_{-\infty}^{\infty}\int_{-\infty}^{\infty}\int_{-\infty}^{\infty} \oint \bar{G}(\bar{x}'', \bar{t} - \bar{t}'; q) \bar{g}(\psi) \bar{f}(\bar{x}', \bar{t}'; q') \, d\bar{x}' d\bar{t}' d\Omega(\theta_{q'}, \phi_{q'})$$

(7.71)

where $\bar{x}'' = \bar{x} - \bar{x}'$, and ψ is the angle between directions q' and $\bar{x}''$. The nondimensional normalized directional Green function is given by

$$\bar{G}(\bar{x}, \bar{t}; q) = \frac{e^{-(1+\bar{b})\bar{t}}}{4\pi \bar{r}^2} \delta(\bar{t} - \bar{r}) H(\bar{t}) \delta_\Omega(\bar{x}; q)$$

(7.72)

where $\bar{r} = |\bar{x}|$.

In addition to the Fourier transform in space and the Laplace transform in time, we use a spherical harmonic series expansion of the solid angle with respect to q (see the Appendix):

$$\hat{\bar{f}}_{lm}(\mathbf{k}, s) = \int_{-\infty}^{\infty}\int_{-\infty}^{\infty}\int_{-\infty}^{\infty} d\bar{x} \, e^{-i\mathbf{k}\bar{x}} \int_0^{\infty} d\bar{t} \, e^{-s\bar{t}} \oint d\Omega(\theta_q, \phi_q) Y_{lm}^*(\theta_q, \phi_q) \bar{f}(\bar{x}, \bar{t}; q)$$

(7.73)

The inverse transform is

$$\bar{f}(\bar{x}, \bar{t}; q) = \frac{1}{(2\pi)^3} \int_{-\infty}^{\infty}\int_{-\infty}^{\infty}\int_{-\infty}^{\infty} d\mathbf{k} \, e^{i\mathbf{k}\bar{x}} \frac{1}{2\pi i} \int_{-\infty i}^{\infty i} ds \, e^{i s} \sum_{l=0}^{\infty} \sum_{m=-l}^{l} Y_{lm}(\theta_q, \phi_q) \hat{\bar{f}}_{lm}(\mathbf{k}, s)$$

(7.74)

Taking the Laplace transform with respect to time and the spherical harmonic series expansion of (7.72),

$$\hat{\bar{G}}_{lm}(\bar{\mathbf{x}},s) = \frac{e^{-(s+1+\bar{b})\bar{r}}}{4\pi\bar{r}^2} Y_{lm}^*(\theta,\phi) \qquad (7.75)$$

where subscripts have the same meaning as when used with $\bar{f}$. The Fourier transform of (7.75) with respect to space coordinates becomes the product of a spherical harmonic function and a function of $k = |\mathbf{k}|$:

$$\hat{\bar{\bar{G}}}_{lm}(\mathbf{k},s) = \int\int\int_{-\infty}^{\infty} d\bar{\mathbf{x}} e^{-i\mathbf{k}\bar{\mathbf{x}}} \hat{\bar{G}}_{lm}(\bar{\mathbf{x}},s) = (-i)^l \overline{\overline{G}}_l(k,s) Y_{lm}^*(\theta_k,\phi_k) \qquad (7.76)$$

where wavenumber vector $\mathbf{k} = (k,\theta_k,\phi_k)$ in spherical coordinates and $\overline{\overline{G}}_l$ is given by (7.46). We used expansion formula (A.7) and the orthogonality of spherical harmonic functions (A.5).

Next, using the addition theorem (A.6), we expand the nondimensional normalized scattering coefficient using spherical harmonic functions as

$$\bar{g}(\psi) = \sum_{l=0}^{\infty} (-i)^l \bar{g}_l Y_{l0}(\psi,0) = \sum_{l=0}^{\infty} \sqrt{\frac{4\pi}{2l+1}} \bar{g}_l \sum_{m=-l}^{l} Y_{lm}(\theta'',\phi'') Y_{lm}^*(\theta_{q'},\phi_{q'}) \qquad (7.77)$$

where $\bar{\mathbf{x}}'' = (\bar{r}'',\theta'',\phi'')$ and $\mathbf{q}' = (1,\theta_{q'},\phi_{q'})$ in spherical coordinates and expansion coefficients $\bar{g}_l$ are real quantities. The lowest term with $l=0$ corresponds to isotropic scattering, where $\bar{g}_0 = 1/\sqrt{4\pi}$.

Substituting (7.77) in the Laplace transform of (7.71),

$$\hat{\bar{f}}(\bar{\mathbf{x}},s;\mathbf{q}) = \hat{\bar{G}}(\bar{\mathbf{x}},s;\mathbf{q})$$
$$+ 4\pi \int\int\int_{-\infty}^{\infty} \hat{\bar{G}}(\bar{\mathbf{x}}'',s;\mathbf{q}) \sum_{l'=0}^{\infty} \sqrt{\frac{4\pi}{2l'+1}} \bar{g}_{l'} \sum_{m'=-l'}^{l'} Y_{l'm'}(\theta'',\phi'') \hat{\bar{f}}_{l'm'}(\bar{\mathbf{x}}',s) d\bar{\mathbf{x}}' \qquad (7.78)$$

Substituting the spherical harmonic expansion of $\hat{\bar{G}}$ with respect to $\mathbf{q}$ in (7.78) and using the orthogonality relationship (A.5),

$$\hat{\bar{f}}_{lm}(\bar{\mathbf{x}},s) = \hat{\bar{G}}_{lm}(\bar{\mathbf{x}},s)$$
$$+ 4\pi \sum_{l'=0}^{\infty} \sqrt{\frac{4\pi}{2l'+1}} \bar{g}_{l'} \sum_{m'=-l'}^{l'} \int\int\int_{-\infty}^{\infty} \hat{\bar{G}}_{lm}(\bar{\mathbf{x}}'',s) Y_{l'm'}(\theta'',\phi'') \hat{\bar{f}}_{l'm'}(\bar{\mathbf{x}}',s) d\bar{\mathbf{x}}' \qquad (7.79)$$

The second term is a convolution integral over $\bar{\mathbf{x}}'$. By using (7.75) and (7.76),

$$\int\int\int_{-\infty-\infty-\infty}^{\infty\ \infty\ \infty} \hat{\bar{G}}_{lm}(\bar{\mathbf{x}},s)Y_{l'm'}(\theta,\phi)e^{-i\mathbf{k}\bar{\mathbf{x}}}d\bar{\mathbf{x}} = \sum_{l''=0}^{\infty}\sum_{m''=-l''}^{l''} (Y_{l''m''})_{l'm'}^{lm}\ \hat{\bar{G}}_{l''m''}(\mathbf{k},s) \qquad (7.80)$$

where $\left(Y_{l''m''}\right)_{l'm'}^{lm}$ can be written explicitly using the definition of the Wigner 3-j symbols given in (A.11) to represent the integral of a product of three spherical harmonic functions. This term vanishes except when the triangular condition described following (A.11) holds. Taking the Fourier transform of (7.79) with respect to space coordinates and substituting (7.80) in the result, we finally obtain a set of simultaneous linear equations for $0 \le l < \infty$ and $-l \le m \le l$:

$$\hat{\bar{f}}_{lm}(\mathbf{k},s) = \hat{\bar{G}}_{lm}(\mathbf{k},s)$$

$$+ 4\pi \sum_{l'=0}^{\infty} \sqrt{\frac{4\pi}{2l'+1}}\bar{g}_{l'} \sum_{m'=-l'}^{l'}\sum_{l''=0}^{\infty}\sum_{m''=-l''}^{l''} (Y_{l''m''})_{l'm'}^{lm}\ \hat{\bar{G}}_{l''m''}(\mathbf{k},s)\ \hat{\bar{f}}_{l'm'}(\mathbf{k},s) \qquad (7.81)$$

We try to find a solution as a product of a spherical harmonic function and a function of k as

$$\hat{\bar{f}}_{lm}(\mathbf{k},s) = (-i)^l\ \bar{f}_l(k,s)\ Y_{lm}{}^*(\theta_k,\phi_k) \qquad (7.82)$$

That is, we may write $\bar{f}$ as a Legendre expansion of angle θ_{xq} between directions $\mathbf{q}$ and $\bar{\mathbf{x}}$ as

$$\bar{f}(\bar{\mathbf{x}},s;\mathbf{q}) = \sum_{l=0}^{\infty}\frac{2l+1}{4\pi}\ P_l(\cos\theta_{xq})\frac{1}{2\pi^2}\int_0^{\infty} j_l(k\bar{r})\bar{f}_l(k,s)k^2 dk \qquad (7.83)$$

The factorization (7.82) means that $\bar{f}(\bar{\mathbf{x}},\bar{t};\mathbf{q})$ is a function of angle θ_{xq}. However, our objective is not to obtain $\bar{f}$ but the lowest order term $\bar{f}_0$, which is the only term needed to describe the energy density, as will be shown below. Multiplying $Y_{lm}(\theta_k,\phi_k)$ by (7.81), integrating over the solid angle in the wavenumber space, and summing up for m from $-l$ to l, we get the equation for $\bar{\bar{f}}_l$

$$\bar{\bar{f}}_l(k,s) = \bar{\bar{G}}_l(k,s) + \frac{4\pi}{2l+1}\sum_{l'=0}^{\infty}\sqrt{\frac{4\pi}{2l'+1}}\bar{g}_{l'}\sum_{m'=-l'}^{l'}$$

$$\times \sum_{l''=0}^{\infty}(-i)^{l'+l''-l}\sum_{m''=-l''}^{l''}\left|(Y_{l'',m''})_{l',m'}^{l,m'+m''}\right|^2\bar{\bar{G}}_{l'}(k,s)\ \bar{\bar{f}}_{l'}(k,s) \qquad (7.84)$$

where we set $m = m'+m''$ for the nonvanishing component according to the selection rule for the addition of angular momenta. Functions $\bar{\bar{f}}_l$ and $\bar{\bar{G}}_l$ are real since the exponent of $-i$, which is $l'+l''-l$, is even for nonvanishing $\left(Y_{l''m''}\right)^{l,m'+m''}_{l'm'}$. In the equation corresponding to (7.84), denominator $2l+1$ was missing in Sato [1995a].

If the scattering coefficient is written as a finite series of spherical harmonic functions, the triangular inequality for the Wigner 3-j symbols makes the right-hand side of (7.84) a finite series, that is, if the highest order of the expansion is l_{Max}, we have to solve the simultaneous linear equations for $l_{Max}+1$ unknowns. For isotropic scattering, solving (7.84) with $\bar{g}_l = \delta_{l0}/\sqrt{4\pi}$, we get $\bar{\bar{f}}_0 = \bar{\bar{G}}_0/(1-\bar{\bar{G}}_0)$, which coincides with (7.8).

Although the mathematical derivation involves expansion of $\bar{\bar{f}}_l$ into spherical harmonics, only the lowest order term of $l=0$ remains in the energy density (7.67), given as an integral over the solid angle, since other terms vanish. Thus, we obtain

$$\bar{E}(\bar{x},\bar{t}) = \oint d\Omega\left(\theta_q,\phi_q\right)\bar{f}(\bar{x},\bar{t};\mathbf{q})$$

$$= \frac{1}{\bar{r}(2\pi)^2}\int_{-\infty}^{\infty}\int dk\,d\omega\,e^{-ik\bar{r}-i\omega\bar{t}}\left[\frac{ik}{2\pi}\bar{\bar{f}}_0(k\rightarrow -k, s\rightarrow -i\omega)\right] \tag{7.85}$$

With no intrinsic absorption $b=0$, taking the limit $k\rightarrow 0$, we can prove that the total energy given by the space integral of $\bar{E}(\bar{x},\bar{t})$ is conserved.

7.4.2 Simulation

Now we will show a method for simulating seismogram envelopes in media with nonisotropic scattering using the method developed in Section 7.4.1. We formally decompose each of $\bar{\bar{f}}_0$ and $\bar{E}$ into three terms corresponding to the direct, the single scattering, and the multiple scattering term with order greater than or equal to two as (7.12–13)

$$\bar{\bar{f}}_0(k,s) = \bar{\bar{G}}_0(k,s) + \bar{\bar{f}}_0^1(k,s) + \bar{\bar{f}}_0^M(k,s) \tag{7.86}$$

$$\bar{E}(\bar{x},\bar{t}) = \bar{G}(\bar{x},\bar{t}) + \bar{E}^1(\bar{x},\bar{t}) + \bar{E}^M(\bar{x},\bar{t}) \tag{7.87}$$

In practice, we use integral (7.85) only to evaluate the multiple scattering term, which converges quickly.

Substituting $\bar{\bar{G}}_l$ in place of $\bar{\bar{f}}_l$ on the right-hand side of (7.84) for $l=0$, we get

$$\bar{\bar{f}}_0^1(k,s) = 4\pi\sum_{l'=0}^{\infty}\sqrt{\frac{4\pi}{2l'+1}}\bar{g}_{l'}\sum_{m'=-l'}^{l'}\sum_{l''=0}^{\infty}(-i)^{l'+l''}\left|\left(Y_{l'',-m'}\right)^{0,0}_{l',m'}\right|^2\bar{\bar{G}}_{l''}(k,s)\,\bar{\bar{G}}_{l'}(k,s)$$

$$= 2\sum_{l'=0}^{\infty}(-1)^{l'}\sqrt{\pi(2l'+1)}\bar{g}_{l'}\,\bar{\bar{G}}_{l'}(k,s)^2 \tag{7.88}$$

where we used the explicit representation of the Wigner 3-j symbols. Substituting (7.88) in (7.86), where $\bar{\bar{f}}_0$ is the solution of simultaneous linear equations (7.84), we get $\bar{\bar{f}}_0{}^M$. Substituting it in (7.85), we numerically calculate $\bar{E}^M(\bar{x},\bar{t})$ by using a 2-D FFT.

To calculate $\bar{E}^1(\bar{x},\bar{t})$, we directly integrate the single nonisotropic scattering term in space. Substituting $\bar{G}_0$ in $\bar{f}$ in the second term of the right-hand side of (7.71), we get the single scattering term $\bar{f}^1(\bar{x},\bar{t};q)$, whose integral over the solid angle gives $\bar{E}^1(\bar{x},\bar{t})$. We take the source and the receiver as the foci of the dimensionless prolate spheroidal coordinates (w,v,ϕ) defined in (3.13), where we replace $z \to \bar{x}'$, $r \to \bar{r}$, $r_a \to \bar{r}'$ and $r_b \to \bar{r}''$. Then,

$$
\bar{E}^1(\bar{x},\bar{t}) = \oint \bar{f}^1(\bar{x},\bar{t};q)\,d\Omega\big(\theta_q,\phi_q\big)
$$

$$
= 4\pi \oint \int\int\int\int\int \oint \bar{G}(\bar{x}'',\bar{t}-\bar{t}';q)\bar{g}(\psi)\bar{G}(\bar{x}',\bar{t}';q')\,d\bar{x}'d\bar{t}'d\Omega\big(\theta_{q'},\phi_{q'}\big)\,d\Omega\big(\theta_q,\phi_q\big)
$$

$$
= 4\pi \oint \int\int\int\int\int \oint \frac{1}{4\pi\bar{r}''^2}e^{-(1+\bar{b})(\bar{t}-\bar{t}')}\delta(\bar{t}-\bar{t}'-\bar{r}'')H(\bar{t}-\bar{t}')\delta_\Omega(\bar{x}'';q)\bar{g}(\psi)
$$

$$
\times \frac{1}{4\pi\bar{r}'^2}e^{-(1+\bar{b})\bar{t}'}\delta(\bar{t}'-\bar{r}')H(\bar{t}')\delta_\Omega(\bar{x}';q')\,d\bar{x}'d\bar{t}'d\Omega\big(\theta_{q'},\phi_{q'}\big)\,d\Omega\big(\theta_q,\phi_q\big)
$$

$$
= \frac{e^{-(1+\bar{b})\bar{t}}}{4\pi}\int\int\int \frac{1}{\bar{r}''^2\,\bar{r}'^2}\delta(\bar{t}-\bar{r}'-\bar{r}'')\bar{g}(\psi)\,d\bar{x}'
$$

$$
= \frac{e^{-(1+\bar{b})\bar{t}}}{4\pi}\int_1^\infty dv\int_{-1}^1 dw\int_0^{2\pi} d\phi\,\frac{1}{\bar{r}''^2\bar{r}'^2}\delta(\bar{t}-\bar{r}'-\bar{r}'')\bar{g}[\psi(w,v)]\frac{\bar{r}\bar{r}'\bar{r}''}{2}
$$

$$
= \frac{e^{-(1+\bar{b})\bar{t}}}{\bar{r}^2}H\!\left(\frac{\bar{t}}{\bar{r}}-1\right)\int_{-1}^1 \frac{\bar{g}\!\left[\psi\!\left(w,\dfrac{\bar{t}}{\bar{r}}\right)\right]}{\left(\dfrac{\bar{t}}{\bar{r}}\right)^2 - w^2}\,dw \tag{7.89}
$$

The single scattering term given by the surface integral over the isochronal scattering shell corresponding to $v = \bar{t}/\bar{r}$ is written as an integral over the prolate spheroidal coordinate w [Sato, 1982c]. Then, ψ becomes the scattering angle between directions $\bar{x}'$ and $\bar{x}''$, which is explicitly written as a function of w and $v = \bar{t}/\bar{r}$ as in (3.17). We have to numerically integrate (7.89) in general; however, we can analytically integrate it using (3.19a) and get (7.17) for isotropic scattering $\bar{g} = 1/4\pi$. Thus, adding the three terms representing the direct, single scattered, and multiple scattered energy, we get the spatiotemporal distribution of the normalized energy density $\bar{E}(\bar{x},\bar{t})$.

One-Parameter Model

To examine the mathematical structure of the formulation, we take the following nonisotropic scattering model having one parameter $\bar{g}_1$:

$$\bar{g}(\psi) = \bar{g}_0 Y_{00}(\psi,0) - i\bar{g}_1 Y_{10}(\psi,0)$$

$$= \frac{1}{4\pi} + \bar{g}_1 \sqrt{\frac{3}{4\pi}} \cos\psi \tag{7.90}$$

This is the simplest model of the spherical harmonic series (7.77). There is a limit on the range of $\bar{g}_1$ in this one-parameter model since we require that $\bar{g}(\psi) > 0$. We show $\bar{g}(\psi)$ for various values of $\bar{g}_1$ in Figure 7.23. By using the explicit representation of the Wigner 3-j symbols, we write the first two linear equations of (7.84) as

$$\bar{\bar{f}}_0 = \overline{\overline{G}}_0 + \sqrt{4\pi}\,\bar{g}_0 \overline{\overline{G}}_0 \bar{\bar{f}}_0 - 2\sqrt{3\pi}\,\bar{g}_1 \overline{\overline{G}}_1 \bar{\bar{f}}_1$$

$$\bar{\bar{f}}_1 = \overline{\overline{G}}_1 + \sqrt{4\pi}\,\bar{g}_0 \overline{\overline{G}}_1 \bar{\bar{f}}_0 + 2\sqrt{\pi/3}\,\bar{g}_1 \overline{\overline{G}}_0 \bar{\bar{f}}_1 - 4\sqrt{\pi/3}\,\bar{g}_1 \overline{\overline{G}}_2 \bar{\bar{f}}_1 \tag{7.91}$$

Substituting $\overline{\overline{G}}_l$ in $\bar{\bar{f}}_l$ in the right-hand side of the first equation, we get the single scattering term in (7.86) as

$$\bar{\bar{f}}_0^1 = \sqrt{4\pi}\,\bar{g}_0 \overline{\overline{G}}_0^2 - 2\sqrt{3\pi}\,\bar{g}_1 \overline{\overline{G}}_1^2 \tag{7.92}$$

Subtracting $\overline{\overline{G}}_0$ and $\bar{\bar{f}}_0^1$ given by (7.92) from the solution $\bar{\bar{f}}_0$ of (7.91) and using $\sqrt{4\pi}\,\bar{g}_0 = 1$, we get $\bar{\bar{f}}_0^M$ using (7.86). Then, substituting it in (7.85), we numerically calculate $\overline{E}^M(\overline{x},\bar{t})$ by using a 2-D FFT, where we use the explicit representation of $\overline{\overline{G}}_l$ for $l = 0$, 1, and 2, as given by (7.47, 7.48 and 7.50).

We plot temporal traces and the spatial sections of the normalized energy density in Figures 7.24a and b, respectively, for $\bar{g}_1 = 0.08$ and no intrinsic absorption. Forward scattering is stronger than backward scattering, as illustrated in Figure 7.23. In Figure 7.24a, comparison of the solid lines showing the result for nonisotropic scattering with the broken lines, which give the results for the isotropic scattering model ($\bar{g}_1 = 0$), reveals that the temporal traces are very close to the isotropic scattering case for short distances. At slightly longer distances,

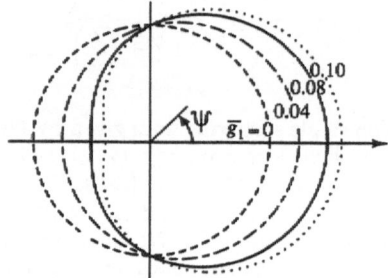

FIGURE 7.23. Nonisotropic scattering pattern of one-parameter model (7.90) for different values of $\bar{g}_1$, where the broken line is for isotropic scattering ($\bar{g}_1 = 0$).

such as $\bar{r}$=0.8 and 1.6, strong forward scattering increases the energy density immediately after the direct arrival, and weak backward scattering decreases the later coda energy density compared with the isotropic scattering case. Nonisotropic scattering has a strong effect even at times when multiple scattering dominates the envelope. At longer distances, such as $\bar{r}$=3.2, a broad secondary peak appears after the direct arrival and occurs much earlier than in the isotropic scattering case. If forward scattering is stronger than in the case investigated here, the secondary peak will appear closer to the direct arrival. Thus, we may infer that both the direct and the secondary peaks constitute a broadened envelope due to multiple forward scattering as will be discussed in Chapter 8 for the observed broadening of S-wave envelopes [Sato, 1989; Scherbaum and Sato, 1991; Obara and Sato, 1995].

In Figure 7.24b, the difference in the spatial distribution of energy density between the isotropic and nonisotropic scattering cases is apparent even at short lapse times, as illustrated by the top spatial section. Most of the energy propagates outward for the nonisotropic case, but the spatial section has a Gaussian-like peak at the hypocenter for isotropic scattering. As lapse time increases, the spatial section for nonisotropic scattering also becomes a Gaussian-like curve because of multiple scattering, but the maximum energy density at the hypocenter is smaller than that for isotropic scattering. At lapse times $\bar{t}$=0.96 and 1.92, spatial sections for the nonisotropic scattering case look similar to those derived using the energy-flux model in Section 3.2.2 and shown in Figure 3.9. When scattering is much stronger in the forward direction and described by a higher order series expansion than used in (7.90), we will expect more uniform spatial distribution of energy density around the hypocenter for longer lapse times than in Figure 7.24b.

The coda-normalization method for attenuation measurements and site factor estimation is based on the hypothesis that coda energy is uniformly distributed within some volume whose size depends on lapse time. Typically, a lapse time of 50 s is chosen as appropriate for a 100 km region surrounding the source for S-wave velocity 4km/s. Normalized lapse time $\bar{t}$=2 for g_0=0.01 km^{-1} corresponds to a 50 s lapse time. Normalized distance $\bar{r}$=1 corresponds to 100 km in this case. The results derived from our simple case study, shown in Figure 7.24b, show that energy is relatively uniformly distributed within a 100 km radius surrounding the source at 50 s lapse time. This provides some theoretical support for the coda-normalization method as applied in regional seismology. We have already discussed a fractal distribution of scatterers in Section 7.1.3 which also predicts the uniform distribution of coda energy even though isotropic scattering is assumed. Both nonisotropic scattering and fractal distribution of scatterers are physically important.

Gusev and Abubakirov [1987] and Hoshiba [1995] used Monte Carlo simulations to study the envelope formation due to nonisotropic multiple scattering with much stronger forward scattering than modeled in this Section. To model these stronger forward scattering cases using the method developed in this Section, we would have to include additional higher order terms of the spherical harmonic series with respect to the solid angle.

The MLTW method for finding the contributions to total attenuation of scattering attenuation and intrinsic absorption based on the multiple isotropic scattering model was presented in Section 7.2.2. As found in the above simulation, large forward scattering increases the amplitude just after the direct-wave arrival and decreases the amplitude of the late S-coda compared to an isotropic scattering model with the same total scattering coefficient. The isotropic scattering MLTW analysis using the single station method described by (7.39) can easily be extended to nonisotropic scattering.

For the practical study of seismogram envelopes, it is necessary to develop a method to simultaneously account for multiple nonisotropic scattering as developed in this section along with nonspherical radiation from a point shear-dislocation

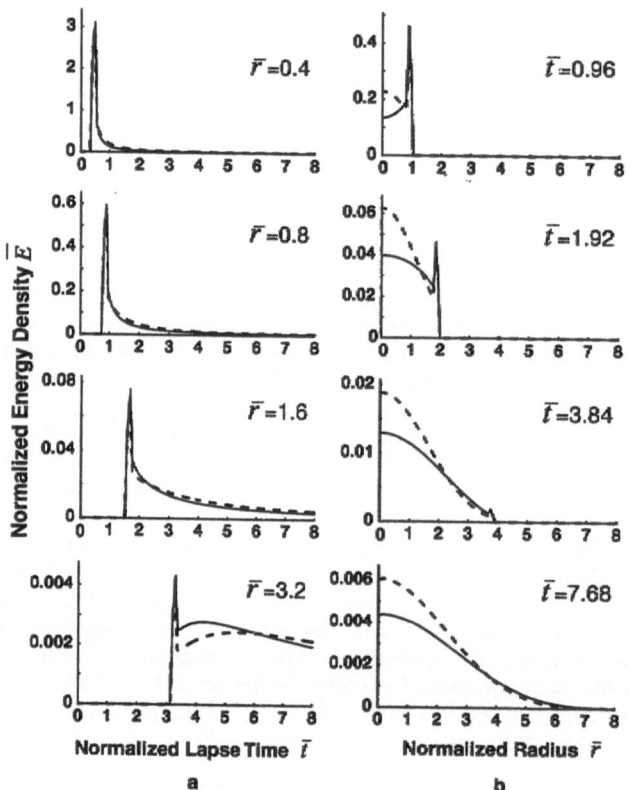

FIGURE 7.24. Plots of normalized energy density resulting from the one-parameter model of the nonisotropic scattering process given in (7.90) for $\bar{g}_1 = 0.08$ and no intrinsic absorption: (a) temporal traces, where numerals show the normalized hypocentral distance at which the temporal plots are made; (b) spatial sections, where numerals show the normalized lapse times at which the spatial distributions are plotted. The broken curves show normalized energy density distribution for isotropic scattering for reference (see Fig. 7.3). For the calculation, the 2-D FFT was done over 200x200 points for ((0–16), (0–16)) in the normalized space-time $(\bar{r}, \bar{t})$. The source duration time is taken as 0.16 and is two samples long.

source as discussed in Section 7.3, since nonisotropic scattering impacts both the direct arrival and the early coda. Watanabe et al. [1996] mathematically formulated a multiple scattering process based on the radiative transfer theory, and derived a coupling relationship between the spherical harmonic expansion coefficients of the source radiation pattern and the nonspherical scattering coefficients by using the Wigner 3-j symbols.

7.5 WHOLE SEISMOGRAM ENVELOPE: ISOTROPIC SCATTERING INCLUDING CONVERSIONS BETWEEN P- AND S-WAVES

Up to this point in this chapter, we have been studying the radiative transfer theory for a single-wave mode. However, the shape of a seismogram envelope beginning with the P-wave onset and continuing through the S-coda contains a great deal of information about seismic propagation effects. Seismogram envelopes have been used to classify earthquakes into groups by their hypocentral location, focal depth, and seismotectonic province [Tsujiura, 1988]. Three-component seismograms of a local microearthquake observed in Kanto, Japan are shown in Figure 7.25a. In addition to the large excitation of S-coda as discussed in Chapter 3, seismograms of local earthquakes have incoherent wave trains, called P-coda, which have nearly stationary amplitude between the direct P- and S-wave arrivals. From timing considerations, the P-coda must consist of a combination of P-to-P, P-to-S and S-to-P scattered waves, as discussed in Chapter 6. The existence of P-coda implies the importance of scattering with conversions between P and S. Figure 7.25b shows examples of high-frequency seismograms in the 50–500 Hz

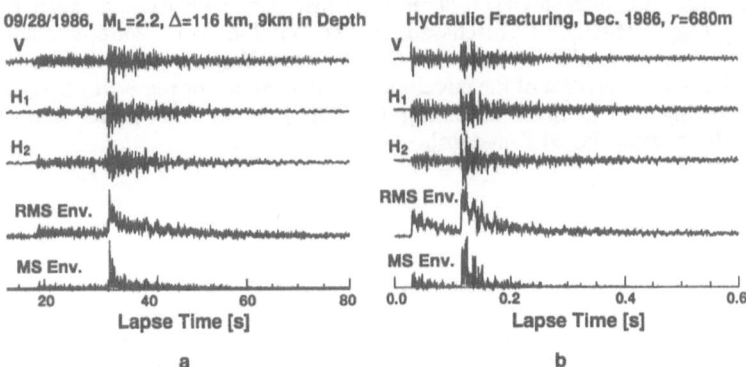

FIGURE 7.25. (a) Velocity seismograms of a local microearthquake in Kanto, Japan, recorded at NIED borehole station IWT [Data courtesy of K. Obara]. (b) High-frequency velocity seismograms of a microearthquake induced by water injection at the Fenton Hill hot dry rock geothermal site, New Mexico, U. S. A. The bottom two traces are RMS and MS envelopes. Note the differences in time scale between the natural and induced events.

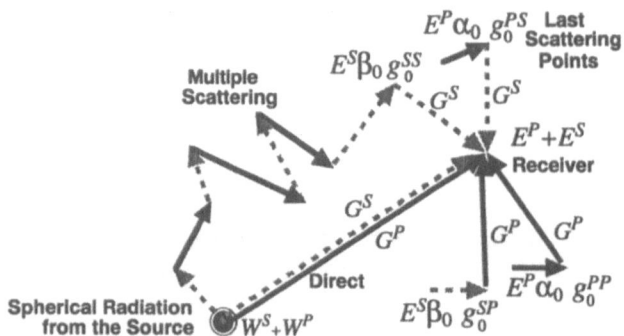

FIGURE 7.26. Configuration of the source, receiver, and last scattering points for multiple isotropic scattering including P–S conversions.

band from microearthquakes induced by water injection into a borehole drilled in crystalline rock [Fehler and Phillips, 1991]. Although the frequencies of the induced microearthquakes are much higher than those from natural earthquakes, without time scales in the plots, it would not be possible to distinguish which seismograms came from the natural sources and which came from the induced events. The bottom trace of each figure shows the smoothed trace of the square sum of the three components of the particle velocity, the MS envelope, which is linearly proportional to energy density.

Now we will extend the multiple isotropic scattering model presented in Section 7.1 to include conversion scattering between P- and S-wave modes [Sato, 1994a; Zeng, 1993] to arrive at a model to explain the entire seismogram envelope. Conversion scattering is nonisotropic in general as shown in Chapter 4. However, isotropic scattering is the lowest order term of the spherical harmonic expansion of the scattering coefficient as discussed in Section 7.4. This isotropic-scattering analysis including mode conversions will help us to obtain a better understanding of the basic characteristics of the envelope of seismograms of regional earthquakes. The following derivation is according to Sato [1994a, with permission from Blackwell Science, United Kingdom].

7.5.1 Formulation

We suppose an impulsive spherical radiation of P- and S-wave energies W^P and W^S from a source located at the origin, where two types of wave modes are considered. Scattering is isotropic for four scattering modes: P-to-P, P-to-S, S-to-P, and S-to-S scattering. The distribution of point-like scatterers is random and uniform in 3-D space. Scattering strengths of the four modes are characterized by four total scattering coefficients g_0^{PP}, g_0^{PS}, g_0^{SP} and g_0^{SS}. The energy density is written as a sum of P- and S-wave energy densities:

$$E(\mathbf{x},t) = E^P(\mathbf{x},t) + E^S(\mathbf{x},t) \tag{7.93}$$

Using an approach similar to that presented in Section 7.1, P- and S-wave energy densities can be expressed as sums of coherent wave energy densities and contributions from last scattering points, as shown schematically in Figure 7.26:

$$E^P(\mathbf{x},t) = W^P G^P(\mathbf{x},t)$$

$$+ \int\int\int\int_{-\infty}^{\infty} \{E^P(\mathbf{x}',t')\alpha_0 g_0^{PP} + E^S(\mathbf{x}',t')\beta_0 g_0^{SP}\} G^P(\mathbf{x}-\mathbf{x}',t-t') dt' d\mathbf{x}'$$

$$E^S(\mathbf{x},t) = W^S G^S(\mathbf{x},t)$$

$$+ \int\int\int\int_{-\infty}^{\infty} \{E^S(\mathbf{x}',t')\beta_0 g_0^{SS} + E^P(\mathbf{x}',t')\alpha_0 g_0^{PS}\} G^S(\mathbf{x}-\mathbf{x}',t-t') dt' d\mathbf{x}'$$

(7.94)

where α_0 and β_0 are the P- and S-wave velocities in the background medium, respectively. We choose Green functions for coherent wave energies that include the effects of geometric spreading, attenuation due to scattering and intrinsic mechanisms, and causality as given by

$$G^P(\mathbf{x},t) = \frac{1}{4\pi r^2 \alpha_0} H(t)\delta\left(t - \frac{r}{\alpha_0}\right) e^{-(\alpha_0 g_0^{PP} + \alpha_0 g_0^{PS} + b)t}$$

$$G^S(\mathbf{x},t) = \frac{1}{4\pi r^2 \beta_0} H(t)\delta\left(t - \frac{r}{\beta_0}\right) e^{-(\beta_0 g_0^{SS} + \beta_0 g_0^{SP} + b)t}$$

(7.95)

where $r = |\mathbf{x}|$ and b is the parameter for intrinsic absorption which is chosen to be the same for P- and S-waves. Since scatterers are assumed to be point-like, the scattering field is exactly described by the above three equations (7.93–7.95). This is a natural extension of the formulation for a single wave-propagation mode given by (7.1–7.2).

To solve for the energy distribution as a function of space and time from equation (7.94), we use the Fourier transform in space and the Laplace transform in time, as done in Section 7.1. Here we keep the dimensional quantities. We first take the Fourier–Laplace transform of the Green functions as

$$\hat{G}^P(k,s) = \int\int\int_{-\infty}^{\infty} d\mathbf{x}\, e^{-i\mathbf{k}\mathbf{x}} \int_0^{\infty} dt\, e^{-st} G^P(\mathbf{x},t) = \frac{1}{\alpha_0 k}\tan^{-1}\frac{\alpha_0 k}{s + \alpha_0 g_0^{PP} + \alpha_0 g_0^{PS} + b}$$

$$\hat{G}^S(k,s) = \int\int\int_{-\infty}^{\infty} d\mathbf{x}\, e^{-i\mathbf{k}\mathbf{x}} \int_0^{\infty} dt\, e^{-st} G^S(\mathbf{x},t) = \frac{1}{\beta_0 k}\tan^{-1}\frac{\beta_0 k}{s + \beta_0 g_0^{SS} + \beta_0 g_0^{SP} + b}$$

(7.96)

where the argument is $k = |\mathbf{k}|$ since there is no specific orientation.

Taking the Fourier-Laplace transform of (7.94) and rearranging,

$$\hat{E}^P(k,s) = \frac{W^P \hat{\tilde{G}}^P\left(1 - \beta_0 g_0^{SS}\hat{\tilde{G}}^S\right) + W^S \hat{\tilde{G}}^S \beta_0 g_0^{SP}\hat{\tilde{G}}^P}{\left(1 - \alpha_0 g_0^{PP}\hat{\tilde{G}}^P\right)\left(1 - \beta_0 g_0^{SS}\hat{\tilde{G}}^S\right) - \alpha_0 g_0^{PS}\beta_0 g_0^{SP}\hat{\tilde{G}}^P\hat{\tilde{G}}^S}$$

$$\hat{E}^S(k,s) = \frac{W^S \hat{\tilde{G}}^S\left(1 - \alpha_0 g_0^{PP}\hat{\tilde{G}}^P\right) + W^P \hat{\tilde{G}}^P \alpha_0 g_0^{PS}\hat{\tilde{G}}^S}{\left(1 - \alpha_0 g_0^{PP}\hat{\tilde{G}}^P\right)\left(1 - \beta_0 g_0^{SS}\hat{\tilde{G}}^S\right) - \alpha_0 g_0^{PS}\beta_0 g_0^{SP}\hat{\tilde{G}}^P\hat{\tilde{G}}^S}$$

(7.97)

As a check on our formulation, now we investigate the temporal dependence of the total energy in the P- and S-wavefields. The total energies in the P- and S-wavefields are defined as space integrals of the P- and S-energy densities, each determined from the Fourier transform at wavenumber $k = 0$:

$$\int\!\!\!\int\!\!\!\int_{-\infty}^{\infty} dx\, E^P(\mathbf{x},t) = \left[\frac{\beta_0 g_0^{SP}\left(W^P + W^S\right)}{\alpha_0 g_0^{PS} + \beta_0 g_0^{SP}} + \frac{W^P \alpha_0 g_0^{PS} - W^S \beta_0 g_0^{SP}}{\alpha_0 g_0^{PS} + \beta_0 g_0^{SP}} e^{-\left(\alpha_0 g_0^{PS} + \beta_0 g_0^{SP}\right)t}\right] e^{-bt}$$

$$\int\!\!\!\int\!\!\!\int_{-\infty}^{\infty} dx\, E^S(\mathbf{x},t) = \left[\frac{\alpha_0 g_0^{PS}\left(W^P + W^S\right)}{\alpha_0 g_0^{PS} + \beta_0 g_0^{SP}} + \frac{W^S \beta_0 g_0^{SP} - W^P \alpha_0 g_0^{PS}}{\alpha_0 g_0^{PS} + \beta_0 g_0^{SP}} e^{-\left(\alpha_0 g_0^{PS} + \beta_0 g_0^{SP}\right)t}\right] e^{-bt}$$ (7.98)

$$\int\!\!\!\int\!\!\!\int_{-\infty}^{\infty} dx\, E(\mathbf{x},t) = \left(W^P + W^S\right) e^{-bt}$$

The last term in (7.98) shows that energy is removed from the total wavefield only by intrinsic attenuation; when $b = 0$, the total energy in the wavefield is conserved.

The partition of energy into P- and S-wave modes for large lapse times is controlled by the ratio of conversion scattering power per time between P and S, which is the first term on the right-hand side of the first two equations in (7.98) since the second terms vanish as lapse time increases. If we take $g_0^{SP} = g_0^{PS}/2\gamma_0^2$, as given by reciprocity between modes (4.53), the ratio of P-wave energy to S-wave energy asymptotically approaches $\beta_0 g_0^{SP}/\alpha_0 g_0^{PS} = 1/2\gamma_0^3 \approx 0.1$.

7.5.2 Analytical Representation of the Single Scattering Term

By using the inverse Fourier–Laplace transform of (7.97), we can formally solve for the space-time distribution of energy density. This must be done numerically; however, the convergence of the numerical integration is slow since the integral kernel oscillates rapidly for large wavenumbers. A procedure for performing the numerical integration that overcomes the rapid oscillation is described by Zeng [1993]. Here, we take another method along the line discussed in the case of single propagation mode in Section 7.1. The direct energy density can be expressed as a delta function and the single scattering energy density diverges logarithmically at the wave front in the time domain. Therefore, we formally decompose the energy density in the Fourier–Laplace domain into three terms: the direct energy density E^0, the single scattered energy density E^1, and the energy density for multiple scattering of order greater than or equal to two E^M:

$$\hat{E}(k,s) = \hat{E}^0(k,s) + \hat{E}^1(k,s) + \hat{E}^M(k,s) \tag{7.99}$$

where

$$\hat{E}^0(k,s) = W^P \hat{\tilde{G}}^P + W^S \hat{\tilde{G}}^S \tag{7.100a}$$

$$\hat{E}^1(k,s) = W^P \hat{\tilde{G}}^P \left(\alpha_0 g_0^{PP} \hat{\tilde{G}}^P + \alpha_0 g_0^{PS} \hat{\tilde{G}}^S \right) + W^S \hat{\tilde{G}}^S \left(\beta_0 g_0^{SS} \hat{\tilde{G}}^S + \beta_0 g_0^{SP} \hat{\tilde{G}}^P \right) \tag{7.100b}$$

$$
\begin{aligned}
\hat{E}^M(k,s) = {} & \frac{1}{\left(1 - \alpha_0 g_0^{PP} \hat{\tilde{G}}^P\right)\left(1 - \beta_0 g_0^{SS} \hat{\tilde{G}}^S\right) - \alpha_0 g_0^{PS} \beta_0 g_0^{SP} \hat{\tilde{G}}^P \hat{\tilde{G}}^S} \\
& \times \Bigg\langle W^P \hat{\tilde{G}}^P \left\{ \alpha_0 g_0^{PS} \hat{\tilde{G}}^S \left[\beta_0 g_0^{SS} \hat{\tilde{G}}^S \left(1 - \alpha_0 g_0^{PP} \hat{\tilde{G}}^P\right) + \beta_0 g_0^{SP} \hat{\tilde{G}}^P \left(1 + \alpha_0 g_0^{PS} \hat{\tilde{G}}^S\right) \right] \right. \\
& \qquad + \alpha_0 g_0^{PP} \hat{\tilde{G}}^P \left[\alpha_0 g_0^{PP} \hat{\tilde{G}}^P \left(1 - \beta_0 g_0^{SS} \hat{\tilde{G}}^S\right) + \alpha_0 g_0^{PS} \hat{\tilde{G}}^S \left(1 + \beta_0 g_0^{SP} \hat{\tilde{G}}^P\right) \right] \Bigg\} \\
& + W^S \hat{\tilde{G}}^S \left\{ \beta_0 g_0^{SP} \hat{\tilde{G}}^P \left[\alpha_0 g_0^{PP} \hat{\tilde{G}}^P \left(1 - \beta_0 g_0^{SS} \hat{\tilde{G}}^S\right) + \alpha_0 g_0^{PS} \hat{\tilde{G}}^S \left(1 + \beta_0 g_0^{SP} \hat{\tilde{G}}^P\right) \right] \right. \\
& \qquad \left. + \beta_0 g_0^{SS} \hat{\tilde{G}}^S \left[\beta_0 g_0^{SS} \hat{\tilde{G}}^S \left(1 - \alpha_0 g_0^{PP} \hat{\tilde{G}}^P\right) + \beta_0 g_0^{SP} \hat{\tilde{G}}^P \left(1 + \alpha_0 g_0^{PS} \hat{\tilde{G}}^S\right) \right] \right\} \Bigg\rangle
\end{aligned}
\tag{7.100c}
$$

As shown in Figure 5.3, the ratio of P- to S-attenuation observed for frequencies higher than 1 Hz in the lithosphere is in the range $Q_P^{-1}/Q_S^{-1} = 0.7 - 2$ [Yoshimoto et al. 1993]. Therefore, for mathematical simplicity, we assume that the total attenuation, equal to the sum of scattering attenuation and intrinsic absorption, is the same for P- and S-waves

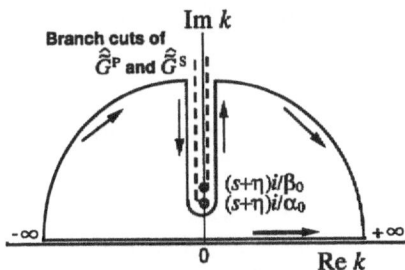

FIGURE 7.27. Contour of integral in the complex k-plane for the single scattering term where mode conversion occurs. Broken lines are branch cuts. [From Sato, 1994a, with permission from Blackwell Science, United Kingdom.]

$$\alpha_0 g_0^{PP} + \alpha_0 g_0^{PS} + b = \beta_0 g_0^{SS} + \beta_0 g_0^{SP} + b = \eta \tag{7.101}$$

Using (7.101) allows us to solve analytically for the single scattering term. The inverse Fourier transform is written as an integral along the real k-axis. For example, the second term of (7.100b) gives the integral kernel for the energy density due to single PS-scattering:

$$
\begin{aligned}
\hat{E}^{1PS}(\mathbf{x},s) &= \frac{W^P \alpha_0 g_0^{PS}}{(2\pi)^3} \int_{-\infty}^{\infty}\int_{-\infty}^{\infty}\int_{-\infty}^{\infty} \hat{\tilde{G}}^P(k,s)\hat{\tilde{G}}^S(k,s)e^{i k x}\,d\mathbf{k} \\
&= \frac{W^P \alpha_0 g_0^{PS}}{(2\pi)^2 ir} \int_{-\infty}^{\infty} \hat{\tilde{G}}^P(k,s)\hat{\tilde{G}}^S(k,s)e^{ikr}k\,dk
\end{aligned}
\tag{7.102}
$$

where $\hat{\tilde{G}}^{P,S}(k,s) = \hat{\tilde{G}}^{P,S}(-k,s)$. We use the technique of residue and branch cut integration by closing the contour of integration at infinity in the complex k-plane. The branch points of $\hat{\tilde{G}}^P$ and $\hat{\tilde{G}}^S$ in the complex k-plane are $(s+\eta)i/\alpha_0$ and $(s+\eta)i/\beta_0$, respectively, as illustrated in Figure 7.27. Branch cuts are taken from these branch points to infinity on the imaginary k-axis. We take the integral contour as two quarter-circles and along two branch cuts on the imaginary axis. The integral around the two quarter-circles vanishes as discussed in Section 7.1, and there remains an integral between the branch points of $\hat{\tilde{G}}^P$ and $\hat{\tilde{G}}^S$ and that between the branch point of $\hat{\tilde{G}}^S$ and infinity on the imaginary k-axis:

$$
\begin{aligned}
\hat{E}^{1PS}(\mathbf{x},s) &= \frac{W^P \alpha_0 g_0^{PS}}{(2\pi)^2 ir} \int_{\substack{\text{Around} \\ \text{Branch Cuts}}} \hat{\tilde{G}}^P(k,s)\,\hat{\tilde{G}}^S(k,s)e^{ikr}k\,dk \\
&= \frac{W^P \gamma_0 g_0^{PS}}{4\pi\alpha_0 r}\left[\int_1^{\gamma_0}\frac{1}{u}\left(\tanh^{-1}\frac{u}{\gamma_0}\right)e^{-(s+\eta)\frac{r}{\alpha_0}u}\,du \right. \\
&\qquad \left. + \int_{\gamma_0}^{\infty}\frac{1}{u}\left(\tanh^{-1}\frac{1}{u} + \tanh^{-1}\frac{\gamma_0}{u}\right)e^{-(s+\eta)\frac{r}{\alpha_0}u}\,du \right] \\
&= \frac{W^P \gamma_0 g_0^{PS}}{4\pi r^2}\int_0^{\infty} K_C\left(\frac{\alpha_0 t}{r}\right)H\left(\frac{\alpha_0 t}{r}-1\right)e^{-(\eta+s)t}\,dt
\end{aligned}
\tag{7.103}
$$

where $\gamma_0 \equiv \alpha_0/\beta_0 > 1$, and we used $\tan^{-1}iz = i\tanh^{-1}z$ for $|z| < 1$ and $\tan^{-1}iz = i\tanh^{-1}(1/z) \pm \pi/2$ for $|z| > 1$ on the right/left side of the imaginary axis. The result is written in the form of a Laplace transform. We introduce a function [Sato, 1977b, 1994a]

$$K_C(x) = \begin{cases} \dfrac{1}{x}\tanh^{-1}\dfrac{x}{\gamma_0} = \dfrac{1}{2x}\ln\dfrac{\gamma_0+x}{\gamma_0-x} & \text{for } 1<x<\gamma_0 \\[3mm] \dfrac{1}{x}\left(\tanh^{-1}\dfrac{1}{x}+\tanh^{-1}\dfrac{\gamma_0}{x}\right) = \dfrac{1}{2x}\ln\dfrac{(x+1)(x+\gamma_0)}{(x-1)(x-\gamma_0)} & \text{for } x>\gamma_0 \end{cases} \tag{7.104}$$

Thus, the integral kernel of (7.103) gives the solution in time:

$$E^{1PS}(\mathbf{x},t) = \frac{W^P\gamma_0 g_0^{PS}}{4\pi r^2}K_C\left(\frac{\alpha_0 t}{r}\right)H\left(\frac{\alpha_0 t}{r}-1\right)e^{-\eta t} \tag{7.105a}$$

In a similar way, for single SP-scattering,

$$E^{1SP}(\mathbf{x},t) = \frac{W^S g_0^{SP}}{4\pi r^2}K_C\left(\frac{\alpha_0 t}{r}\right)H\left(\frac{\alpha_0 t}{r}-1\right)e^{-\eta t} \tag{7.105b}$$

In the case of PP- and SS-single scattering, using the integral around one branch cut as derived in Section 7.1, we obtain

$$E^{1PP}(\mathbf{x},t) = \frac{W^P g_0^{PP}}{4\pi r^2}K\left(\frac{\alpha_0 t}{r}\right)H\left(\frac{\alpha_0 t}{r}-1\right)e^{-\eta t} \tag{7.105c}$$

$$E^{1SS}(\mathbf{x},t) = \frac{W^S g_0^{SS}}{4\pi r^2}K\left(\frac{\beta_0 t}{r}\right)H\left(\frac{\beta_0 t}{r}-1\right)e^{-\eta t} \tag{7.105d}$$

where K is given by (3.19a). Functions $K(x)$ and $K_C(x)$ diverge logarithmically at $x=1$ and $x=\gamma_0$, respectively, as shown in Figure 7.28. These functions show that single PP- and SS-scattering transfer energy in the volume behind the direct

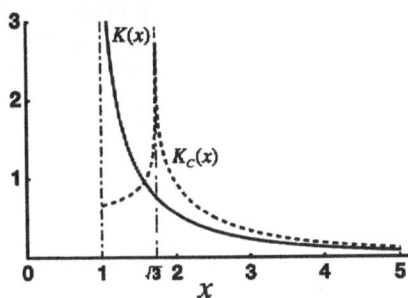

FIGURE 7.28. Plots of $K(x)$ in (3.19a) (solid) and $K_C(x)$ in (7.104) (broken), where $\gamma_0 = \sqrt{3}$. [From Sato, 1977b, with permission from Center for Academic Publications Japan, Tokyo, Japan.]

wave arrivals; however, both single PS and SP scattering have peaks at the S-wave arrival and scattered energy appears later than the P-arrival and earlier and later than the S-arrival.

7.5.3 Time Trace of the Total Energy Density

We can numerically evaluate the multiple scattering term E^M in (7.100c) in a manner similar to that in Section 7.1 by using an FFT instead of an inverse Laplace transform along the imaginary s-axis. Adding the multiple scattering term to the direct terms (7.95) and single scattering terms (7.105a–d), we get the total energy density in space-time. We numerically synthesize time traces of energy density, where we choose $\gamma_0 = \sqrt{3}$, no intrinsic absorption (b=0), and use two choices of total scattering coefficients. So far, our model assumes spherical radiation from the source, but we suppose that the ratio of P- to S-wave radiation is that of a point shear-dislocation, $W^S / W^P = 1.5\gamma_0^5$, as given by (6.10). We normalize time, length and energy density by η, α_0 and the radiated energy $W^P + W^S$ respectively:

$$\bar{x} = \frac{\eta}{\alpha_0}x, \ \bar{t} = \eta t \ \text{ and } \ \bar{E} = \left(\frac{\alpha_0}{\eta}\right)^3 \frac{E}{W^P + W^S} \qquad (7.106)$$

A 2-D FFT is done over 256×256 points for $((0\text{–}8.), (0\text{–}8.))$ in normalized space-time $(\bar{r}, \bar{t})$ where $\bar{r} = |\bar{x}|$. The single scattering term diverges logarithmically at $\bar{t} = \bar{r}$ and $\bar{t} = \gamma_0\bar{r}$ because of singularities in $K(x)$ and $K_C(x)$. However, the integrals over the windows containing the singularities are finite. The first example is chosen for mathematical simplicity: $\alpha_0 g_0^{PP} = \alpha_0 g_0^{PS} = \beta_0 g_0^{SS} = \beta_0 g_0^{SP} = \eta/2$. The cross-term containing $\hat{\bar{G}}^P \, \hat{\bar{G}}^S$ in the denominator of (7.97) vanishes. Figure 7.29a shows the temporal evolution of the normalized energy density $\bar{E}$ at three distances. The arrival time of the P-wave corresponds to the scaled hypocentral distance. P-coda is excited by a combination of PP- and conversion scattering. The P-coda level is initially small but gradually increases approaching the S-arrival. The multiple scattering contribution, shown by a broken curve, gradually increases with increasing lapse time. The direct wave energy rapidly decreases with increasing travel distance due to scattering. The second example is for $\alpha_0 g_0^{PP} = 2\eta/3$, $\alpha_0 g_0^{PS} = \eta/3$, $\beta_0 g_0^{SS} = 9\eta/10$, $\beta_0 g_0^{SP} = \eta/10$. This example has a smaller SP scattering coefficient than PS-scattering coefficient, in agreement with the result of elastic wave scattering (4.53). The time traces in Figure 7.29b show that the P-coda maintains a rather stable amplitude irrespective of lapse time until just before the direct S-wave arrival. These two simulations give a good qualitative explanation of the stable or gradual increase in amplitude of the P-coda that is often observed (see envelopes in Figures 2.29, 7.25a and b).

In this chapter, the radiative transfer equation in the form of a convolution integral has been used. This is a phenomenological approach consistent with causality in a background medium having a constant velocity and including geometrical spreading. These assumptions are generally acceptable; however, the only link with the wave equation approach is the use of total scattering coefficients. There have been attempts to derive the transport equation for energy density from the wave equation. Introducing the Wigner distribution, Ryzhik et al. [1996] derived the energy transport equation for elastic waves when the wavelength is much smaller than the characteristic scale of heterogeneity. Their formulation also correctly modeled S-wave polarization and showed the diffusive behavior of energy at long lapse time and large source–receiver distance.

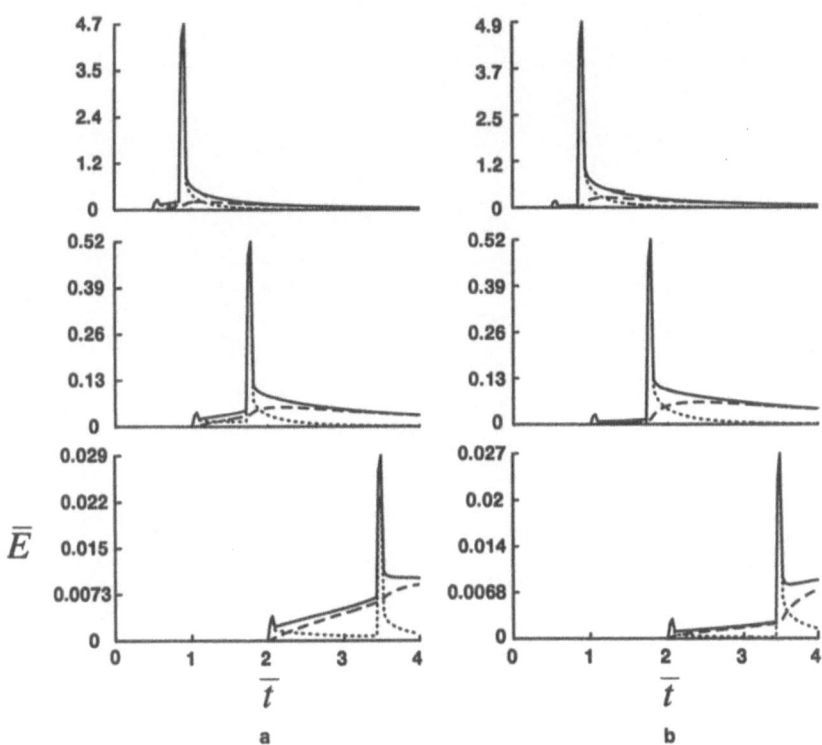

FIGURE 7.29. Temporal evolution of energy density at three distances for various scattering orders and two sets of total scattering coefficients: (a) $\alpha_0 g_0^{PP} = \alpha_0 g_0^{PS} = \beta_0 g_0^{SS} = \beta_0 g_0^{SP} = \eta / 2$; (b) $\alpha_0 g_0^{PP} = 2\eta / 3$, $\alpha_0 g_0^{PS} = \eta / 3$, $\beta_0 g_0^{SS} = 9\eta / 10$, and $\beta_0 g_0^{SP} = \eta / 10$. Solid line shows total normalized energy density, dotted lines show analytical solution for the single scattering and direct energy terms, and the broken curves show the contributions due to the multiple scattering term obtained by a 2-D FFT. Source duration is two samples, 1/16 in normalized time. [From Sato, 1994a, with permission from Blackwell Science, United Kingdom.]

CHAPTER 8

Diffraction and Broadening of Seismogram Envelopes

The inhomogeneity of the earth's lithosphere has a broad spectrum compared with the wavelength of regional seismic waves. We have presented models for wave propagation through inhomogeneous media, in which scattering by inhomogeneities of characteristic scales shorter than or equal to the wavelength dominates S-coda wave excitation. For the evaluation of scattering loss, as discussed in Chapter 5, subtraction of the travel-time fluctuation due to long-wavelength components of velocity inhomogeneities gave us a model that is consistent with measurement practice and in agreement with observed frequency characteristics. We will now present a model for wave propagation through media having long-wavelength components of velocity inhomogeneity, that is, we focus on diffraction and multiple forward-scattering effects.

Coherence measurements based on seismic array data have been used to detect variations in the phase and amplitude of seismic waves with increasing travel distance. These measurements have helped quantify the inhomogeneity of the lithosphere under the seismic arrays. As shown in Section 2.4.3, the time width of the S-wave portion of seismogram envelopes recorded at hypocentral distances of a few hundred kilometers is much longer than the source duration time estimated from earthquake magnitude. This phenomenon can be interpreted as resulting from strong diffraction.

Starting from the mathematical study of the parabolic wave equation through random media, we investigate the amplitude and phase correlations of scalar waves. We will use the concept of the ensemble average of these correlations and show how to statistically characterize wave propagation. Quantitative measurements of phase and amplitude correlations will be presented and interpreted. Next, we introduce a method to synthesize mean square envelopes for quasi-monochromatic waves based on the Markov approximation. We discuss the application of this formalism to observed field data and show quantitative measurements of the broadening of high-frequency S-wave seismogram envelopes. Finally, we briefly introduce how the phase-screen (split-step Fourier) method can be used to model wave propagation in inhomogeneous media.

8.1 AMPLITUDE AND PHASE DISTORTIONS OF SCALAR WAVES

8.1.1 Parabolic Wave Equation

We assume an inhomogeneous medium with a large correlation length compared with the wavelength of the propagating waves. As found by the Born approximation in Section 4.1.2 and shown in Figure 4.4b, scattering occurs into a small angle around the forward direction in such a case. Then we may neglect the gradient of the velocity inhomogeneity, conversion scattering between P- and S-waves and backscattering [see Richards, 1974]. Therefore, we can study the main propagation characteristics of P- or S-waves through inhomogeneous media using the wave equation for a scalar wavefield $u(\mathbf{x}, t)$:

$$\left(\Delta - \frac{1}{V(\mathbf{x})^2} \partial_t^2 \right) u(\mathbf{x}, t) = 0 \tag{8.1}$$

We will study the line-of-sight propagation through a randomly inhomogeneous medium extending over $z > 0$ of a plane wave incident along the z-axis, where the medium is homogeneous for $z < 0$. The receiver is located at $z = Z > 0$, and the propagation distance is long compared with the correlation distance, $Z \gg a$. Inhomogeneous structure in velocity is written as $V(\mathbf{x}) = V_0(1 + \xi(\mathbf{x}))$, where the fractional fluctuation is assumed to be small: $|\xi| \ll 1$. Then, (8.1) can be written

$$\left(\Delta - \frac{1}{V_0^2} \partial_t^2 \right) u + \frac{2}{V_0^2} \xi \partial_t^2 u = 0 \tag{8.2}$$

Introducing transverse coordinates $\mathbf{x}_\perp = (x, y)$ orthogonal to the mean propagation direction, the z-direction, we write the scalar wavefield using the Fourier transform with respect to time as

$$u(\mathbf{x}_\perp, z, t) = \frac{1}{2\pi} \int_{-\infty}^{\infty} d\omega\, U(\mathbf{x}_\perp, z, \omega) e^{i(kz - \omega t)} \tag{8.3}$$

where wavenumber $k = \omega / V_0$ and $U(\mathbf{x}_\perp, z, \omega)$ is the amplitude of a harmonic wave. Substituting (8.3) in (8.2), we get

$$\partial_z^2 U + 2ik \partial_z U + \Delta_\perp U - 2k^2 \xi U = 0 \tag{8.4}$$

where $\Delta_\perp \equiv \partial_x^2 + \partial_y^2$ is Laplacian in the transverse plane. When

$$ak \gg 1 \tag{8.5}$$

the amplitude changes very slowly; therefore, we may neglect the first term in (8.4) and we obtain the parabolic equation

$$2ik\partial_z U + \Delta_\perp U - 2k^2\xi U = 0 \tag{8.6}$$

8.1.2 Transverse Correlations of Amplitude and Phase Fluctuations

We solve (8.6) using a perturbation method known as the Rytov method, in which amplitude in (8.3) is written as

$$U(\mathbf{x}_\perp, z, \omega) = \exp\left[\Xi(\mathbf{x}_\perp, z, \omega)\right] = \exp(\Delta \ln A_0 + i\Delta\varphi) \tag{8.7}$$

where Ξ is a first-order small quantity, $\Delta \ln A_0$ is the log-amplitude fluctuation, and $\Delta\varphi$ is the phase fluctuation for the mean propagation direction in the z-direction. $\Delta \ln A_0 = \Delta\varphi = 0$ for an incident plane wave of unit amplitude for $z \le 0$. This exponential expression represents wave propagation better than the algebraic series of the Born approximation for line-of-sight propagation problems [see Ishimaru, 1978, p.346]. Neglecting $(\nabla_\perp \Xi)^2$ because Ξ is considered smooth in the transverse plane, the wave equation for Ξ is

$$2ik\,\partial_z\Xi + \Delta_\perp\Xi = 2k^2\xi \tag{8.8}$$

Using the Fourier transform in the transverse plane

$$\Xi(\mathbf{x}_\perp, z, \omega) = \frac{1}{(2\pi)^2} \int\limits_{-\infty}^{\infty}\int\limits_{-\infty}^{\infty} d\mathbf{m}_\perp\, e^{i\mathbf{m}_\perp \mathbf{x}_\perp}\, \breve{\Xi}(\mathbf{m}_\perp, z, \omega) \tag{8.9}$$

where $\mathbf{m}_\perp$ is the wavenumber vector in the transverse plane, the parabolic equation (8.6) becomes

$$2ik\,\partial_z\breve{\Xi} - m_\perp^2\,\breve{\Xi} = 2k^2\breve{\xi} \tag{8.10}$$

where $m_\perp \equiv |\mathbf{m}_\perp|$ and $\breve{\xi}$ is the Fourier transform of ξ in the transverse plane. By using the Green function

$$\breve{G}(\mathbf{m}_\perp, z, \omega) = \frac{-i}{2k}\exp\left[-i\frac{m_\perp^2 z}{2k}\right]H(z) \tag{8.11}$$

that satisfies

$$\left(2ik\partial_z - m_\perp^2\right)\breve{G}(\mathbf{m}_\perp, z, \omega) = \delta(z) \tag{8.12}$$

we can solve (8.10) as

$$\breve{\Xi}(\mathbf{m}_\perp, Z, \omega) = 2k^2 \int_0^Z dz\, \breve{G}(\mathbf{m}_\perp, Z - z, \omega) \breve{\xi}(\mathbf{m}_\perp, z) \qquad (8.13)$$

where the convolution integral between 0 and Z means that we have neglected the backward scattering contribution from the region $z > Z$. Taking the inverse Fourier transform,

$$\Xi(\mathbf{x}_\perp, Z, \omega) = \frac{2k^2}{(2\pi)^2} \int_0^Z dz \int\int_{-\infty}^{\infty} d\mathbf{m}_\perp\, e^{i m_\perp x_\perp} \breve{G}(\mathbf{m}_\perp, Z - z, \omega) \breve{\xi}(\mathbf{m}_\perp, z) \quad (8.14)$$

Now, imagine an ensemble of random media $\{\xi(\mathbf{x})\}$, where $\xi(\mathbf{x})$ is a spatially uniform and isotropic random function of coordinate $\mathbf{x}$, and $\langle \xi(\mathbf{x}) \rangle = 0$. We will investigate wavefield statistics for this ensemble. The characteristic of the ensemble is statistically given by the ACF or the PSDF of $\xi(\mathbf{x})$. We summarize the statistical relations which will be used in the following. The PSDF of the random fluctuation given in (2.6) can be written

$$P(\mathbf{m}_\perp, m_z) = \int_{-\infty}^{\infty} dz\, e^{-i m_z z} \int\int_{-\infty}^{\infty} d\mathbf{x}_\perp\, e^{-i m_\perp x_\perp}\, R(\mathbf{x}_\perp, z) = \int_{-\infty}^{\infty} dz\, e^{-i m_z z}\, \breve{R}(\mathbf{m}_\perp, z) \quad (8.15)$$

The double integral term in (8.15) is the Fourier transform of the ACF in the transverse plane. We define the longitudinal integral of the ACF along the mean propagation direction on the z-axis as

$$A(\mathbf{x}_\perp) = A(r_\perp) = \int_{-\infty}^{\infty} dz\, R(\mathbf{x}_\perp, z) \qquad (8.16)$$

where we may write the argument as scalar $r_\perp = |\mathbf{x}_\perp|$ because of the assumed isotropy of the inhomogeneity. Using (8.15), the Fourier transform of A in the transverse plane is written using the isotropic PSDF:

$$\breve{A}(\mathbf{m}_\perp) = \breve{A}(m_\perp) = \int\int_{-\infty}^{\infty} d\mathbf{x}_\perp\, e^{-i m_\perp x_\perp}\, A(\mathbf{x}_\perp)$$

$$= \int_{-\infty}^{\infty} dz\, \breve{R}(\mathbf{m}_\perp, z) = P(\mathbf{m}_\perp, m_z = 0) \qquad (8.17)$$

where we may write the argument as scalar $m_\perp = |\mathbf{m}_\perp|$. Because of the homogeneity of the randomness,

$$\left\langle \breve{\xi}(m_{\perp}',z')\breve{\xi}(m_{\perp}'',z'')\right\rangle = (2\pi)^2 \delta(m_{\perp}'+m_{\perp}'')\tilde{R}(m_{\perp}',z'-z'') \qquad (8.18)$$

Using

$$\breve{\xi}^*(m_{\perp},z)=\breve{\xi}(-m_{\perp},z) \quad \text{and} \quad \tilde{G}(-m_{\perp},z,\omega)=\tilde{G}(m_{\perp},z,\omega) \qquad (8.19)$$

we get the complex conjugate of (8.14) as

$$\Xi^*(x_{\perp},Z,\omega)=\frac{2k^2}{(2\pi)^2}\int_0^Z dz \int_{-\infty}^{\infty}\int dm_{\perp} e^{im_{\perp}x_{\perp}}\,\tilde{G}^*(m_{\perp},Z-z,\omega)\breve{\xi}(m_{\perp},z) \qquad (8.20)$$

The following derivation is due to Chernov [1960] and Ishimaru [1978]. From (8.14) and (8.20),

$$\Delta \ln A_0\left(x_{\perp},Z,\omega\right)=\frac{-k}{(2\pi)^2}\int_0^Z dz \int_{-\infty}^{\infty}\int dm_{\perp} e^{im_{\perp}x_{\perp}}\sin\frac{(Z-z)m_{\perp}^{\ 2}}{2k}\breve{\xi}(m_{\perp},z)$$

$$\Delta \varphi\left(x_{\perp},Z,\omega\right)=\frac{-k}{(2\pi)^2}\int_0^Z dz \int_{-\infty}^{\infty}\int dm_{\perp} e^{im_{\perp}x_{\perp}}\cos\frac{(Z-z)m_{\perp}^{\ 2}}{2k}\breve{\xi}(m_{\perp},z) \qquad (8.21)$$

Taking an average over the ensemble, we get the correlation of log-amplitude fluctuations on the transverse plane as

$$\left\langle \Delta \ln A_0\left(x_{\perp}',Z,\omega\right)\Delta \ln A_0\left(x_{\perp}'',Z,\omega\right)\right\rangle = \frac{k^2}{(2\pi)^2}\int_0^Z dz' \int_0^Z dz''$$

$$\times \int_{-\infty}^{\infty}\int dm_{\perp} e^{im_{\perp}(x_{\perp}'-x_{\perp}'')}\sin\frac{(Z-z')m_{\perp}^{\ 2}}{2k}\sin\frac{(Z-z'')m_{\perp}^{\ 2}}{2k}\tilde{R}(m_{\perp},z'-z'') \qquad (8.22)$$

where we used (8.18). To evaluate integrals in (8.22), we introduce the center-of-mass and difference coordinates as $z_c=(z'+z'')/2$, $z_d=z'-z''$, $x_{\perp c}=(x_{\perp}'+x_{\perp}'')/2$, and $x_{\perp d}=x_{\perp}'-x_{\perp}''$. We note that $\tilde{R}(m_{\perp},z)$ is very small for $m_{\perp}>1/a$. Therefore, it is sufficient to study only $m_{\perp}<1/a$. For a travel distance of the order of a, the diffraction effect given by $\sin(Z-z)m_{\perp}^{\ 2}/2k$ is small for $m_{\perp}<1/a$ since $m_{\perp}^{\ 2}a/k<1/ak<<1$. We note that $\sin(Z-z)m_{\perp}^{\ 2}/2k$ is a slowly varying function of z. Therefore, we may discard argument z_d in the diffraction term. For $Z>>a$ we may write (8.22) as

$$\left\langle \Delta \ln A_0\left(\mathbf{x}_\perp{'},Z,\omega\right)\Delta \ln A_0\left(\mathbf{x}_\perp{''},Z,\omega\right)\right\rangle$$

$$\approx \frac{k^2}{(2\pi)^2}\int\limits_{-\infty}^{\infty}\int\limits_{-\infty}^{\infty} dm_\perp e^{im_\perp x_{\perp d}}\int\limits_0^Z dz_c \sin^2\frac{(Z-z_c)m_\perp{}^2}{2k}\int\limits_{-\infty}^{\infty} dz_d \breve{R}\left(m_\perp,z_d\right)$$

$$\approx \frac{k^2 Z}{8\pi^2}\int\limits_{-\infty}^{\infty}\int\limits_{-\infty}^{\infty} dm_\perp e^{im_\perp x_{\perp d}}\left(1-\frac{k}{Zm_\perp{}^2}\sin\frac{Zm_\perp{}^2}{k}\right)\breve{A}(m_\perp) \qquad (8.23a)$$

$$=\frac{k^2 Z}{4\pi}\int\limits_0^\infty dm_\perp m_\perp J_0\left(r_{\perp d}m_\perp\right)\left(1-\frac{k}{Zm_\perp{}^2}\sin\frac{Zm_\perp{}^2}{k}\right)\breve{A}(m_\perp)$$

This correlation depends only on the transverse distance $r_{\perp d}\equiv\left|\mathbf{x}_{\perp d}\right|$ and is insensitive to the fluctuation in the mean propagation direction. In the same way,

$$\left\langle\Delta\varphi'\,\Delta\varphi''\right\rangle \approx \frac{k^2 Z}{4\pi}\int\limits_0^\infty dm_\perp m_\perp J_0\left(r_{\perp d}m_\perp\right)\left(1+\frac{k}{Zm_\perp{}^2}\sin\frac{Zm_\perp{}^2}{k}\right)\breve{A}(m_\perp)$$

$$\left\langle\Delta\ln A_0{'}\,\Delta\varphi''\right\rangle \approx \frac{k^2 Z}{4\pi}\int\limits_0^\infty dm_\perp m_\perp J_0\left(r_{\perp d}m_\perp\right)\frac{\sin^2\dfrac{Zm_\perp{}^2}{2k}}{\dfrac{Zm_\perp{}^2}{2k}}\breve{A}(m_\perp) \qquad (8.23b)$$

where $'$ and $''$ abbreviate functions whose arguments depend on transverse coordinates $\mathbf{x}_\perp{'}$ and $\mathbf{x}_\perp{''}$, respectively.

We introduce the scattering strength parameter Φ [after Flatté et al., 1979, p. 92], whose square is known as the optical distance [Ishimaru, 1978, p. 119]. The square of this parameter is the MS of the phase fluctuation in the geometrical optic region for $Z \gg a$:

$$\Phi^2 = \omega^2\left\langle\left(\int\limits_0^Z \frac{dz'}{V(z')}-\frac{Z}{V_0}\right)^2\right\rangle = k^2\left\langle\left(\int\limits_0^Z dz'\,\xi\left(\mathbf{x}_\perp=0,z'\right)\right)^2\right\rangle$$

$$= k^2\int\limits_0^Z dz'\int\limits_0^Z dz''R\left(\mathbf{x}_\perp=0,z'-z''\right)\approx k^2\int\limits_0^Z dz_c\int\limits_{-\infty}^{\infty} dz_d R\left(\mathbf{x}_\perp=0,z_d\right)=k^2 Z A(0) \qquad (8.24)$$

This function increases linearly with increasing travel distance. We introduce the wave parameter D following Chernov [1960, p. 74]:

$$D=\frac{4Z}{a^2 k} \qquad (8.25)$$

Diffraction effects begin to be significant when the radius of the first Fresnel zone $\sqrt{\pi Z/k}$ exceeds the correlation distance a. The wave parameter is the square of the ratio of the first Fresnel zone to the scale of inhomogeneities, so it characterizes the order of the diffraction effect.

Gaussian ACF

For the Gaussian ACF (2.7), from (8.16-8.17),

$$A(x_\perp) = A(r_\perp) = \sqrt{\pi}\,\varepsilon^2 a \exp\left(-r_\perp^2 / a^2\right)$$
$$\tilde{A}(m_\perp) = \tilde{A}(m_\perp) = \sqrt{\pi^3}\,\varepsilon^2 a^3 \exp\left(-a^2 m_\perp^2 / 4\right) \tag{8.26}$$

Then, the square of scattering strength parameter is given by

$$\Phi^2 = \sqrt{\pi}\,\varepsilon^2 a k^2 Z \tag{8.27}$$

At $x_{\perp d} = 0$, substituting (8.26) and (8.27) into (8.23a–b), we get

$$\left\langle (\Delta \ln A_0)^2 \right\rangle = \frac{\sqrt{\pi}\varepsilon^2 a\, k^2 Z}{2}\left(1 - \frac{a^2 k}{4Z}\tan^{-1}\frac{4Z}{a^2 k}\right) = \frac{\Phi^2}{2}\left(1 - \frac{1}{D}\tan^{-1} D\right)$$

$$\left\langle (\Delta\varphi)^2 \right\rangle = \frac{\Phi^2}{2}\left(1 + \frac{1}{D}\tan^{-1} D\right) \tag{8.28}$$

$$\left\langle \Delta \ln A_0\, \Delta\varphi \right\rangle = \frac{\Phi^2}{4D}\ln\left(1 + D^2\right)$$

We plot the correlation functions (8.28) normalized by Φ^2 against D in Figure 8.1.

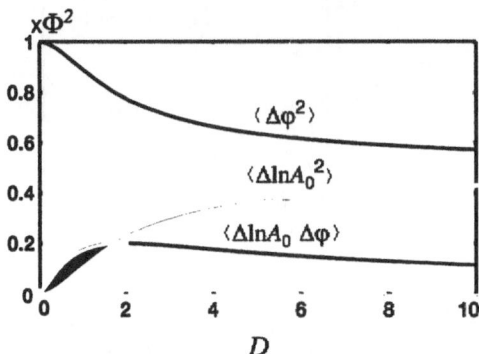

FIGURE 8.1. Phase and log-amplitude fluctuations (8.28) against wave parameter D for the Gaussian ACF.

The phase fluctuation is of the order of Φ^2. This tends to (5.26) as D goes to zero.

Numerically simulating wave propagation through a random medium, Frankel and Clayton [1986] measured the travel-time fluctuation of waves with $ak \gg 1$ for a fixed travel distance with differing values of a. They determined the standard deviation of the travel times measured for media characterized by a given a by using several realizations of media having the given value. They found that the standard deviation of the measured travel time at a fixed propagation distance increases with increasing a. Their measured standard deviation agrees well with the prediction of (8.28) for a Gaussian ACF and even for an exponential ACF; however, their measured standard deviation is half the theoretical prediction for a von Kármán ACF. When the wavelength is much shorter than the correlation distance, the travel time at large distances in a random medium can be estimated using a ray theoretical approach. Müller et al. [1992] and Roth et al. [1993] numerically confirmed that the effective propagation velocity is higher than the average propagation velocity because of diffraction around low-velocity regions, that is, waves prefer fast paths. These numerical experiments confirm the validity of the parabolic approximation for Gaussian and exponential ACF media and the breakdown for von Kármán ACF media which are rich in short-wavelength inhomogeneities.

Ratios of the correlations are independent of Φ^2 and uniquely determined by wave parameter D as

$$\sqrt{\frac{\langle(\Delta \ln A_0)^2\rangle}{\langle(\Delta\varphi)^2\rangle}} = \sqrt{\frac{D - \tan^{-1} D}{D + \tan^{-1} D}}$$

$$\frac{\langle\Delta \ln A_0 \, \Delta\varphi\rangle}{\sqrt{\langle(\Delta \ln A_0)^2\rangle\langle(\Delta\varphi)^2\rangle}} = \frac{\ln(1 + D^2)}{2\sqrt{D^2 - (\tan^{-1} D)^2}}$$

(8.29)

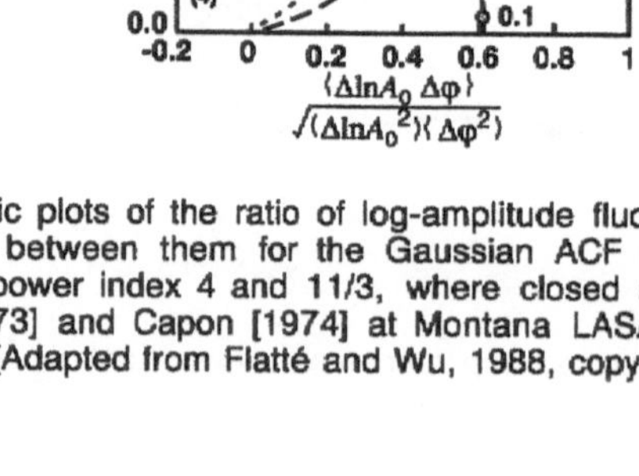

FIGURE 8.2. Parametric plots of the ratio of log-amplitude fluctuation to phase fluctuation vs. correlation between them for the Gaussian ACF (solid curve) and the power-law PSDF with power index 4 and 11/3, where closed symbols are observations made by Aki [1973] and Capon [1974] at Montana LASA and the large open circle is for NORSAR. [Adapted from Flatté and Wu, 1988, copyright by the American Geophysical Union.]

Figure 8.2 shows the ratio of log-amplitude fluctuation to phase fluctuation vs. correlation between log-amplitude and phase fluctuations, where a solid curve labeled with D values is for the Gaussian ACF.

When $D \ll 1$, in the geometrical optics region, taking the lowest order term of the expansion of the diffraction term in (8.23),

$$\langle \Delta \ln A_0' \, \Delta \ln A_0'' \rangle \approx \frac{\sqrt{\pi}\varepsilon^2 a^3 k^2 Z}{4} \int_0^\infty dm_\perp m_\perp J_0(r_{\perp d} m_\perp)$$

$$\times \frac{1}{6} \left(\frac{Zm_\perp^2}{k} \right)^2 \exp(-a^2 m_\perp^2 / 4)$$

$$= \frac{8\sqrt{\pi}\varepsilon^2}{3} \left(\frac{Z}{a} \right)^3 {}_1F_1 \left[3,1; -\left(\frac{r_{\perp d}}{a} \right)^2 \right] = \frac{\Phi^2 D^2}{6} {}_1F_1 \left[3,1; -\left(\frac{r_{\perp d}}{a} \right)^2 \right]$$

(8.30)

$$\langle \Delta \varphi' \, \Delta \varphi'' \rangle \approx \Phi^2 \exp\left(-\frac{r_{\perp d}^2}{a^2} \right)$$

where ${}_1F_1$ is the confluent hypergeometric function.

For $D \gg 1$, in the diffraction region, we may discard the highly oscillating diffraction factor $\sin Zm_\perp^2 / k$ in integral (8.23). Then,

$$\langle \Delta \ln A_0' \, \Delta \ln A_0'' \rangle = \langle \Delta \varphi' \, \Delta \varphi'' \rangle$$

$$\approx \frac{\sqrt{\pi}\varepsilon^2 a^3 k^2 Z}{4} \int_0^\infty dm_\perp m_\perp J_0(r_{\perp d} m_\perp) \exp(-a^2 m_\perp^2 / 4) \quad (8.31)$$

$$\approx \frac{\Phi^2}{2} \exp\left(-\frac{r_{\perp d}^2}{a^2} \right) = \frac{\sqrt{\pi}\varepsilon^2 a \, k^2 Z}{2} \exp\left(-\frac{r_{\perp d}^2}{a^2} \right)$$

We plot these correlations against the transverse separation in Figure 8.3. They decrease with increasing separation in the transverse direction. The behavior of the transverse correlation for intermediate values of D is discussed in Chernov [1960].

The range of applicability of the parabolic approximation has implicit conditions

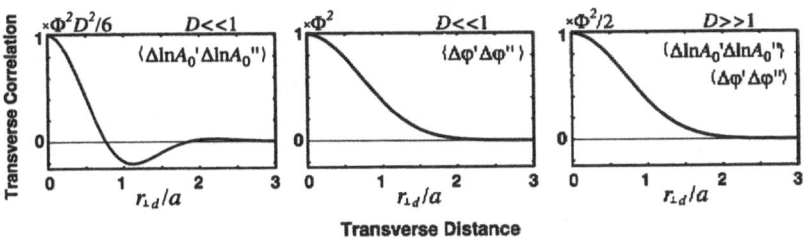

FIGURE 8.3. Transverse correlation functions of log-amplitude and phase fluctuations for extreme cases of D.

in addition to condition $ak \gg 1$ given by (8.5). Here we briefly discuss the application conditions for a Gaussian ACF [Rytov et al., 1989]. The parabolic approximation models strong forward scattering for large wavenumbers, where the characteristic scattering angle is of the order of $1/ak$, that is, $r_\perp/Z \approx 1/ak$, as discussed in Section 4.1.2. The phase term in the exponent of the spherically outgoing scattered wave is written as $k\sqrt{Z^2 + r_\perp^2} \approx kZ + kr_\perp^2/2Z - kr_\perp^4/8Z^3$. The second term is of the order of the wave parameter D. The parabolic approximation corresponds to neglecting the third term, which is written as $Z/a^4k^3 \ll 1$, that is,

$$D \ll (ak)^2 \tag{8.32}$$

The diffraction effect is accounted for in the parabolic approximation. The other limiting condition for the approximation is due to the neglect of backscattering in (8.13). The effect of backscattering can be found by the corrected Born approximation as given by (5.20). For the Gaussian ACF, scattering attenuation corresponding to the backward half-space is given by

$$^{TSc}Q^{-1} = \frac{k^3}{2\pi} \int_{\pi/2}^{\pi} \varepsilon^2 \sqrt{\pi^3}\, a^3 e^{-k^2 a^2 \sin^2\frac{\psi}{2}} \sin\psi \, d\psi$$
$$= \sqrt{\pi}\varepsilon^2 ake^{-k^2 a^2/2}\left(1 - e^{-k^2 a^2/2}\right) \tag{8.33}$$

for cutoff scattering angle $\psi_C = \pi/2$. Under condition (8.5), the condition for a small backscattering loss for propagation distance Z is written as

$$^{TSc}Q^{-1}kZ \approx \sqrt{\pi}\varepsilon^2 ak^2 Ze^{-k^2 a^2/2} \ll 1 \tag{8.34}$$

Therefore, the neglect of backscattering is not a severe restriction for the parabolic approximation.

8.1.3 Measurements of Amplitude and Phase Fluctuations

P-Wave Turbidity Coefficient in the Lithosphere

For a plane wave propagating a long travel distance through a randomly inhomogeneous medium, the MS of log-amplitude fluctuation increases with travel distance as given by (8.28). The ratio of $\langle (\Delta \ln A_0)^2 \rangle$ to travel distance Z characterizes the randomness or turbidity of the heterogeneous medium [Chernov, 1960; Nikolaev, 1975, p. 16]:

$$g_F \equiv \langle (\Delta \ln A_0)^2 \rangle / Z \tag{8.35}$$

The turbidity coefficient g_F can be considered related to the total scattering coefficient. The P-wave turbidity coefficient of the crust and the upper mantle was extensively measured by Russian investigators in the 1960s by analyzing the amplitude fluctuation of P-wave first motions from explosions and natural earthquakes [see Nikolaev, 1975, p. 113]. In these measurements the turbidity coefficient is a rather phenomenological parameter that includes spatial variations of site factors and attenuation strength, and the ensemble average was replaced with a spatial average. The measurements were made on horizontally propagating P waves from the Zhalanash–Kurgan 200 km long deep seismic sounding profile in southern Kazakhstan in which the sensor interval was 250 m and the predominant frequency was 7–8 Hz. The depth profile of turbidity coefficient g_F was made by estimating the contribution to amplitude fluctuations of the different layers in which seismic rays traveled. The value of g_F was as high as 2×10^{-3} km^{-1} near the surface and dropped rapidly as depth increased to 10 km. A layer of high "transparency" with $g_F \approx 10^{-5}$ km^{-1} was found at depths between 10–25 km [Nikolajev and Tregub, 1970]. Between depths of 28–54 km, g_F is constant of the order of 10^{-4} km^{-1}. Estimates of g_F were also made using data from active seismic experiments conducted at sea. In the Tatar Strait, g_F for the crust was 0.0025 km^{-1}. Values of 0.002 km^{-1} for the crust and 0.0007 km^{-1} were found for the upper mantle beneath the Kuril Islands (Iturup and Ishishir Islands); 0.001 km^{-1} for the crust and 0.0005 km^{-1} for the upper mantle east of Iturup island and east of southern Kamchatka; 0.002 km^{-1} for the crust under the Sea of Okhotsk; 0.0013 km^{-1} in the crust and 0.00015 km^{-1} in the upper mantle under the Black Sea. Nikolaev [1975] concluded that g_F for 5 Hz P-waves in the crust to the upper mantle is 0.0001–0.0025 km^{-1} with an error of about a factor of two.

Amplitude and Phase Correlations of Teleseismic P-Waves

Aki [1973] first analyzed array recordings of teleseismic P-waves made at the Large Aperture Seismic Array (LASA) in Montana. The aperture of this array is about 200 km and waveforms contained frequencies centered on about 0.6 Hz. He measured transverse correlation functions of teleseismic P-waves arriving from near vertical incidence and estimated a=10 km. He found a positive correlation between log-amplitude and phase fluctuations that agreed with the theoretical prediction given by (8.29). From plots of the ratio of RMS log-amplitude to RMS phase fluctuations against the correlation between log-amplitude and phase fluctuations as given by Figure 8.2, he estimated D=5, Z=60 km and ε^2=0.0016 (ε=4%). He predicted $g_F \approx 0.008$ km^{-1}, which is much larger than the turbidity coefficient estimated for horizontally traveling rays by Nikolaev and his colleagues, as discussed above. The difference between the estimated turbidities might be attributed to the difference in ray directions and the departure of the real randomness from the assumed isotropic randomness. The difference may also be due to differences in tectonic settings of the regions where the measurements were made or due to differences in the frequency band used for the measurements.

Comparing 0.8 Hz band P-wave slowness fluctuations across subarrays at LASA with those for the whole array, Capon [1974] estimated D=6 and a=12 km (see Figure 8.2), which are similar to Aki's [1973] results. However, Capon's estimates of $\epsilon^2 = 0.00036$ ($\epsilon = 1.9\%$) and $Z = 136$ km are quite different from Aki's since it is difficult to separate the product $\epsilon^2 Z$ into ϵ^2 and Z in Φ^2.

Analyzing travel-time fluctuations of teleseismic P-waves of dominant frequency near 1 Hz observed in southern California, Powell and Meltzer [1984] inferred that $a = 25$ km with $\epsilon^2 = 0.001$ ($\epsilon = 3.26\%$) to depths of at least 119 km. Examining array data at both Montana LASA and a large seismic array NORSAR in Norway, Berteussen et al. [1975] pointed out difficulties in uniquely determining the parameters characterizing the randomness because of the finite aperture of the seismic array. Haddon and Husebye [1978] tried to explain both amplitude and travel-time anomalies observed for P-waves at NORSAR in terms of deterministic velocity inhomogeneities in the upper mantle.

Flatté and Wu [1988] measured the transverse correlation of log-amplitude and phase fluctuations of more than one hundred teleseismic P-wave beams with 2 Hz center frequency recorded at NORSAR. They also introduced the concept of angular correlation functions, which are based on measurements of two rays with differ-

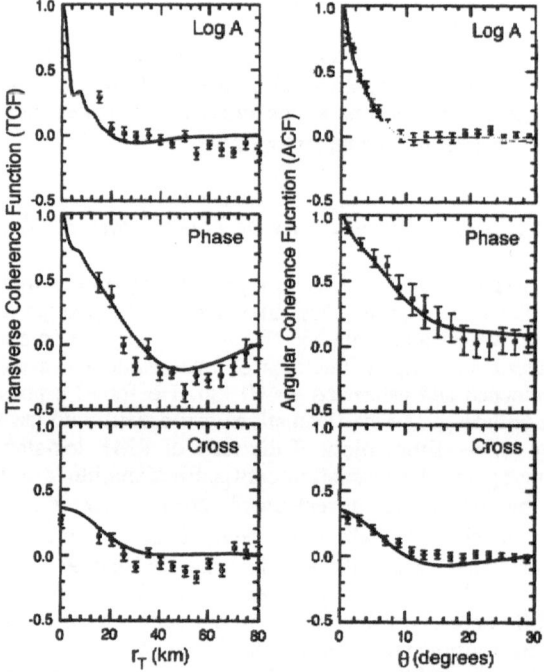

FIGURE 8.4. Transverse correlation functions and angular correlation functions measured at NORSAR, where closed circles and vertical bars are observed teleseismic P-wave data and standard deviations, respectively. Solid curves are predictions for a two-layer model. [From Flatté and Wu, 1988, copyright by the American Geophysical Union.]

ent incident angles. The rigorous derivation of angular correlation functions was done by Wu and Flatté [1990] and later by Chen and Aki [1991]. Flatté and Wu [1988] suggested that the von Kármán type ACF is more appropriate than the Gaussian ACF for modeling the NORSAR data. Figure 8.4 shows both the transverse and angular correlation functions at NORSAR. Investigating the depth dependence of random inhomogeneity, Flatté and Wu [1988] proposed a model for lithospheric and asthenospheric inhomogeneities beneath NORSAR that consists of two overlapping layers. The spectra of inhomogeneities in both layers are band-limited between the wavelengths of 5.5 and 110 km. The upper layer has a flat PSDF, $P(m) \sim m^0$, extending from the surface to about the 200 km depth. The lower layer has $P(m) \sim m^{-4}$ extending from 15 to 250 km. The latter spectrum corresponds to an exponential ACF having a longer scale than the observation aperture of 110 km and ε in the range 1–4%. The difference between the power spectra means that there are more small scale inhomogeneities near the surface compared with the deeper portions.

Mori and Frankel [1992] correlated observed relative amplitudes with travel-time residuals of teleseismic P-waves in southern California. Their data show an amplitude increase of about factor of two for a 1 s increase in travel time. The simplest interpretation of this result is that velocity inhomogeneity causes both amplitude and travel-time variations. Figure 8.5 shows the average amplitude measured at a number of stations vs. the station average time residual. Stations are separated into two groups based on surface geology; hard sites are located on bedrock and soft sites on alluvium or soil. Travel times to the soft sites are longer on average than the average on hard sites. Soft sites have larger amplitudes. The soft sites also show a stronger correlation between the time residuals and amplitudes than the hard sites. Mori and Frankel [1992] pointed out the importance of surface geology for travel-time fluctuations. Their interpretation is in general agreement with the larger velocity fluctuation and the shorter correlation distance at shallower depth that Flatté and Wu [1988] obtained from studying data at NORSAR.

The studies we have described led to the interpretation of observed amplitude and phase fluctuations of P-waves using random media models. The model of a homogeneous random medium first assumed was later expanded to include a depth dependence for the randomness. The power-law spectra corresponding to the exponential ACF fits data better than the Gaussian ACF. It would be appropriate to ex-

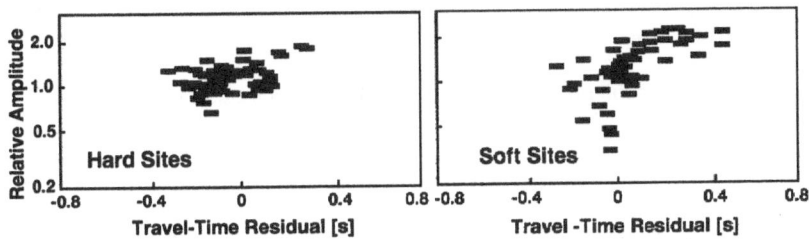

FIGURE 8.5. Relative amplitude versus travel-time residual for stations in southern California. [Modified from Mori and Frankel, 1992, published by the American Geophysical Union.]

pand the model presented here by including an anisotropic inhomogeneity and then reanalyze the amplitude and phase fluctuations of S-waves in the lithosphere.

8.2 MARKOV APPROXIMATION FOR PREDICTING THE MS ENVELOPE DUE TO DIFFRACTION

For seismograms of regional earthquakes, the time difference between the S-wave onset and the time that the S-wave envelope has its maximum amplitude is much larger than the source duration, as introduced in Section 2.4.3. Therefore, we infer that the observed delay or broadening of the S-wave energy packet is caused by the effects of wave propagation through randomly inhomogeneous media. Pulse broadening in random media has been an important subject in fields other than seismology [Tatarskii, 1971; Rytov et al., 1989]. For example, an impulsive source emitted from a ruby laser whose duration is on the order of nanoseconds broadens more than 100 times after passing through a typical fog with a fluctuating refractive index [Ishimaru, 1978, p. 325]. Lee and Jokipii [1975a, b] studied the broadening of radio waves by turbulent plasma. Pulse broadening of acoustic waves in the ocean is caused by wave-speed fluctuation through internal waves [Flatté et al., 1979].

Now we will describe a model for wave propagation in randomly inhomogeneous media where the wavelength is much smaller than the correlation distance. Using the parabolic approximation, we will develop a model that predicts that an impulsive wave packet at the source stretches out in time and reduces amplitude with increasing travel distance. First, we introduce the concepts of coherent wavefield, mutual coherence function, and two-frequency mutual coherence function. These are necessary to describe how wavefields lose their original form with increasing travel distance. Then we derive the equation for each quantity based on the Markov approximation. Solving the master equation for the two-frequency mutual coherence function for quasi-monochromatic waves, we will get the time trace of the intensity of the wavefield, which corresponds to the mean square envelope of the wavefield in a frequency band. We will discuss the relationship between the characteristics of the time trace intensity and the stochastic characteristics of the random medium.

8.2.1 Coherent Wavefield

First, we study the propagation of the coherent wavefield $\langle U \rangle$ in (8.3). Taking the ensemble average of (8.6),

$$2ik\partial_z \langle U \rangle + \Delta_\perp \langle U \rangle - 2k^2 \langle \xi U \rangle = 0 \qquad (8.36)$$

We follow the simple derivation according to Lee and Jokipii [1975a] to evaluate the last term of the left-hand side. From (8.6), we can write the wavefield at z in an integral form by using the wavefield at $z - \Delta z$, where $\Delta z > 0$:

$$U(\mathbf{x}_\perp,z,\omega)=U(\mathbf{x}_\perp,z-\Delta z,\omega)$$
$$+\frac{i}{2k}\int_{z-\Delta z}^{z}dz'\big[\Delta_\perp U(\mathbf{x}_\perp,z',\omega)-2k^2\xi(\mathbf{x}_\perp,z')U(\mathbf{x}_\perp,z',\omega)\big] \qquad (8.37)$$

We suppose the existence of an intermediate scale Δz, which is larger than the correlation distance a but smaller than the scale of variation of U. Then, we can write

$$U(\mathbf{x}_\perp,z,\omega)\approx U(\mathbf{x}_\perp,z-\Delta z,\omega)+\frac{i}{2k}\Delta z\,\Delta_\perp U(\mathbf{x}_\perp,z-\Delta z,\omega)$$
$$-ikU(\mathbf{x}_\perp,z-\Delta z,\omega)\int_{z-\Delta z}^{z}dz'\xi(\mathbf{x}_\perp,z') \qquad (8.38)$$

Multiplying $\xi(\mathbf{x}_\perp',z)$ and taking the ensemble average, we have

$$\langle\xi(\mathbf{x}_\perp',z)U(\mathbf{x}_\perp,z,\omega)\rangle\approx-ik\int_{z-\Delta z}^{z}dz'\langle\xi(\mathbf{x}_\perp',z)\xi(\mathbf{x}_\perp,z')\rangle\langle U(\mathbf{x}_\perp,z-\Delta z,\omega)\rangle$$
$$\approx-ik\int_{0}^{\infty}dz_d R(\mathbf{x}_\perp-\mathbf{x}_\perp',z_d)\langle U(\mathbf{x}_\perp,z,\omega)\rangle \qquad (8.39)$$
$$=-\frac{i}{2}k\,A(\mathbf{x}_\perp-\mathbf{x}_\perp')\langle U(\mathbf{x}_\perp,z,\omega)\rangle$$

Since the variation of U is small, $\langle U(\mathbf{x}_\perp,z-\Delta z,\omega)\rangle\approx\langle U(\mathbf{x}_\perp,z,\omega)\rangle$, and there is no contribution of the inhomogeneity at z to the wavefield at $z-\Delta z$, $\langle\xi(\mathbf{x}_\perp',z)U(\mathbf{x}_\perp,z-\Delta z,\omega)\rangle=0$. This means that we neglect backward scattering. Substituting (8.39) with $\mathbf{x}_\perp'=\mathbf{x}_\perp$ in (8.36),

$$2ik\partial_z\langle U\rangle+\Delta_\perp\langle U\rangle+ik^3A(0)\langle U\rangle=0 \qquad (8.40)$$

The derivation of this stochastic equation is called the Markov approximation [Tatarskii, 1971] since the last term in (8.40) depends only on the local value of z. Alternative derivations based on a functional formulation are given in Tatarskii [1971], Ishimaru [1978] and Rytov et al. [1989]. The range of application conditions for this approximation is discussed in detail by Barabanenkov et al. [1971]. The above derivation was not based on the assumption of isotropy of the ACF in 3-D space. The longitudinal integral of ACF is taken along the mean propagation direction and the resultant function A describes the correlation of media on the transverse plane.

Under the initial condition $U(\mathbf{x}_\perp,z=0,\omega)=1$,

$$\langle U(\mathbf{x}_\perp,Z,\omega)\rangle=e^{-A(0)k^2Z/2}=e^{-\Phi^2/2} \qquad (8.41)$$

The coherent component decays exponentially with increasing propagation distance characterized by the square of the scattering strength parameter.

8.2.2 Mutual Coherence Function

We define the mutual coherence function of wavefield U measured at different locations on the transverse plane at distance z and angular frequency ω as

$$\Gamma_1(\mathbf{x}_\perp', \mathbf{x}_\perp'', z, \omega) \equiv \left\langle U(\mathbf{x}_\perp', z, \omega) U(\mathbf{x}_\perp'', z, \omega)^* \right\rangle \tag{8.42}$$

Multiplying U by (8.6) and taking the ensemble average, we obtain

$$2ik\partial_z \Gamma_1 + (\Delta_\perp' - \Delta_\perp'')\Gamma_1 - 2k^2 \left\langle (\xi - \xi'')U' U''^* \right\rangle = 0 \tag{8.43}$$

where U' and U'' mean that their arguments are at $\mathbf{x}_\perp'$ and $\mathbf{x}_\perp''$, respectively. Using the same procedure as for the derivation of (8.39),

$$\left\langle (\xi - \xi'')U' U''^* \right\rangle = -ik\left[A(0) - A(r_{1d}) \right]\Gamma_1 \tag{8.44}$$

where $r_{1d} = |\mathbf{x}_\perp' - \mathbf{x}_\perp''|$. Then, the equation for Γ_1 is

$$2ik\partial_z \Gamma_1 + (\Delta_\perp' - \Delta_\perp'')\Gamma_1 + 2ik^3 \left[A(0) - A(r_{1d}) \right]\Gamma_1 = 0 \tag{8.45}$$

Introducing center-of-mass and difference coordinates in the transverse plane $\mathbf{x}_\perp' = \mathbf{x}_{1c} + \mathbf{x}_{1d}/2$ and $\mathbf{x}_\perp'' = \mathbf{x}_{1c} - \mathbf{x}_{1d}/2$, we write Laplacians in the transverse plane as

$$\Delta_\perp' = \Delta_{1d} + \frac{1}{4}\Delta_{1c} + \nabla_{1c}\nabla_{1d} \quad \text{and} \quad \Delta_\perp'' = \Delta_{1d} + \frac{1}{4}\Delta_{1c} - \nabla_{1c}\nabla_{1d} \tag{8.46}$$

Because of the statistical homogeneity of the random media, Γ_1 is independent of the center-of-mass coordinates: $(\Delta_\perp' - \Delta_\perp'')\Gamma_1 = 0$. Then, (8.45) becomes

$$\partial_z \Gamma_1 + k^2 \left[A(0) - A(r_{1d}) \right]\Gamma_1 = 0 \tag{8.47}$$

Under the initial condition $\Gamma_1(\mathbf{x}_\perp', \mathbf{x}_\perp'', z = 0, \omega) = 1$, the solution for the mutual coherence function is given by

$$\Gamma_1(\mathbf{x}_\perp', \mathbf{x}_\perp'', Z, \omega) = \exp\left\{ -k^2 \left[A(0) - A(r_{1d}) \right]Z \right\} \tag{8.48}$$

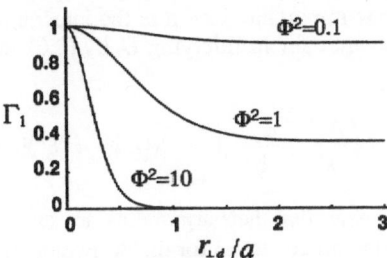

FIGURE 8.6. Decay of the mutual coherence function for the Gaussian ACF with increasing travel distance for different values of Φ^2.

For the Gaussian ACF,

$$\Gamma_1(\mathbf{x}_\perp',\mathbf{x}_\perp'',Z,\omega)=\exp\left\{-\Phi^2\left[1-\exp(-r_{1d}^{\,2}/a^2)\right]\right\} \tag{8.49}$$

The asymptotic behavior is given by

$$\begin{aligned}\Gamma_1(\mathbf{x}_\perp',\mathbf{x}_\perp'',Z,\omega)&\approx(1-\Phi^2)+\Phi^2\exp(-r_{1d}^{\,2}/a^2) && \text{for } \Phi^2\ll1 \\ &\approx\exp(-\Phi^2 r_{1d}^{\,2}/a^2)=\exp(-r_{1d}^{\,2}/a_\perp^2) && \text{for } \Phi^2\gg1\end{aligned} \tag{8.50}$$

where we introduce the transverse correlation distance at large travel distance:

$$a_\perp\equiv\frac{a}{\Phi}=\frac{a}{\sqrt{\sqrt{\pi}\varepsilon^2 aZk^2}} \tag{8.51}$$

Mutual coherence on a transverse plane decreases with increasing travel distance, as shown in Figure 8.6. This means that the coherence parallel to an isochron becomes smaller as the propagation distance increases. This characteristic will be used in the first-order expansion of A on the transverse plane in the following. Parameter Φ^2 characterizes not only the decay of the coherent wavefield itself, as shown in (8.41), but also the decay of mutual coherence in the transverse plane.

8.2.3 Two-Frequency Mutual Coherence Function

We define the two-frequency mutual coherence function at distance z as the correlation between different locations on the transverse plane and different angular frequencies at ω' and ω'' [Hong and Ishimaru, 1976] as

$$\Gamma_2(\mathbf{x}_\perp',\mathbf{x}_\perp'',z,\omega',\omega'')\equiv\left\langle U(\mathbf{x}_\perp',z,\omega')U(\mathbf{x}_\perp'',z,\omega'')^*\right\rangle \tag{8.52}$$

This coherence function is important since it is the integral kernel of the wavefield intensity in the frequency domain. Multiplying U by (8.6) and taking the ensemble average, we obtain

$$2i\partial_z\Gamma_2 + \left(\frac{\Delta_\perp{}'}{k'} - \frac{\Delta_\perp{}''}{k''}\right)\Gamma_2 - 2\langle(k'\xi'-k''\xi'')U'U''^*\rangle = 0 \tag{8.53}$$

where U' and U'' now mean that their arguments are $(\mathbf{x}_\perp{}',\omega')$ and $(\mathbf{x}_\perp{}'',\omega'')$, respectively. Using the same procedure as for the derivation of (8.39) and (8.44),

$$\langle(k'\xi'-k''\xi'')U'U''^*\rangle = -\frac{i}{2}\left[(k'^2+k''^2)A(0) - 2k'k''A(r_{\perp d})\right]\Gamma_2 \tag{8.54}$$

We finally obtain the master equation for the two-frequency mutual coherence function

$$2i\partial_z\Gamma_2 + \left(\frac{\Delta_\perp{}'}{k'} - \frac{\Delta_\perp{}''}{k''}\right)\Gamma_2 + i\left[(k'^2+k''^2)A(0) - 2k'k''A(r_{\perp d})\right]\Gamma_2 = 0 \tag{8.55}$$

We have to solve this differential equation to get the temporal change in the intensity of the wavefield; however, it is difficult to analytically solve (8.55) in general. It is possible to solve this equation under the condition that the waves are quasi-monochromatic.

8.2.4 Master Equation for Quasi-Monochromatic Waves

Since the random media are statistically homogenous, Γ_2 depends on the difference coordinates in the transverse plane, and we may put $\Delta_\perp{}'=\Delta_\perp{}''=\Delta_{\perp d}$ in (8.55). We study quasi-monochromatic waves having angular frequency centered around ω_c following Lee and Jokipii [1975a, b]. We introduce center-of-mass and difference coordinates in the wavenumber space as $k_c = (k'+k'')/2$ and $k_d = k'-k''$. Then, $k'^2+k''^2 \approx 2k_c^2 + k_d^2/2$, $k'k'' \approx k_c^2 - k_d^2/4$, and $1/k'-1/k'' \approx -k_d/k_c^2$. Corresponding coordinates for angular frequency will also be used. When we study propagation over a long travel distance, contributions to variations in Γ_2 come from only short offsets in the transverse plane, as shown in Figure 8.6. Substituting these coordinates in (8.55), we get

$$\partial_z\Gamma_2 + i\frac{k_d}{2k_c^2}\Delta_{\perp d}\Gamma_2 + k_c^2\left[A(0)-A(r_{\perp d})\right]\Gamma_2 + \frac{k_d^2}{2}A(0)\Gamma_2 = 0 \tag{8.56}$$

Lerche [1979] discussed the validity of the above equation for a wide spectral range.

We may factor Γ_2 into the following product

$$\Gamma_2 = {}_0\Gamma_2 \, e^{-k_d^2 A(0)z/2} \tag{8.57}$$

The master equation for ${}_0\Gamma_2$ is written as

$$\partial_z {}_0\Gamma_2 + i\frac{k_d}{2k_c^2}\Delta_{\perp d} \, {}_0\Gamma_2 + k_c^2\left[A(0) - A(r_{\perp d})\right]_0\Gamma_2 = 0 \tag{8.58}$$

We will examine the meaning of the above factorization in the following.

Intensity

We define the intensity of wavefield at distance z and time t as

$$
\begin{aligned}
I(z,t) &\equiv \left\langle u(\mathbf{x}_\perp, z, t)u^*(\mathbf{x}_\perp, z, t)\right\rangle \\
&= \frac{1}{(2\pi)^2}\int_{-\infty}^{\infty} d\omega' \int_{-\infty}^{\infty} d\omega'' \left\langle U(\mathbf{x}_\perp, z, \omega')U(\mathbf{x}_\perp, z, \omega'')^*\right\rangle e^{-i(\omega'-\omega'')(t-z/V_0)} \\
&= \frac{1}{(2\pi)^2}\int_{-\infty}^{\infty} d\omega_c \int_{-\infty}^{\infty} d\omega_d \, \Gamma_2\left(\mathbf{x}_{\perp d}=0, z, \omega', \omega''\right) e^{-i\omega_d(t-z/V_0)} \\
&= \frac{1}{2\pi}\int_{-\infty}^{\infty} d\omega_c \, \hat{I}(z,t;\omega_c)
\end{aligned} \tag{8.59}
$$

The intensity spectral density $\hat{I}$ is written as the integral over the difference angular frequency as

$$\hat{I}(z,t;\omega_c) = \frac{1}{2\pi}\int_{-\infty}^{\infty} d\omega_d \, \Gamma_2\left(\mathbf{x}_{\perp d}=0, z, \omega', \omega''\right) e^{-i\omega_d(t-z/V_0)} \tag{8.60}$$

where the integral kernel is the two-frequency mutual coherence function.

The intensity spectral density corresponding to only the exponential term of (8.57) is written as

$$\frac{1}{2\pi}\int_{-\infty}^{\infty} d\omega_d \, e^{-A(0)k_d^2 z/2} \, e^{-i\omega_d(t-z/V_0)} = \frac{\omega_c}{\sqrt{2\pi\Phi}} e^{-\frac{\omega_c^2(t-z/V_0)^2}{2\Phi^2}} \tag{8.61}$$

which represents the phase fluctuation. This term does not correspond to the broadening of the individual wave packet but shows the wandering effect from the statistical averaging of the phase fluctuations of different rays on the transverse plane at distance z [Lee and Jokipii, 1975b]. To study envelope broadening through a single realization, we need to find $_0\Gamma_2$ that solves (8.58). The intensity spectral density (8.60) for this term is given by

$$\hat{I}(z,t;\omega_c) = \frac{1}{2\pi} \int_{-\infty}^{\infty} d\omega_d \; _0\Gamma_2\left(\mathbf{x}_{\perp d} = 0, z, \omega', \omega''\right) e^{-i\omega_d(t-z/V_0)} \qquad (8.62)$$

This corresponds to the mean square of a bandpass-filtered trace, that is, the MS envelope.

For plane wave incidence $U = 1$, we have $_0\Gamma_2 = 1$ and $\hat{I}(z,t;\omega_c) = \delta(t - z/V_0)$ for $z < 0$. We may put the initial condition as

$$_0\Gamma_2\left(\mathbf{x}_{\perp d} = 0, z = 0, \omega', \omega''\right) = 1 \qquad (8.63)$$

Energy Conservation

Integrating (8.62) over time,

$$\int_{-\infty}^{\infty} dt\, \hat{I}(z,t;\omega_c) = \int_{-\infty}^{\infty} d\omega_d \; _0\Gamma_2\left(\mathbf{x}_{\perp d} = 0, z, \omega', \omega''\right) \delta(\omega_d) e^{+i\omega_d z/V_0}$$

$$= {_0\Gamma_2}\left(\mathbf{x}_{\perp d} = 0, z, \omega_d = 0\right) = 1 \qquad (8.64)$$

since $\partial_{z_0}\Gamma_2\left(\mathbf{x}_{\perp d} = 0, z, \omega_d = 0\right) = 0$ from (8.58) and initial condition (8.63). This confirms that energy is conserved in each angular frequency band at any distance $z > 0$.

8.2.5 MS Envelope

Using the fact that the contribution to the MS envelope at a long travel distance comes from a small transverse distance, as shown by the study of the mutual coherence given by (8.50) and Figure 8.6, we solve (8.58). We will then derive the MS envelope.

Gaussian ACF

For the Gaussian ACF, we may expand A given by (8.26) for a small transverse distance as

$$A(\mathbf{x}_\perp) = A(r_\perp) \approx \sqrt{\pi}\varepsilon^2 a\left[1 - (r_\perp/a)^2\right] \quad \text{for} \quad r_\perp \ll a \qquad (8.65)$$

Substituting this in (8.58),

$$\partial_{z0}\Gamma_2 + i\frac{k_d}{2k_c^2}\Delta_{\perp d0}\Gamma_2 + \frac{\Phi^2}{Z}\left(\frac{r_{\perp d}}{a}\right)^2 {}_0\Gamma_2 = 0 \tag{8.66}$$

We introduce the nondimensional transverse distance χ and longitudinal distance τ scaled by the transverse correlation distance $a_\perp$, as defined by (8.51), where k is replaced with k_c, and travel distance Z as

$$z = Z\tau \quad \text{and} \quad r_{\perp d} = a_\perp\chi \tag{8.67}$$

We define the characteristic wavenumber and characteristic time as

$$k_M \equiv \frac{2k_c^2 a_\perp^2}{Z} = \frac{2a}{\sqrt{\pi}\varepsilon^2 Z^2} \quad \text{and} \quad t_M \equiv \frac{1}{V_0 k_M} = \frac{\sqrt{\pi}\varepsilon^2 Z^2}{2V_0 a} = \frac{D\Phi^2}{8\omega_c} \tag{8.68}$$

Then, we may write (8.66) in nondimensional form as

$$\partial_{\tau 0}\Gamma_2 + i\left(\frac{k_d}{k_M}\right)\left(\partial_\chi^2 + \frac{1}{\chi}\partial_\chi\right)_0\Gamma_2 + \chi^2{}_0\Gamma_2 = 0 \tag{8.69}$$

We solve this differential equation following Sreenivasiah et al. [1976]. Under the initial condition ${}_0\Gamma_2(\tau = 0, \chi) = 1$, we want to find ${}_0\Gamma_2(\tau = 1, \chi = 0)$. First, we assume that the solution has the following form:

$$_0\Gamma_2(\tau, \chi) = \frac{e^{v(\tau)\chi^2}}{w(\tau)} \tag{8.70}$$

Then, (8.69) reduces to

$$\left[\frac{dv}{d\tau} + 4i\left(\frac{k_d}{k_M}\right)v^2 + 1\right]\chi^2 + \left[4i\left(\frac{k_d}{k_M}\right)v - \frac{1}{w}\frac{dw}{d\tau}\right] = 0 \tag{8.71}$$

Each term in brackets in (8.71) must be zero to satisfy the equation regardless of χ. The differential equation for $v(\tau)$ is a Riccati equation. Using the initial condition, $v(0) = 0$ and $w(0) = 1$,

$$v(\tau) = -v_0\tanh\frac{\tau}{v_0} \quad \text{and} \quad w(\tau) = \cosh\frac{\tau}{v_0} \tag{8.72}$$

where $v_0 = \frac{1}{2\sqrt{k_d/k_M}}e^{-3\pi i/4}$. Finally we obtain

$$_0\Gamma_2(\tau,\chi) = \dfrac{\exp\left(-\dfrac{e^{-3\pi i/4}\chi^2}{2\sqrt{k_d/k_M}}\tanh\left(2e^{3\pi i/4}\sqrt{k_d/k_M}\,\tau\right)\right)}{\cosh\left(2e^{3\pi i/4}\sqrt{k_d/k_M}\,\tau\right)} \tag{8.73}$$

At $\tau=1$ and $\chi=0$, it becomes

$$_0\Gamma_2(1,0) = \operatorname{sech}\left(2e^{3\pi i/4}\sqrt{k_d/k_M}\right) \tag{8.74}$$

Substituting (8.74) in (8.62), we get the intensity spectral density as

$$\hat{I}(Z,t;\omega_c) = \frac{1}{2\pi}\int_{-\infty}^{\infty} d\omega_d\,\operatorname{sech}\left(2e^{3\pi i/4}\sqrt{\frac{\omega_d}{V_0 k_M}}\right)e^{-i\omega_d(t-Z/V_0)} \tag{8.75}$$

To evaluate this integral, we use the following expansion [see Gradshteyn and Ryzhik, 1994, p. 44]:

$$\operatorname{sech} x = \sum_{n=0}^{\infty}\frac{(-1)^n(2n+1)\pi}{x^2+\left[(2n+1)\dfrac{\pi}{2}\right]^2} \tag{8.76}$$

There are poles on the lower half of the complex ω_d plane. We close the integral contour by following the upper or lower semicircle according to $t-Z/V_0 < 0$ or > 0. By using the residue integral, we get the following analytical solution

$$\hat{I}(Z,t;\omega_c) = H\left(t-\frac{Z}{V_0}\right)\frac{\pi}{4t_M}\sum_{n=0}^{\infty}(-1)^n(2n+1)e^{-\left(\frac{2n+1}{4}\pi\right)^2\frac{(t-Z/V_0)}{t_M}} \tag{8.77}$$

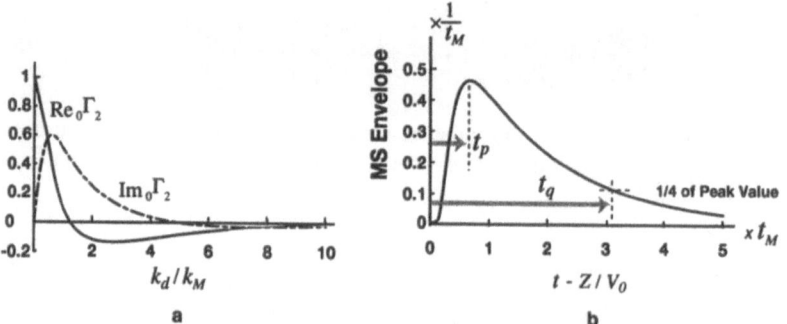

FIGURE 8.7. (a) Real and imaginary parts of $_0\Gamma_2(\tau=1,\chi=0)$. (b) Temporal evolution of the intensity spectral density corresponding to the MS envelope.

The time integral of the intensity spectral density is conserved at any travel distance since

$$\int_{-\infty}^{\infty} dt\, \hat{I}(Z,t;\omega_c) = \frac{4}{\pi}\sum_{n=0}^{\infty}(-1)^n \frac{1}{2n+1} = 1 \qquad (8.78)$$

[see Gradshteyn and Ryzhik, 1994, p. 9].

We plot the real and imaginary parts of $_0\Gamma_2(\tau=1, \chi=0)$ in Figure 8.7a and the resultant temporal trace of the intensity spectral density $\hat{I}$ in Figure 8.7b. This is the MS envelope. The time delay between the direct wave onset and the peak arrival t_p and the duration t_q, defined as the time between the onset and when the MS envelope amplitude decreases to one quarter of its peak value, are given by

$$t_p \approx 0.67 t_M \quad \text{and} \quad t_q \approx 3.11 t_M \qquad (8.79)$$

where t_M is defined as in (8.68). We note that t_q corresponds to the time when the RMS trace reaches half its maximum height.

General Form for the Longitudinal Integral of the ACF

Here we study the envelope shape for a more general form of the longitudinal integral of the autocorrelation function A as an extension of (8.65). We introduce the following functional form characterizing the transverse correlation of random media, as shown in Figure 8.8:

$$A(\mathbf{x}_\perp) = A(r_\perp) = \sqrt{\pi}\,\varepsilon^2 a \exp\left[-(r_\perp/a)^p\right]$$
$$\approx \sqrt{\pi}\,\varepsilon^2 a\left[1-(r_\perp/a)^p\right] \quad \text{for} \quad r_\perp \ll a \qquad (8.80)$$

As parameter p decreases, function A drops off more rapidly for small distances in a manner similar to an exponential and the corresponding PSDF in the transverse plane becomes richer in short-wavelength components compared with the Gaussian ($p=2$). We focus on the correlation or power spectra of random media on the transverse plane with respect to the mean propagation direction, where we do not insist on isotropic randomness. The master equation (8.58) becomes

$$\partial_{z0}\Gamma_2 + i\frac{k_d}{2k_c^2}\Delta_{\perp d}\,_0\Gamma_2 + \frac{\Phi^2}{Z}\left(\frac{r_{\perp d}}{a}\right)^p\,_0\Gamma_2 = 0 \qquad (8.81)$$

From the comparison with (8.66), it is necessary to introduce a correlation scale in the transverse plane appropriate to (8.80) instead of (8.51). We normalize the longitudinal distance and the transverse distance by using the transverse correlation distance redefined as an extension of (8.51):

$$z = Z\tau \text{ and } r_{Ld} = a_\perp \chi \tag{8.82}$$

where $a_\perp \equiv a\Phi^{-2/p} = a\left(\sqrt{\pi}\varepsilon^2 aZk_c^2\right)^{-1/p}$. The transverse correlation distance decreases with increasing travel distance, where the power of travel distance depends on $-1/p$. The characteristic wavenumber and the characteristic time are written as

$$k_M \equiv \frac{2k_c^2 a_\perp^2}{Z} = \frac{2k_c^2 a^2}{Z\left(\sqrt{\pi}\varepsilon^2 aZk_c^2\right)^{2/p}} \text{ and } t_M \equiv \frac{1}{V_0 k_M} = \frac{Z\left(\sqrt{\pi}\varepsilon^2 aZk_c^2\right)^{2/p}}{2V_0 k_c^2 a^2} \tag{8.83}$$

The characteristic time is proportional to the power $4/p$ of the fractional fluctuation, that is, scattering becomes stronger for a smaller p value. The dependence of characteristic time on wavenumber is power $(4/p) - 2$. This means there is no frequency dependence of MS envelope for the Gaussian ($p=2$), but the frequency dependence increases as the p value decreases. The non-dimensional master equation becomes

$$\partial_{\tau\,0}\Gamma_2 + i\left(\frac{k_d}{k_M}\right)\left(\partial_\chi^2 + \frac{1}{\chi}\partial_\chi\right)_0\Gamma_2 + \chi^p{}_0\Gamma_2 = 0 \tag{8.84}$$

Solving this equation numerically and substituting the solution at $\tau = 1$ and $\chi=0$ in (8.62), we can numerically simulate the intensity spectral density for different p values as illustrated in Figure 8.9 [Sato, 1989]. The peak becomes sharper and the duration decreases as the p value decreases.

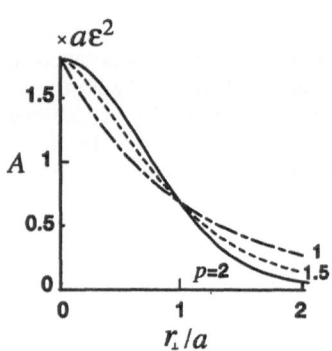

FIGURE 8.8. Plots of A for different p values, as defined in (8.80).

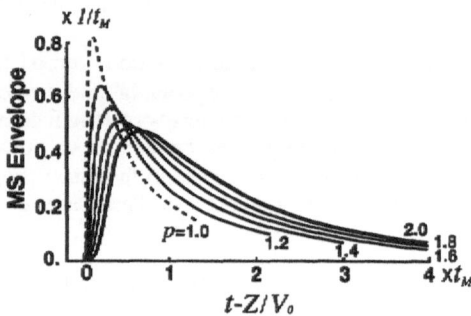

FIGURE 8.9. Temporal evolution of the intensity spectral density corresponding to the MS envelope for different p values. [From Sato, 1989, copyright by the American Geophysical Union.]

8.3 OBSERVED BROADENING OF S-WAVE SEISMOGRAM ENVELOPES

High-frequency seismogram envelopes around direct P- and S-arrivals have been characterized and modeled. While more observations have been reported of the S-wave portion of the seismogram, there have been some studies of the P-wave portion. The complexity of direct P-wave and P coda of nuclear explosions in 0.5–2.25 Hz band recorder at epicentral distances of 27°–44° was pointed out by Douglas et al. [1973]. They discussed two possibilities: greater absorption of the direct P-wave relative to the following P-coda and multipathing. On teleseismic P-waves recorded in the French Massif Central, Ritter et al. [1997] found a delay of high-frequency components (2–4 Hz) of about 2–4 s and longer duration compared with lower-frequency components. They interpreted the delay of the high-frequency wavelets to be caused by P-to-S scattering that takes place mostly in the strongly heterogeneous lower crust. Examining the long propagation distance (greater than 3000 km) P-wave signals from an explosion recorded at the NORSAR array, McLaughlin and Anderson [1987] found that 5 Hz-band signals arrive later than those in the 1 Hz band. Comparison with numerical simulations of P-wave propagation through a random medium characterized by a Gaussian ACF, they interpreted this observed velocity dispersion to be caused by randomness having multiple correlation distances. Analyzing array seismograms of crustal earthquakes in southern California and Nevada, Wagner [1997] reported that the P- and S-wave trains are composed predominantly of forward scattered waves with relatively little mode conversions.

Envelope broadening phenomena are more prominent in high-frequency S-wave seismograms recorded at long travel distances. Figure 2.30 shows horizontal NS-component velocity seismograms and their bandpass filtered RMS traces of magnitude 2.6 and 3.6 earthquakes recorded distances at 91 km and 175 km in southeastern Japan. The lithosphere in this region is thought to be very heterogeneous since the Pacific plate is subducting beneath the Eurasian plate. The strong heterogeneity is revealed from the microearthquake distribution and 3-D velocity tomography investigations of this region [Ishida and Hasemi, 1988]. The recording station ASO is located in northern Kanto, Japan (see Figure 8.10a), and has a 1 Hz natural frequency seismometer installed on hard chert. The time difference between the arrival of the maximum peak and the S-wave onset is much larger than the source duration time, which is thought to be much less than 1 s for earthquakes as small as magnitude 2.6 or 3.6. The delay in arrival time of the maximum amplitude signal relative to the direct arrival must be due to a path effect. Broadening of S waveforms is also found on continents. Atkinson [1993] reported an increase in the duration of S-wave trains with increasing travel distance in the distance range from 10 to 500 km in eastern North America. She defined the time duration of the initial wave packet as the interval containing 90% of the S-wave energy of 1 to 10 Hz seismograms. She reported that the duration T_d[s] is proportional to travel distance although the scatter of data is large. She suggested that the observed duration time is related to the reciprocal of the corner frequency f_C[Hz] of the earthquake source spectrum and propagation distance r [km] by the linear relation: $T_d = 1/f_C + 0.05r$.

Here, applying the model developed in Section 8.2 for the envelope broadening to observed S-wave data, we will estimate the strength and the scale of the long wavelength component of the inhomogeneity in the lithosphere.

8.3.1 Envelope Broadening Observed in Kanto, Japan

First, we examine the broadening of S-wave seismogram envelopes (NS component) observed at station ASO in Kanto, Japan [Sato, 1989]. Figure 8.10a shows epicenters of 103 earthquakes with local magnitudes from 2 to 4.5 and hypocentral distances ranging from 80 km to about 300 km used for the analysis. Crustal earthquakes with focal depths shallower than 30 km were excluded from the data set, since the Moho head wave might be larger than the direct S-wave through the crust for shallow events. As shown by two examples in Figure 2.30, there is a time lag between the S-wave onset and the maximum amplitude arrival. We can interpret the time lag as caused by the contribution of diffraction and multiple forward scattering due to long-wavelength components of velocity inhomogeneity. We define two quantities that are independent of the absolute amplitude as illustrated in Figure 2.30: time lag t_p [s], equal to the difference in time between the arrival of the

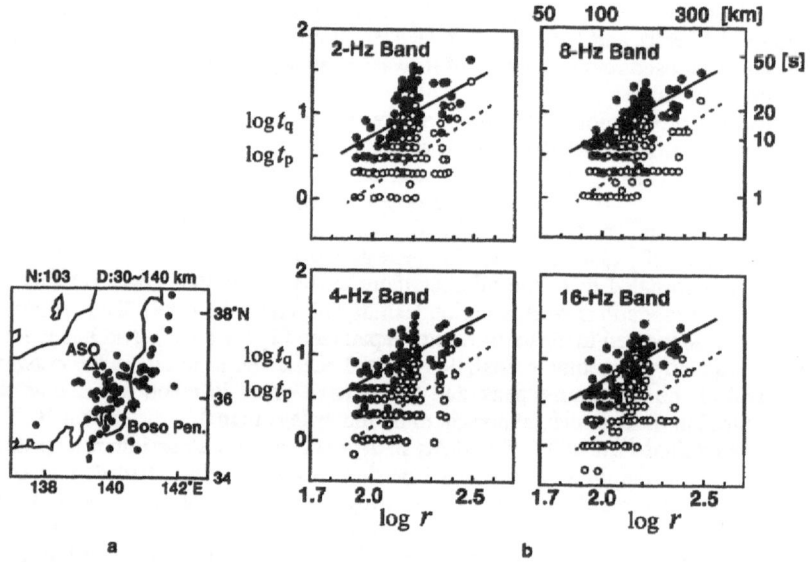

FIGURE 8.10. (a) Station ASO and epicenter distribution of earthquakes having focal depths between 30–140 km in SE Honshu, Japan that were used in the analysis of S-wave envelope characteristics. (b) Plots of $\log t_p$ (open circles) and $\log t_q$ (closed circles) against $\log r$. Solid and broken lines are predicted by the model (8.85) using the Gaussian ACF and the sum of intrinsic and scattering attenuation parameterized by $Q_s^{-1} \approx 0.014 f^{-1}$. [From Sato, 1989, copyright by the American Geophysical Union.]

maximum RMS amplitude and the direct S-wave arrival, and the envelope duration time t_q[s], which is the time between the direct S-wave onset and the time when the envelope has decreased to half the maximum RMS amplitude (the quarter-maximum power spectrum). These times were read from RMS traces in octave-width frequency bands. Figure 8.10b shows the measured times plotted against hypocentral distance r [km] for four frequency bands. Even though there is considerable scatter, a positive correlation appears without any correction for radiation pattern. The slopes of regression lines for $\log t_p$ against $\log r$ range from 1.35 to 1.59, and those for $\log t_q$ range from 1.68 to 1.82. The delay of the arrival time of the peak amplitude increases and the envelope broadens with increasing hypocentral distance. We note that t_p and t_q are much longer than the source duration empirically predicted from local magnitude.

Analysis Using the Gaussian ACF and Q_S^{-1}

When we model the observations shown in Figure 8.10 with the theory developed in Section 8.2 using the parabolic approximation for a plane wave incident on a random medium characterized by a Gaussian ACF, distance Z should be replaced with hypocentral distance r. As given by (8.79), both t_p and t_q are proportional to t_M, which is proportional to the square of r from (8.68). However, plots of data in Figure 8.10b show that the observed power of r is less than two.

Recall that we neglected attenuation in the formulation of envelope synthesis. To include attenuation, we multiply $\exp[-Q_S^{-1}\omega t]$ by the intensity spectral density (8.77), where we interpret the attenuation as the sum of intrinsic absorption and scattering attenuation due to short-wavelength components of inhomogeneities. This means that our model is appropriate for only the long-wavelength components of the real earth medium. Based on the analytical solution, we show in Figure 8.11

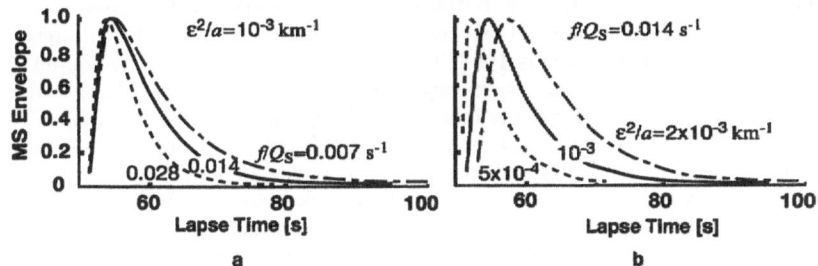

FIGURE 8.11. Temporal character of the synthesized intensity spectral density corresponding to the MS envelope at r=200 km, where lapse time is measured from the origin time. Each trace is normalized by its maximum value. (a) For different values of attenuation and a fixed character of inhomogeneity. (b) For different characterizations of inhomogeneity and a fixed attenuation. All results are for a Gaussian ACF. [From Sato, 1991a, with permission from Elsevier Science - NL, Sara Burgerhartsraat 25, 1055 KV, Amsterdam, The Netherlands.]

how inhomogeneity and attenuation control the temporal trace of the intensity spectral density in random media characterized by a Gaussian ACF. As expected, attenuation decreases the envelope duration time but does not affect the arrival time of the envelope peak. The amount and scale of inhomogeneity changes both the peak delay and the envelope duration. Choosing $Q_S^{-1} \approx 0.014 f^{-1}$ as observed in Kanto, Japan [Sato, 1984a], we numerically synthesize the intensity spectral density and estimate both t_p and t_q as functions of ratio ε^2 / a and travel distance r:

$$\log t_p = -0.73 + 0.86 \log \frac{\varepsilon^2}{a} + 1.71 \log r$$
$$\log t_q = \quad 0.05 + 0.68 \log \frac{\varepsilon^2}{a} + 1.36 \log r$$

(8.85)

where $V_0 = 4$ km/s for S-waves. The introduction of attenuation reduces the predicted coefficient of $\log r$ to be less than two, which is more consistent with the observed relationship between time and travel distance shown in Figure 8.10b.

Applying (8.85) to observed data plotted in Figure 8.10b, we estimate the ratio ε^2 / a. The values of ε^2 / a found by fitting t_p and t_q are nearly the same and independent of frequency band: $\varepsilon^2 / a \approx 10^{-2.98 \pm 0.32}$ km^{-1}. The frequency independence means that the Gaussian ACF is a reasonable choice to statistically characterize the inhomogeneity.

Full Envelope Inversion for the Character of Inhomogeneity and Q_S^{-1}

Scherbaum and Sato [1991] removed the a priori assumption of the specific frequency dependence of attenuation Q_S^{-1} and modeled the entire SH seismogram envelopes of earthquakes in southeastern Honshu using the theory outlined in Section 8.2. They used t_M, Q_S^{-1}, the onset time of the S-arrival and the gain factor as model parameters and applied an inversion scheme based on the Marquardt–Levenberg method [Marquardt, 1963] to model the observed MS seismogram envelopes in four octave-width frequency bands (see Figure 8.12a). The Gaussian ACF was used to characterize the inhomogeneities. The time window for analysis was extended beyond the arrival time of the maximum peak amplitude by up to half the travel time of the direct S-wave. S-wave attenuation obtained from the inversion agrees well with the result of the earlier attenuation study in the same region, $Q_S^{-1} = 0.014 f^{-1}$. The estimated f / Q_S shows a strong scatter for distances smaller than 150–200 km, where $t_M < 2$ s; however, the scatter reduces rapidly with increasing distance, that is, increasing characteristic time $t_M \gg 2$ s. They estimated $\varepsilon^2 / a \approx 10^{-3.27 \pm 0.32}$ km^{-1} (see Figure 8.12b), which is smaller than the previous estimate. The smaller estimated value may have resulted because they discarded some complex seismograms from earthquakes that took place beneath the Boso Peninsula.

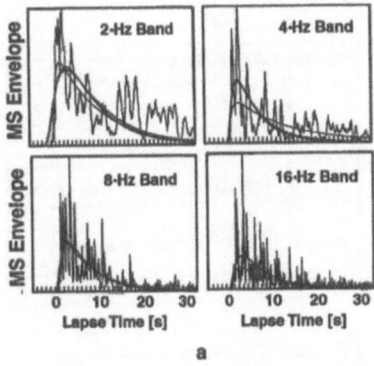

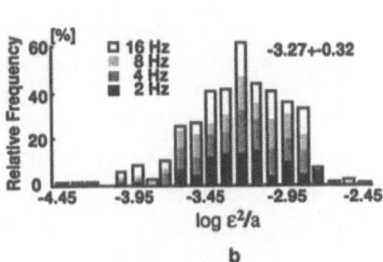

a

b

FIGURE 8.12. (a) Example of the inversion of full MS seismogram envelopes re-corded at station ASO, where time is measured from the S-wave onset. Smoothed curves correspond to differing constraints used in the inversion. (b) Histogram of $\log \varepsilon^2 / a[\text{km}^{-1}]$ estimated for SE Honshu, Japan from inversion of many MS enve-lope traces. All results are for a Gaussian ACF. [From Scherbaum and Sato, 1991, copyright by the American Geophysical Union.]

8.3.2 Differences of Random Inhomogeneities across the Volcanic Front in the Kanto–Tokai District, Japan

On the island arc of Japan, the sharp boundary defined by the distribution of volcanoes running from north to south on the fore-arc side of the island is called the volcanic front (VF) [Sugimura, 1960]. The VF is located at the projection onto the surface of the 110 km isodepth contour of seismicity associated with the subduct-ing Pacific plate [Tatsumi, 1986] and is shown by a bold broken line in Figure 8.13a. The distribution of heat flow is quite different on either side of the front [Yuhara, 1973]. The upper mantle above the descending plate on the back-arc side is an aseismic region called a mantle wedge. Seismogram envelope broadening in this region was studied by Obara and Sato [1995], who found differences in wave-form characteristics depending upon which side of the VF the propagation path followed. Earthquakes occurring along the subducting Pacific plate with depths ranging from 80 to 500 km were observed at 73 stations of the NIED seismic ob-servation network and were used in this analysis (see Figure 8.13a).

Figure 8.13b shows seismogram envelopes recorded at two stations for event E in Figure 8.13a that took place at a depth of 201 km beneath the southern end of the Izu Peninsula. The top panel of Figure 8.13b shows the observed EW component seismogram for station KGN and the RMS traces of the seismogram after band-pass filtering using central frequencies of 1, 2, 4, 8 and 16 Hz. The lower panel shows those observed at station KIB located near the Pacific coast. At KIB, enve-lope shapes are impulsive at all frequencies, and the envelope broadening is inde-pendent of frequency, as seen at ASO (see Figure 8.12a). On the other hand, at sta-tion KGN, which is located west of the VF, we find a frequency dependence of en-velope broadening: envelopes at this station are impulsive at low frequencies like

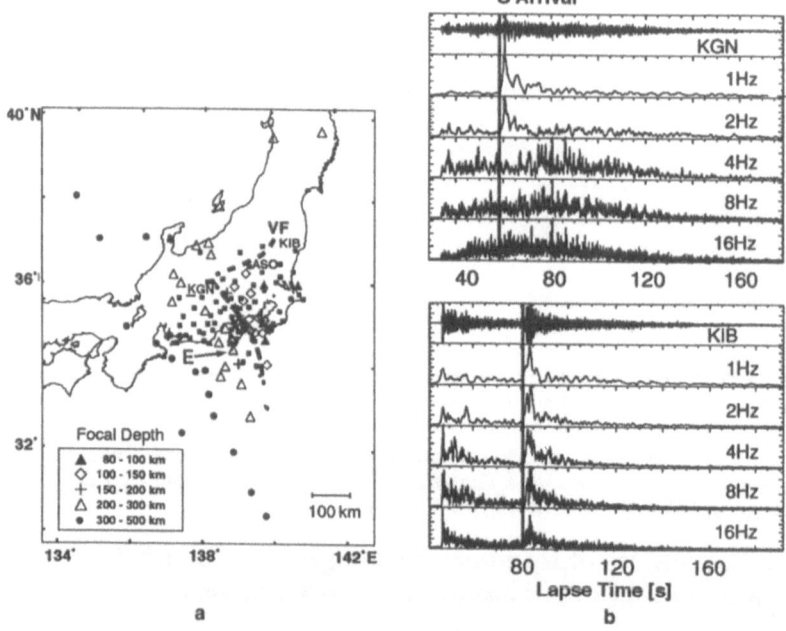

FIGURE 8.13. (a) Epicenter distribution of 58 earthquakes in the Kanto–Tokai area, Japan used in the analysis and 73 seismic stations (solid squares) of the NIED network. (b) Horizontal component (EW) seismograms and their RMS octave-width bandpass-filtered traces at two stations KGN and KIB for event E at a focal depth of 201 km beneath the location indicated on the map. Each trace is normalized by its maximum RMS amplitude. Vertical bars indicate S-wave onsets. [From Obara and Sato, 1995, copyright by the American Geophysical Union.]

the 1- and 2-Hz band, however, they look spindle-like at frequencies higher than 4 Hz. The frequency dependence of the observed envelope broadening is different from station to station even for the same earthquake.

Distance Dependence and Frequency Dependence

Obara and Sato [1995] read the differences between the S-wave onset and the arrival time of the maximum amplitude t_p and the time when the envelope had decayed to half of the maximum amplitude t_q in each frequency band. Figure 8.14 shows plots of log t_p (cross) and log t_q (open circle) against log hypocentral distance r for each frequency band at stations KGN and KIB. Solid lines and broken lines are linear regression lines calculated for log t_p and log t_q, respectively:

$$\log t_p = A_{pR}^{obs} + B_{pR}^{obs} \log r$$
$$\log t_q = A_{qR}^{obs} + B_{qR}^{obs} \log r \qquad (8.86)$$

At station KIB the correlation coefficients are small. The slope of the regression line for KGN is steep and the correlation coefficient is large. Data for the 1 Hz band from most stations studied have considerable scatter; however, the data scatter becomes smaller with increasing frequency. From plots at 73 stations, Obara and Sato [1995] found that the slope of the regression shows a clear increase from east to west indicating that the envelope width has a stronger dependence on hypocentral distance as observation location moves from east to west. The hypocentral distance dependence is not clear east of the VF.

Figure 8.15 shows the range of time lag t_p and envelope width t_q obtained using data from many source-receiver measurements, where each measurement is normalized by the value at 2 Hz for the same source–receiver combination. Normalizing by the measurements at 2 Hz has removed the distance dependence shown in Figure 8.14. These plots show a frequency dependence of envelope broadening that is independent of travel distance. At stations KIB and KTU, the frequency dependence of the normalized time lags is weak. At stations YMK and ENZ located near the VF, normalized time lags increase slightly with increasing frequencies. At JIZ, HDA, HTN, KGN, and GER, all located west of the VF, the frequency dependence becomes strong. To compare the frequency dependence of t_p and t_q at each station, taking 2 Hz as the reference frequency, we calculate the

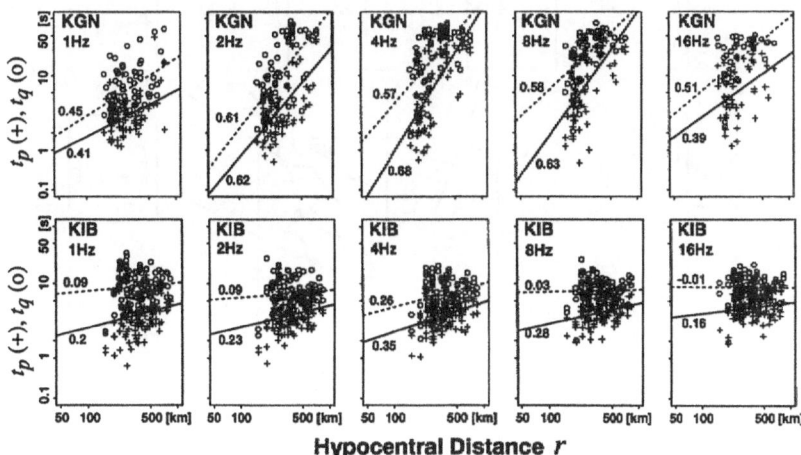

Hypocentral Distance r

FIGURE 8.14. Log-log plots of t_p (cross) and t_q (open circle) against hypocentral distance r at stations KGN and KIB. Solid and broken lines are linear regression lines for t_p and t_q, respectively. Numerals are correlation coefficients. [From Obara and Sato, 1995, copyright by the American Geophysical Union.]

linear regressions for t_q and t_q against frequency f in Hz:

$$\log \frac{t_p(f)}{t_p(2)} = A_{pf}^{obs} + B_{pf}^{obs} \log f$$

$$\log \frac{t_q(f)}{t_q(2)} = A_{qf}^{obs} + B_{qf}^{obs} \log f$$

(8.87)

We plot regression coefficients B_{pf}^{obs} and B_{qf}^{obs} at each station in Figure 8.16, where the size of symbol indicates the value of the regression coefficient. At stations along the Pacific coastline, the envelope broadening is independent of frequency. However, frequency dependence increases from east to west across the network with the VF acting as a sharp boundary.

Figure 8.17 schematically illustrates the characteristics of the envelope width on the west-east section. The frequency dependence of the envelope broadening is not observed at all stations when the earthquake occurs east of the VF. However, the frequency dependence becomes stronger at stations near and west of the VF, when the earthquakes occurs on the back-arc (west) side. Therefore, the frequency dependence of the envelope broadening suggests that the velocity structure of the mantle wedge west of the VF is more inhomogeneous than that to the east, because the envelope broadening is more conspicuous for seismic rays that have traveled through the western region.

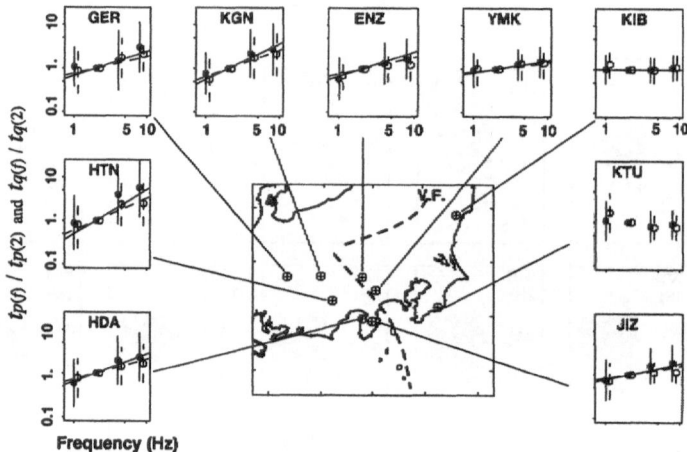

FIGURE 8.15. Frequency dependence of t_p (solid circle) and t_q (open) obtained using data from all events shown in Figure 8.13a. Results are normalized by values at 2 Hz. Vertical bars indicate standard deviation. Solid and dashed slant lines are best fit theoretical frequency dependence for t_p and t_q, respectively, for estimated p defined in (8.80). [From Obara and Sato, 1995, copyright by the American Geophysical Union.]

Estimation of the Spectral Structure of the Inhomogeneity

Using the theory outlined in Section 8.2.5, we can interpret the observed differences in the frequency dependence of envelope broadening as being caused by a
difference in the spectral structure of the velocity inhomogeneity. Obara and Sato
[1995] removed the a priori assumption that the Gaussian ACF is appropriate to
explain the velocity inhomogeneity. They assumed a general form for A the longitudinal integral of the autocorrelation function, using parameter p defined in (8.80).
As p decreases, function A becomes sharper at zero lag-distance and the PSDF
becomes richer in short-wavelength components. Replacing Z by r in (8.83), we
get the characteristic time $t_M \propto \varepsilon^{4/p} f^{4/p-2} r^{2/p+1}$. We know that $t_M \propto \varepsilon^2 f^0 r^2$ for the
Gaussian case (p=2). On the other hand, $t_M \propto \varepsilon^4 f^2 r^3$ for p=1. Thus, a smaller
value of p gives a stronger frequency dependence for envelope broadening and
more envelope broadening in general. The theoretical prediction of more envelope
broadening and stronger frequency dependence is in good qualitative agreement
with the observed data. Figure 8.18 shows numerical simulations of RMS envelopes for different frequency bands at a distance of 200 km for $Q_S^{-1} = 0.014 f^{-1}$.
For the Gaussian case (p=2), the envelope is the same for all frequency bands;
however, envelopes for higher frequencies are broadened more than lower frequency envelopes as p becomes smaller.

In Figure 8.19 we show the value of p at each station that was determined
from the frequency dependence of t_p and t_q shown in Figure 8.16. In general, at
stations located east of the VF, p is close to 2 and function A is Gaussian. Near the
VF, parameter p takes a value ranging from 1.6 to 1.8. West of the VF, p is

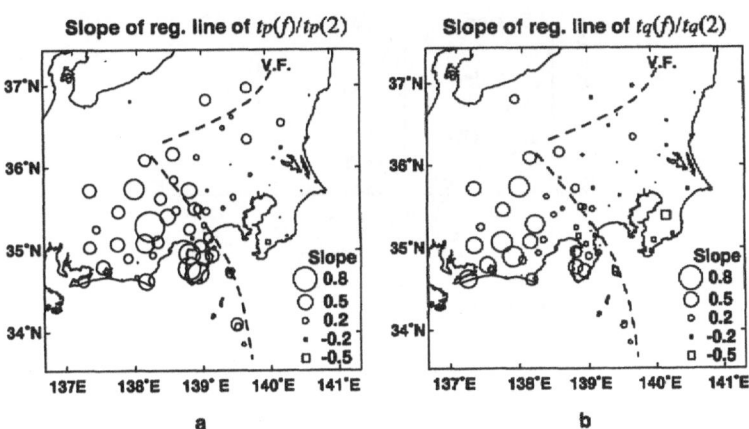

FIGURE 8.16. Frequency dependence (a) B_M^{obs} and (b) B_q^{obs} defined in (8.87) measured at stations in the Kanto–Tokai region, Japan. Larger circles indicate stronger
frequency dependence and squares show negative gradients. [From Obara and
Sato, 1995, copyright by the American Geophysical Union.]

smaller than 1.6 and function A is exponential-like. This means that the random inhomogeneity west of the VF is relatively richer in short-wavelength components compared to that east of the VF. East of the VF, Obara and Sato [1995] estimated $\varepsilon^2 / a \approx 10^{-3.0} \sim 10^{-3.8}$ km^{-1} and $\varepsilon^2 \approx 10^{-2.0} \sim 10^{-2.8}$ assuming that $a=10$ km. This estimate of the fractional fluctuation is larger than obtained by analysis of teleseismic P-waves by Aki [1973] and Capon [1974] at Montana LASA, who estimated $\varepsilon^2 \approx 10^{-2.8} \sim 10^{-3.44}$ using $a=10$ km. The differences may reflect the differences in tectonic settings. The smaller p value found west of the VF qualitatively agrees with power-law spectra found under NORSAR [Flatté and Wu, 1988]. Beneath the back-arc region of the VF, the existence of partially molten mantle diapirs has been proposed by Tatsumi [1989]. Such diapirs, which are partially liquid inclusions, might appear as strong short-wavelength components of the inhomogeneity.

In the study of Obara and Sato [1995], the spectral characteristics of the inhomogeneity were mainly determined from the frequency dependence of the envelope broadening. The predicted travel-distance dependence of envelope width calculated for the p value west of the VF agrees well with the observed distance dependence. On the other hand, there are few observations of the distance dependence of envelope width and they do not agree with predictions for stations along the Pacific coast and east of the VF. The large observed envelope broadening on the back-arc side of the VF means that the randomness may be strong enough to violate the region of validity of the parabolic approximation, that is, the forward scattering approximation. The theory developed in Section 8.2 is based on the assumption that a plane wave was incident on a half-space containing the inhomogeneity whereas we used observations from an earthquake source. Using a method proposed by Williamson [1972, 1975] to model radiation from a point source in a 3-D scattering

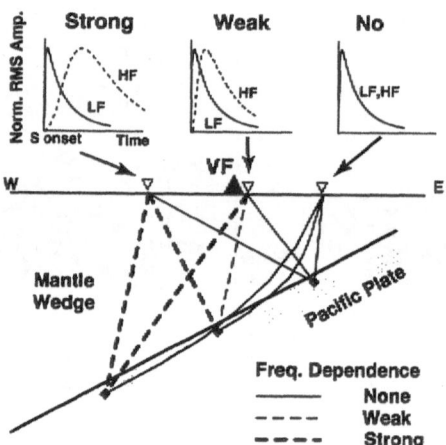

FIGURE 8.17. Schematic illustration of the relationship between tectonic setting and regions showing strong, weak, and no frequency dependence of S-wave envelope broadening in Kanto–Tokai, Japan. [From Obara and Sato, 1995, copyright by the American Geophysical Union.]

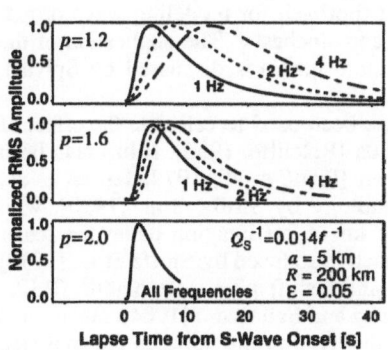

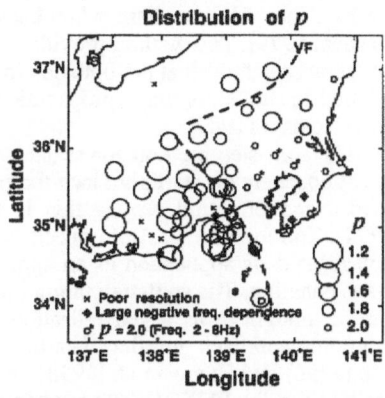

FIGURE 8.18. Numerical simulation of RMS envelopes for different p values in function A in (8.80) (see Figure 8.8). Each trace is normalized by its maximum value. [From Obara and Sato, 1995, copyright by the American Geophysical Union.]

FIGURE 8.19. Plot of p values estimated from the frequency dependence of t_p and t_q. [From Obara and Sato, 1995, copyright by the American Geophysical Union.]

medium using a correct geometrical spreading factor, Gusev and Abubakirov [1995] numerically synthesized seismogram envelopes using Monte Carlo simulations for a wide range of travel distances. The theory outlined in this chapter needs to be extended to use more realistic point-source models. Additional developments should include a method to account for full elastic wave scattering including P-S conversions at wide angles in a 3-D medium with a broad spectrum of inhomogeneities.

8.4 SPLIT-STEP FOURIER METHOD FOR MODELING WAVE PROPAGATION THROUGH AN INHOMOGENEOUS MEDIUM

The split-step Fourier, or phase-screen, method provides a deterministic approach for calculating one-way wave propagation in inhomogeneous media. The method is a marching algorithm, whereby the wavefield is extrapolated along one dimension across thin parallel layers. It can be shown that the wavefield can be propagated in a two-step process. In the first step, the wavefield is propagated from one face of the layer to the next through a constant background medium whose properties are chosen to be the mean of the properties of the medium within the layer. In the second step, the phase of the wavefield is corrected for the effects of the inhomogeneities within the layer. The phase is corrected by collapsing the inhomogeneities onto a screen and calculating the phase change at each point on the layer face. The phase factor is computed by integrating the medium properties along lines normal to the screen. The split-step method provides a discretized version of the path integral formalism when back scattering is neglected [Flatté et al., 1979; Dashen, 1979]. The method provides a natural approach for modeling wave propagation in structures that are both deterministic and stochastic. The application of the method to modeling wave propagation in random media is discussed by Spivack and Uscinski [1989].

The split-step method for scalar waves has been used to calculate the effect of the atmosphere on starlight since the mid-1950s [Ratcliffe, 1956]. It has also been used to model sound propagation in the ocean [Flatté et al., 1979; Jensen et al., 1994]. The method was introduced into seismology by Stoffa et al. [1990], who investigated its application as an approach for migration imaging using the scalar wave equation. The split-step migration approach introduced by Stoffa et al. [1990] can be thought of as a generalization of the phase-shift plus interpolation (PSPI) migration approach introduced earlier by Gazdag and Sguazzero [1984]. Huang and Wu [1996] and Huang et al. [1998] discussed the relationship between the split-step method and the PSPI algorithm when doing migration.

We will briefly develop the method for scalar waves propagating through a region having velocity fluctuations with constant density. The mathematical development is similar to that given in Section 8.1, but we do not use the parabolic equation from the start. We introduce the Fourier transform of wavefield with respect to time as

$$u(\mathbf{x},t) = \frac{1}{2\pi} \int_{-\infty}^{\infty} u(\mathbf{x},\omega) e^{-i\omega t} d\omega \qquad (8.88)$$

We use the scalar wave equation for $u(\mathbf{x},\omega)$ in inhomogeneous media having velocity $V(\mathbf{x}) = V_0[1 + \xi(\mathbf{x})]$ from (8.1):

$$\left[\Delta + \frac{\omega^2}{V(\mathbf{x})^2}\right] u(\mathbf{x},\omega) = 0 \qquad (8.89)$$

In the case of small fluctuation $|\xi| << 1$, we may write the above equation as

$$\left(\Delta + k^2\right) u(\mathbf{x},\omega) = 2k^2 \xi(\mathbf{x}) u(\mathbf{x},\omega) \qquad (8.90)$$

where $k = \omega / V_0$. The wavefield having mean propagation in the z direction can be written as a 2-D Fourier transform on the transverse plane at z as

$$u(\mathbf{x}_\perp, z, \omega) = \frac{1}{(2\pi)^2} \int_{-\infty}^{\infty} \int_{-\infty}^{\infty} \tilde{u}(\mathbf{k}_\perp, z, \omega) e^{i\mathbf{k}_\perp \mathbf{x}_\perp} d\mathbf{k}_\perp \qquad (8.91)$$

where we introduced transverse coordinates $\mathbf{x}_\perp$. We suppose that ξ varies smoothly over distances of the order of a wavelength: $ak >> 1$, where a is the characteristic scale of the inhomogeneity. We try to find a solution at $z + \Delta z$

$$u(\mathbf{x}_\perp, z + \Delta z, \omega) = \frac{1}{(2\pi)^2} \int_{-\infty}^{\infty} \int_{-\infty}^{\infty} \tilde{u}(\mathbf{k}_\perp, z, \omega) e^{i\mathbf{k}_\perp \mathbf{x}_\perp + i\varphi(\mathbf{x}_\perp, z, \Delta z, \mathbf{k}_\perp, \omega)} d\mathbf{k}_\perp \qquad (8.92)$$

where we divide the medium into thin slabs perpendicular to the z-axis and φ is a phase that varies on the transverse plane. This representation assumes that we will be able to calculate the wavefield at $z + \Delta z$ from the 2-D Fourier transform of the wavefield in the transverse plane at z by using phase function φ and is similar to the Rytov approximation in Section 8.1.2.

Substituting (8.92) in (8.90), the equation for φ is

$$\left[i\nabla^2\varphi - (\mathbf{k}_\perp + \nabla_\perp\varphi)^2 - (\partial_z\varphi)^2 + k^2(1 - 2\xi) \right] = 0 \qquad (8.93)$$

Neglecting $\nabla^2\varphi$ and the variation with respect to transverse coordinates $(\nabla_\perp\varphi)^2$ since ξ varies smoothly, we obtain

$$\partial_z\varphi = \sqrt{k^2(1 - 2\xi) - k_\perp^2} \qquad (8.94)$$

Integrating (8.94), we have

$$\varphi(\mathbf{x}_\perp, z, \Delta z, \mathbf{k}_\perp, \omega) = \int_z^{z+\Delta z} \sqrt{k^2[1 - 2\xi(\mathbf{x}_\perp, z')] - k_\perp^2} \, dz' \qquad (8.95)$$

Substituting (8.95) into (8.92), we get the wavefield at $z + \Delta z$

$$u(\mathbf{x}_\perp, z + \Delta z, \omega) = \frac{1}{(2\pi)^2} \int_{-\infty}^{\infty} \int_{-\infty}^{\infty} \tilde{u}(\mathbf{k}_\perp, z, \omega) e^{i\mathbf{k}_\perp \mathbf{x}_\perp + i\int_z^{z+\Delta z} \sqrt{k^2[1 - 2\xi(\mathbf{x}_\perp, z')] - k_\perp^2} \, dz'} d\mathbf{k}_\perp \qquad (8.96)$$

To evaluate (8.96) in an efficient manner, we use the dispersion relation for the background medium $k_z = \sqrt{k^2 - k_\perp^2}$ and expand the square root terms in the integral of (8.95) for $k_\perp \ll k$ to get

$$\varphi(\mathbf{x}_\perp, z, \Delta z, \mathbf{k}_\perp, \omega) \approx k_z\Delta z - k \int_z^{z+\Delta z} \xi(\mathbf{x}_\perp, z') dz' \qquad (8.97)$$

In this case, the term describing the inhomogeneity can be removed from the 2-D Fourier transform and (8.96) becomes

$$u(\mathbf{x}_\perp, z + \Delta z, \omega) = e^{-ik\int_z^{z+\Delta z} \xi(\mathbf{x}_\perp, z') dz'} \frac{1}{(2\pi)^2} \int_{-\infty}^{\infty} \int_{-\infty}^{\infty} \tilde{u}(\mathbf{k}_\perp, z, \omega) e^{i\mathbf{k}_\perp \mathbf{x}_\perp + ik_z\Delta z} d\mathbf{k}_\perp \qquad (8.98)$$

This is an approximation for wave propagation through a thin slab that can be numerically calculated using a 2-D FFT. This is a dual-domain (wavenumber-space) implementation of the split-step method. The interaction between the wavefield and the inhomogeneities in the thin layer is accomplished by multiplication of a phase function with the wavefield in the space domain. The propagation from one thin slab to the next through the background medium is simply a phase shift in the wavenumber domain. Thus all the modeling is done by multiplication; Fourier transforms are used to switch between the wavenumber and space domains. Given a grid of velocity distribution, the medium is equated to a series of phase screens with interval Δz; the wavefield is marched in the forward direction in increments of Δz. When using the split-step method, the increment Δz can be variable depending on the amount of inhomogeneity in the medium. For example, propagation through regions that are homogeneous can be done with very large values of Δz. This means that computational expense can be focused on regions of the model where the greatest inhomogeneity exists.

It should be noted that Claerbout [1985] calls (8.6) a 15° approximation, which means that the equation provides a good approximation to the wave equation for waves that are propagating within 15° of a reference direction. Claerbout [1985] discussed extensions of (8.6) that are also parabolic equations but which have a larger angle of validity than does (8.6). There are several derivations of the split-step Fourier method that use different approximations and have slightly different versions of the propagator equations. The range of validity of three of the methods is discussed by Thomson and Chapman [1983] and they conclude that the propagator that we derived has the widest angular range of reliability of the three methods they discuss. The method derived directly from the parabolic wave equation (8.6), which is given in Jensen et al. [1994, p. 376], has a narrower range of applicability than the one we derived. Our derivation is not completely rigorous. Rigorous derivations are given by Stoffa et al. [1990] and Jensen et al. [1994].

Since the split-step method is an approximate numerical approach for modeling wave propagation, it is important to understand the limitations of the method. Stability and accuracy have been discussed in the oceanic acoustics literature [Jensen et al., 1994] and the seismology literature [Cheng et al., 1996; Huang and Fehler, 1998]. Jensen et al. [1994] show that the method can be applied in a manner in which the dominant terms in the error relate to $(\Delta z)^3$ but also depend on the variation in medium properties in the plane perpendicular to the mean propagation direction. This means that the error can be made smaller by choosing the increment Δz smaller. The method is unconditionally stable, which means that numerical errors when doing computations do not grow with increasing propagation distance to cause the predicted wavefield to blow up [Jensen et al. 1994]. Huang and Fehler [1998] investigated the accuracy of the method as a function of $k_\perp / k$, or the propagation direction relative to the main propagation direction, and the size of the medium inhomogeneity. They found that when the velocity perturbation in the $x - y$ plane is as large as a factor of two, the method is reliable for propagation angles as large as 20° from the main propagation direction. When velocity perturbation is less than 10%, the method is reliable for propagation angles as large as 60° from the main propagation direction provided that Δz is chosen to be much smaller than the wavelength.

Wu et al. [1995] discuss an extension of the split-step method that does not require the small angle approximation to be computationally efficient. The wide angle method is derived using the Born approximation and is reliable for larger propagation angles than the approach developed here [Huang et al., 1998]. Applying the Born approximation within each layer, a solution similar to (8.98) is found:

$$u\left(\mathbf{x}_{\perp}, z + \Delta z, \omega\right) = \frac{1}{(2\pi)^2} \int_{-\infty}^{\infty}\int_{-\infty}^{\infty} \tilde{u}\left(\mathbf{k}_{\perp}, z, \omega\right) e^{i k_{\perp} x_{\perp} + i k_z \Delta z} d\mathbf{k}_{\perp}$$

$$- i k^2 \int_{z}^{z+\Delta z} \xi(\mathbf{x}_{\perp}, z')dz' \frac{1}{(2\pi)^2} \int_{-\infty}^{\infty}\int_{-\infty}^{\infty} \frac{1}{k_z}\tilde{u}\left(\mathbf{k}_{\perp}, z, \omega\right) e^{i k_{\perp} x_{\perp} + i k_z \Delta z} d\mathbf{k}_{\perp}$$

(8.99)

where the first term on the right-hand side is the wavefield after propagation through the background medium and the second term corresponds to the scattered wavefield. When velocity perturbation is small and propagation is in nearly the z direction, (8.99) reduces to (8.98). The approach based on the Born approximation is not guaranteed to be stable in the presence of large velocity contrasts but does extend the range of applicability of the method to larger propagation angles. Huang et al. [1998] present some examples showing the benefits of the enhanced method.

Liu and Wu [1994] compared the synthetic seismograms using the split-step method with seismograms calculated using fourth-order finite difference and eigenfunction expansion methods for scalar wave propagation through 2-D media containing discrete heterogeneities and random fluctuations. Stoffa et al. [1990] tested the method for seismic migration and found that it performed well even in the pres-

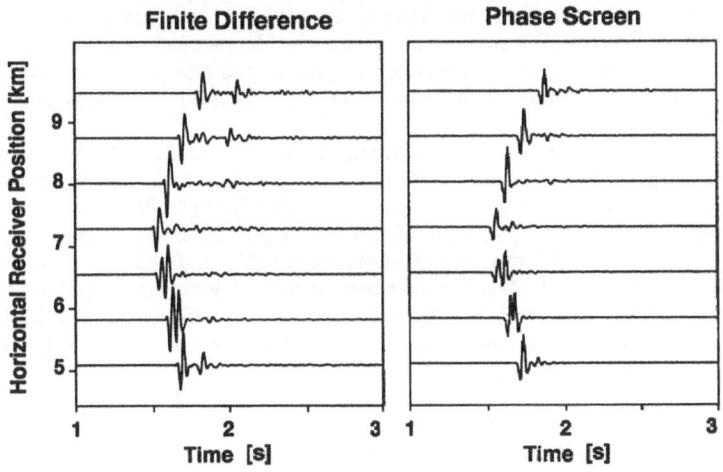

FIGURE 8.20. Waveforms calculated using 2-D finite difference modeling of the wave equation and 2-D split-step Fourier (phase screen) method for the slice of the 3-D model shown in Figure 2.15. The source is located at the surface of the model at horizontal distance 6.9 km. Receivers are located at a depth of 3.65 km at horizontal positions indicated next to the traces. The source is a 20 Hz Ricker wavelet. The finite difference scheme is fourth order in space and second order in time. Grid spacing in both calculations was 6.096 m. The time interval for finite difference and split-step calculations are 0.0005 s and 0.0004 s, respectively. [Courtesy of L.-J. Huang.]

ence of velocity contrasts as large as 2.7 to 4.6 km/s and steeply dipping interfaces like those shown in Figure 2.15. Figure 8.20 shows synthetic seismograms calculated through the 2-D slice of the 3-D model containing a salt body shown in Figure 2.15. The left side shows traces calculated using a fourth-order in space finite difference scheme of the 2-D scalar wave equation for a source located at the surface of the model at horizontal distance 6.9 km in Figure 2.15. The receivers are located near the bottom of the model, at a depth of 3.65 km at the horizontal locations indicated next to the traces. The right side of the figure shows traces calculated using the split-step Fourier method given in (8.98). Comparing the seismograms, we see that the first pulses of the waveforms are similar for the two methods. Later portions of the waveforms, which may be dominated by reverberations in the high-velocity salt body, are different. The difference is due to the limitation of the split-step method to model only forward-scattered energy.

Fisk and McCartor [1991] extended the spit-step method to vector elastic waves assuming the medium is smooth, which allowed them to write a pair of decoupled equations for P- and S-waves. Coupling between P and S was accomplished after phase corrections by decomposing the resulting phase-corrected wavefield back into P- and S-waves. Fisk and McCartor [1991] and Fisk et al. [1992] compared synthetic seismograms calculated by their method for 2-D random elastic media with numerical synthesis based on the finite difference method; however, their formulation does not contain a complete treatment of conversions between P-and S-waves. Wu [1994] formulated one-way elastic wave equations for P- and S-wavefields in inhomogeneous media using the Born approximation. Introducing the parabolic approximation, Wu [1994] arrives at a complex phase-screen for elastic waves instead of a regular phase-screen. He presents comparisons of the elastic phase-screen method with finite difference and exact solutions for simple structures and shows that the method works well for modest velocity and density contrasts.

Wu et al. [1995] extended the split-step method to include the effects of primary reflections. With this approach, it is now possible to model reflection seismograms recorded at the earth's surface. Either (8.98) or (8.99) is used to calculate the forward propagating wavefield. Within each layer, the back-scattered wavefield due to single scattering from inhomogeneities is calculated using a method due to De Wolf [1971]. The forward propagated wavefield interacts at multiple depths within each slab. The wavefield scattered within each layer is propagated back to the surface using the split-step approach. Comparison of synthetic seismograms calculated using this multiple-forward, single backscattering modeling approach with those calculated using finite difference for the 2-D case of a buried cylinder with a 10% velocity perturbation by Wu et al. [1995] showed that the result obtained using the split-step method is in good agreement with the finite difference solution.

CHAPTER 9

Summary and Epilogue

Analysis of high-frequency (>1 Hz) regional seismograms using the scattering approach helps us to obtain a better understanding of the origin of the complexity of observed waveforms. In addition, the scattering approach provides a stochastic method to obtain additional information about the character of the earth's lithosphere beyond that obtained from deterministic methods. Stochastic characterization and deterministic imaging of the lithosphere are complementary to each other. We have introduced a variety of stochastic approaches for modeling high-frequency seismic wave propagation through the randomly inhomogeneous structure of the lithosphere developed during the last two decades. We have emphasized the contribution of the scattering process to the formation of seismogram envelopes. Here, we summarize the state of the art and discuss possible future developments.

9.1 SUMMARY OF METHODS AND OBSERVATIONS

Heterogeneity in the Crust

Geological surveys and well-log data show that the structure of the crust is heterogeneous. The fractional fluctuations of seismic velocity often exceed 10%. Well-log data show an approximate linear correlation between P- and S-wave velocities and mass density. We model the crustal inhomogeneity as a stationary random process in space, which can be statistically characterized using an appropriate ACF or PSDF. For wavenumbers ranging from $10^{-1}\,m^{-1}$ to $10\,m^{-1}$, a power-law PSDF can be used for the crust to match most well-log data. This means that the exponential ACF or the von Kármán ACF, which are rich in short-wavelength components, better explain crustal inhomogeneity than the Gaussian ACF, which contains a relatively smaller short-wavelength component of inhomogeneity.

S-Coda Waves

High-frequency (1–30 Hz) seismograms of local earthquakes contain signals following the direct S-wave that are known as S-coda waves. Array analysis has shown that S-coda consists of incoherent S-waves that have been scattered by distributed random heterogeneities in the lithosphere. Heterogeneities include variations in elastic coefficients and mass density due to changes in rock composition or the presence of cracks or fluids in the cracks. S-coda wave amplitude decreases with increasing lapse time and is independent of epicentral distance and radiation pattern at long lapse times. The epicentral distance independence of S-coda amplitude at long lapse times leads to the coda-normalization method which can be used to reliably estimate the relative strength of source spectra, the site amplification caused by localized structure in the vicinity of a receiver site, and seismic amplitude attenuation with travel distance.

The S-coda envelope decay rate is quantified by a geometric spreading term and a phenomenological exponential decay factor Q_C^{-1}, known as coda attenuation. The parameter Q_C^{-1} has been measured throughout the world and compared with seismotectonic setting. The scatter of worldwide measurements is about a factor of two. The value of Q_C^{-1} is of the order of 10^{-2} at 1 Hz and decreases with increasing frequency to the order of 10^{-3} at 20 Hz. There have been reports of temporal changes in coda characteristics, and some of these changes occurred in advance of earthquakes or volcanic eruptions.

Single Scattering Modeling for S-coda and Seismogram Envelopes

It is difficult to model the entire scattering process using the wave equation. However, phenomenological methods for modeling the single scattering process have been developed. They help to build our intuition about the wave propagation process in inhomogeneous media, where we explicitly suppose the addition of the power of scattered waves. This model predicts that MS S-coda amplitude decays with the inverse square of lapse time when there is no intrinsic attenuation. The S-coda spectrum is a product of the source power spectrum and the total scattering coefficient g_0 that characterizes the S-to-S scattering power per unit volume. The value of g_0 is on the order of 10^{-2} km^{-1} for frequencies of 1–30 Hz in the lithosphere. This corresponds to a mean free path of 100 km and a mean free time of 25 s for S-wave velocity $\beta_0 = 4$ km/s.

Single scattering models based on the Born approximation give S-coda envelope shapes that agree with phenomenological models and with observations for relatively short lapse times. Single scattering modeling using frequency-dependent nonisotropic scattering amplitudes calculated using the Born approximation with travel-time correction for elastic waves that includes the effects of point shear-dislocation source radiation succeeded in explaining variations of three-component seismogram envelopes observed at short source–receiver distances not only for S-coda but also for the P-coda between the direct P- and S-phases. This method for calculating synthetic seismogram envelopes was successfully extended to include reflections on the free surface of the earth.

Multiple Scattering Modeling Based on the Radiative Transfer Theory

The incoherence of scattered waves in the lithosphere means that we can add the power of scattered waves to simulate seismogram envelopes. Models based on the radiative transfer theory have been used to explain the spatiotemporal pattern of the S-wave energy density. The simplest model is based on the assumption of spherically symmetric radiation from an earthquake source and uniform distribution of isotropic scatterers in 3-D space. This model gives predictions compatible with the observed shapes of S-wave envelopes. The formulation of the radiative transfer theory allows us to compare the influence of single and multiple scattering on the distribution of seismic energy density. Inclusion of multiple scattering increases the energy density at long lapse times so the MS envelope decays as the -1.5 th power of lapse time, rather than the -2 power as predicted by the single scattering theory. The slower decay with increasing lapse time is in agreement with observations made in many regions. Models, including nonisotropic scattering and/or nonuniform fractal distribution of scatterers, predict a uniform distribution of coda energy within a volume surrounding the source at long lapse times, agreeing with observations.

The radiative transfer theory provides a method for distinguishing scattering attenuation from intrinsic absorption by analyzing the whole S-wave seismogram. This method is known as the multiple lapse-time window analysis. Reported ratios of scattering loss to total attenuation, known as seismic albedo, range from 0.3 to 0.8 in the frequency band 1-20 Hz using a model based on isotropic scattering. On average g_0 is estimated to be on the order of $10^{-2}\,\mathrm{km}^{-1}$.

The radiative transfer theory has been extended to include the effects of nonspherical radiation from a point shear-dislocation source, nonisotropic scattering, and conversion scattering between P- and S-waves. Although these effects have not all been included in one formalism, the inclusion of individual effects has facilitated investigation of the temporal evolution of the early S-coda, which is sensitive to the source radiation pattern, and of whole seismogram envelopes starting from the P-wave onset to late S-coda.

Scattering Attenuation

Recent observations have shown that S-wave attenuation Q_S^{-1} in the lithosphere is of the order of 10^{-2} at 1 Hz and decreases with increasing frequency to approximately 10^{-3} at 20 Hz. Studies at lower frequencies using surface wave analysis found that attenuation is of the order of 10^{-3} at about 0.05 Hz. Thus, it is expected that Q_S^{-1} has a peak in the 0.5–1 Hz frequency range. Attenuation is considered due to a combination of intrinsic absorption and scattering attenuation. The existence of S-coda suggests that scattering attenuation due to random inhomogeneity in the lithosphere may be significant from the view point of energy conservation.

The first-order Born approximation can be used to estimate the amount of scattering attenuation. To do this, we interpret the ensemble average of the power of singly scattered waves as the energy lost from the incident waves. The ordinary statistical averaging procedure leads to the conclusion that scattering attenuation increases with increasing frequency because it implicitly includes phase changes caused by travel-time fluctuations due to long-wavelength components of velocity inhomogeneity when calculating energy lost by scattering from the incident wave-field. If we modify the procedure by first subtracting the travel-time fluctuations, we obtain scattering attenuation $^{TSc}Q^{-1}$ at high frequency that is constant or decreases with increasing frequency, where prefix "TSc" means the scattering attenuation calculated using travel-time correction. This procedure, known as the travel-time corrected Born approximation, corresponds to neglecting scattered wave energy within a cone around the forward direction when estimating the scattering attenuation.

For a random medium characterized by an exponential ACF with MS fractional fluctuation ε^2 and correlation distance a, the peak value of $^{TSc}Q_S^{-1}$ predicted by the travel-time corrected Born approximation is of the order of ε^2 at frequency $f \approx \beta_0 / \pi a$ and decreases with the reciprocal of frequency for high frequencies. The backscattering coefficient for S-to-S-waves g_π^{SS} is of the order of $1.6\,\varepsilon^2 / a$ for high frequencies. Assuming all attenuation is due to scattering (seismic albedo is 1) and applying the above theoretical results to S-wave attenuation and S-coda excitation measurements, we get that the inhomogeneity has parameters $\varepsilon \approx 10\%$ and $a \approx 2$ km. We also note that this scattering model is consistent with the observed relation $Q_P^{-1} > Q_S^{-1}$ for frequencies higher than 1 Hz. For more precise analysis, we may introduce the von Kármán ACF, which has an additional parameter that controls the power of frequency of $^{TSc}Q_S^{-1}$ for high frequencies. The Gaussian ACF is not appropriate for explaining the excitation of S-coda waves and the frequency dependence of S-wave attenuation for frequencies higher than 1 Hz since it has relatively small short-wavelength components.

Array Observation of Teleseismic P-Waves

Array observation of teleseismic P-waves have been done to investigate the phase and amplitude fluctuations caused by the random inhomogeneities located under the array. Recent measurements focus not only on the change in correlation with transverse distance but also on the incident angle. The correlations of waves have been modeled using the parabolic approximation for waves with wavelengths shorter than the correlation distance of the random inhomogeneity. This approximation includes contributions of forward scattering and diffraction that strongly affect waves near the direct wave arrival time. Combining measurement with theory leads to a larger estimated correlation distance and a smaller MS fractional fluctuation than obtained from S-coda measurements, which are more sensitive to short-wavelength components of inhomogeneities. We also note that there is a contribution of surface waves excited by irregular surface topography in P-coda.

Peak Arrival Delay and Envelope Broadening of S-Waves

The source duration of small earthquakes is very short compared to the duration of observed S-wave first arrival packets at distances of 100–300 km. The packet duration generally increases with increasing travel distance indicating that it is caused by some propagation effect. Increasing packet duration is called envelope broadening. In addition, the arrival of the peak amplitude in the first S-arrival packet often comes well after the S-wave onset. The delay between the arrival time of the peak amplitude and the S-onset time increases with increasing travel distance. Observations in Japan show that the frequency dependence of the peak delay and the S-wave envelope broadening vary with location relative to the volcanic front associated with the subduction zone. On the fore-arc side, the peak delay and envelope broadening are small and frequency independent; however, on the back-arc side, they are larger compared with that in the fore-arc side and become stronger with increasing frequency.

MS seismogram envelopes have been synthesized for the case where strong diffraction and multiple forward scattering alter the envelope shape from impulsive to spindle-like. Stochastic synthesis is done using the Markov approximation for the parabolic wave equation when the wavelength of the seismic wave is smaller than the correlation distance of the random media. Envelope width is frequency-independent for random media described by the Gaussian PSDF, but is frequency-dependent and has a stronger distance dependence if the randomness is richer in short-wavelength components than the Gaussian PSDF. When the simulation is used to model data from Kanto, Japan, the ratio ε^2 / a is estimated to be on the order of $10^{-(3.0-3.3)}$ km^{-1}. The regional differences of the frequency dependence of the peak-arrival delay and the width of the S-wave envelope are explained as due to spectral differences in the random inhomogeneity. Comparison with observed data suggests that the lithosphere beneath the back-arc side of the volcanic front is relatively richer in short-wavelength components of random inhomogeneities than the fore-arc side.

Modeling Wave Propagation Depending on the Description of Heterogeneity

The simplest description of the heterogeneous lithosphere is that it is a locally isotropic inhomogeneous elastic media with Lamé coefficients that are continuous functions of coordinates. Born approximation modeling is based on this description. Full waveform synthesis can be done by an extended version of the split-step Fourier method and the finite difference method when medium properties are smooth. Another description of an inhomogeneous medium is to consider that it is composed of a distribution of cavities or crack inclusions with distinct boundaries. These features are efficient scatterers. The boundary integral method is a powerful method for the synthesis of full waveforms to study coda excitation and amplitude attenuation with travel distance through distributed cracks and/or cavities.

9.2 FUTURE DEVELOPMENTS

Our continuing objective is to improve our models of the earth and seismic wave propagation which allow us to learn more about the structure of the earth and how it works. We would like to formulate an inversion scheme to determine the spatial distribution of intrinsic attenuation and scattering coefficients in addition to the deterministic velocity structure based on the conventional tomography method. Results from such a scheme could provide significant insight into geological processes that shape the earth and important information for resource exploration. Our ability to develop better models will depend on future developments and refinement of mathematics and observations.

Observations and Measurements

Improvements in observation are needed to test the limits of the theories developed and to provide the basis for improved theories. A significant gap in observations of S-wave attenuation exists in the frequency band between the measurements at 0.05 Hz obtained by analysis of surface waves and the measurements at 1 Hz from body waves. The peak in S-wave attenuation in the earth proposed by Aki [1980a] to occur at 0.5 Hz falls within this gap. Measurements of body and surface wave attenuation within this frequency gap are needed to test the validity of Aki's conjecture about the shape of the S-wave attenuation vs. frequency curve. We also note that the number of P-wave attenuation measurements is still too small to allow us to investigate regional differences in P-wave attenuation.

Measurements of scattering characteristics in various tectonic regimes will help to better understand the relationship between scattering and geology. For example, the relationship between the broadening of S-envelopes and the location of the volcanic front was reported only for Japan. It will be interesting to classify envelope characteristics in relation to tectonic setting and examine whether the envelopes have the same characteristics in different tectonic regions.

There have been a few studies on P-wave envelopes. Broadening of P-wave envelopes does not appear to be as significant as for S-waves, and P-waves appear to maintain their original impulsive shape more than S-waves. More observations of P-wave broadening will help us to better understand the differences in scattering of the two wave modes and the character of heterogeneity in the earth.

Broad-band observations of regional seismic waves in deep boreholes will help to clarify the amount of scattering that occurs below the surface weathered layer.

Array observations and analysis will allow us to identify the composition of coda waves and to identify localized regions of high scattering intensity.

Observations have shown that coda attenuation Q_C^{-1} is nearly the same order as direct S-wave attenuation Q_S^{-1} in many regions in the world. It has not been clearly shown whether this coincidence has some physical background. The uniform distribution of coda energy density at long lapse times is the conceptual basis of the conventional coda-normalization method; however, we need more quantitative measurements of the spatial distribution of coda energy density in relation to the geology of receiver sites and tectonic settings.

We have used Birch's law and the similarity of P- and S-wave velocities to reduce the number of independent random functions describing the earth to simplify the mathematical modeling. However, these relationships have been confirmed from only several studies of well-log data. We need to further examine their correlations with in situ field measurements. Some of the more interesting cases of scattering may arise when the simple relationships are not valid. For example, in volcanic regimes, bodies in which the S-wave velocity is nearly zero may exist due to the presence of molten materials. Examination of scattering in media with such bodies may help explain some of the observed complexity of seismograms recorded in volcanic regions and may help to better understand the structure of volcanoes.

Theoretical Developments

We found that the travel-time corrected Born approximation or the neglect of forward scattering resulted in theoretical predictions for the frequency dependence of scattering attenuation that agrees with S-wave attenuation observations; however, the choice of the lower bound of scattering angle is heuristic. The derivation of scattering attenuation based on the Born approximation with travel-time correction is a kind of differential approach, where the energy loss for an infinitesimal path length is calculated and the effect of diffraction is completely excluded. The amplitude change should be theoretically studied for long travel distances.

Three-component envelope syntheses based on the single scattering model by using scattering amplitudes from the Born approximation with travel-time correction have been done; however, near the direct wave arrival-times, the point-source assumption and the Fraunhofer zone assumption may not hold. The predicted envelope time-traces have discontinuities at the direct-wave arrival times. We need to develop models overcoming these difficulties.

The incorporation of both nonisotropic scattering and nonspherical source radiation is important since strong forward scattered energy appears immediately after the direct-wave arrival when the radiation pattern is nonspherical. Models that include conversion scattering between P- and S-waves are also necessary for correct interpretation of whole seismograms starting from the direct P-wave until the end of S-coda. Quantifying the partition of energy density into three components is important but it has been investigated only using models based on the single scattering approximation. It will be possible to formulate a multiple scattering process to model three-component seismogram envelopes by introducing

S-wave polarization parameters into the radiative transfer theory. We also need to incorporate conversion between body waves and surface waves at the irregular free surface or boundaries in the radiative transfer theory.

A complete link between the radiative transfer equation in scattering media and the wave equation in inhomogeneous media has not been demonstrated yet. The radiative transfer formulation presented here has assumed that waves are instantaneously scattered when the incident wave hits a scatterer. However, low-velocity bodies can immediately scatter waves and trap and radiate waves with a time lag depending on the frequency of incident waves and the characteristics of the body. It is necessary to include such resonant scattering in the formulation of the radiative transfer theory.

The radiative transfer theory predicts that coda attenuation Q_C^{-1} is dominated by intrinsic absorption since the theory predicts that the MS coda-envelope decays according to the -1.5th power of lapse time for a "uniform" distribution of "isotropic" scatterers and no intrinsic attenuation. Hence, any coda decay rate larger than $t^{-1.5}$ must be due to intrinsic attenuation. Models using the radiative transfer theory for envelope synthesis need to be investigated for media in which scattering coefficient and intrinsic attenuation are depth dependent or fractally distributed. Ultimately, we would like to formulate an inversion scheme to determine the spatial distribution of intrinsic attenuation and scattering coefficients within the framework of the radiative transfer theory.

Observed S-wave envelope broadening was analyzed based on the parabolic approximation, which is valid only for narrow-angle scattering around the forward direction due to long-wavelength components of velocity inhomogeneity. It might be necessary to incorporate the contribution of large-angle scattering due to short-wavelength components of elastic inhomogeneity. Development of a model that includes wide-angle scattering may help explain the result that the heterogeneity estimated from coda excitation due to large-angle scattering has a larger MS fractional velocity fluctuation and smaller correlation distance than that estimated from forward scattering. Elastic wave theory shows that S-waves are more easily scattered than P-waves; however, a rigorous development of elastic wave theory for envelope broadening is necessary as an extension of the split-step Fourier method.

Deterministic approaches provide the most reliable way to study wave propagation in strongly inhomogeneous media. It is necessary to develop mathematics used for this purpose that has a wider range of applicability than the parabolic approximation. The development of fast codes for computing waveforms is necessary as a complement to the stochastic approaches. Exact methods that include resonance within scatterers and allow the modeling of media containing a large number of scattering bodies are needed. The methods should allow study of wave propagation over long propagation distances. These methods will become more practical as faster computers containing more memory become available.

Spherical Harmonic Functions and Wigner 3-j Symbols

We briefly introduce the definition of the spherical harmonic functions. First we define

$$\Theta_{lm}(\theta) = (-1)^m i^l \sqrt{\frac{(2l+1)}{2}\frac{(l-m)!}{(l+m)!}} P_l^m(\cos\theta) \quad \text{for } m \geq 0$$

$$= i^l \sqrt{\frac{(2l+1)}{2}\frac{(l-|m|)!}{(l+|m|)!}} P_l^{|m|}(\cos\theta) \quad \text{for } m < 0 \tag{A.1}$$

where P_l^m are associated Legendre polynomials and the phase factor i^l is after Landau and Lifshitz [1989]. This definition is different from the usual one; however, this choice is the most natural from the viewpoint of the theory of the addition of angular momenta. Define

$$\Phi_m(\phi) = \frac{1}{\sqrt{2\pi}} e^{im\phi} \tag{A.2}$$

The normalized spherical harmonic function is given by

$$Y_{lm}(\theta,\phi) = \Theta_{lm}(\theta)\Phi_m(\phi) \tag{A.3}$$

where we note that

$$Y_{l,m}^*(\theta,\phi) = (-1)^{l-m} Y_{l,-m}(\theta,\phi) \tag{A.4}$$

The normalized spherical harmonic functions with different l or m are orthonormal:

$$\oint Y_{l_1 m_1}^*(\theta,\phi) Y_{l_2 m_2}(\theta,\phi) d\Omega(\theta,\phi) = \delta_{l_1 l_2} \delta_{m_1 m_2} \tag{A.5}$$

where $d\Omega(\theta,\phi) = \sin\theta d\theta d\phi$. The addition theorem holds for Legendre polynomials:

$$P_l\big(\cos\theta\cos\theta' +\sin\theta\sin\theta'\cos(\phi-\phi')\big)$$

$$=\frac{4\pi}{2l+1}\sum_{m=-l}^{l}Y_{lm}(\theta,\phi)Y_{lm}^{*}(\theta',\phi')=\frac{4\pi}{2l+1}\sum_{m=-l}^{l}Y_{lm}^{*}(\theta,\phi)Y_{lm}(\theta',\phi') \tag{A.6}$$

There is an expansion formula

$$e^{ikr[\cos\theta\cos\theta'+\sin\theta\sin\theta'\cos(\phi-\phi')]}=4\pi\sum_{l=0}^{\infty}i^{l}j_{l}(kr)\sum_{m=-l}^{l}Y_{lm}(\theta,\phi)Y_{lm}^{*}(\theta',\phi') \tag{A.7}$$

where the spherical Bessel function is given by

$$j_l(z)=\sqrt{\frac{\pi}{2z}}J_{l+\frac{1}{2}}(z) \tag{A.8}$$

We note that $j_l(-z)=(-1)^l j_l(z)$. From (A.6), the spherical harmonic closure relation [Arfken and Weber, 1995] gives the delta function for the solid angle as

$$\delta_\Omega(\mathbf{x};\mathbf{x}')=\delta_\Omega(\theta,\phi;\theta',\phi')=\sum_{l=0}^{\infty}\sum_{m=-l}^{l}Y_{lm}(\theta,\phi)Y_{lm}^{*}(\theta',\phi')$$

$$=\sum_{l=0}^{\infty}\sum_{m=-l}^{l}Y_{lm}^{*}(\theta,\phi)Y_{lm}(\theta',\phi') \tag{A.9}$$

where vectors $\mathbf{x}=(r,\theta,\phi)$ and $\mathbf{x}'=(r',\theta',\phi')$ are in spherical coordinates. We note that

$$\oint\delta_\Omega(\theta,\phi;\theta',\phi')d\Omega(\theta',\phi')=1 \quad\text{and}$$

$$\oint f(\theta',\phi')\delta_\Omega(\theta,\phi;\theta',\phi')d\Omega(\theta',\phi')=f(\theta,\phi) \tag{A.10}$$

for any function $f(\theta,\phi)$.

The integral of three spherical harmonic functions is written using the Wigner 3-j symbols [see Landau and Lifshitz, 1989]:

$$\big(Y_{lm}\big)_{l''m''}^{l'm'}\equiv\oint Y_{l'm'}^{*}(\theta,\phi)Y_{lm}(\theta,\phi)Y_{l''m''}(\theta,\phi)d\Omega(\theta,\phi)$$

$$=(-1)^{m'}i^{l-l'+l''}\begin{pmatrix}l' & l & l''\\-m' & m & m''\end{pmatrix}\begin{pmatrix}l' & l & l''\\0 & 0 & 0\end{pmatrix}\sqrt{\frac{(2l'+1)(2l+1)(2l''+1)}{4\pi}} \tag{A.11}$$

The above integral vanishes except when $m'=m+m''$ and $|l-l''|\le l'\le l+l''$ (the triangular condition) according to the selection rules corresponding to the addition of angular momenta. Then, $l+l'+l''$ is even. Wigner 3-j symbols are pure real. There are published programs for the numerical computation of the Wigner 3-j symbols [Shriner and Thompson, 1993; Wolfram 1991].

Glossary of Symbols

a	Correlation distance.
$a_\perp$	Transverse correlation distance in a plane perpendicular to the mean propagation direction.
A	Longitudinal integral of R along the mean propagation direction.
b	Intrinsic absorption parameter.
B	Radiation pattern of seismic waves due to a point shear-dislocation.
B_0	Seismic albedo, the ratio of scattering attenuation to total attenuation.
D	Wave parameter characterizing the diffraction effect.
D_C	Diffusivity.
D_S, D_A	Fractal dimension for the distribution of scatterers or attenuation bodies.
$d\sigma/d\Omega$	Differential scattering cross section of a scatterer.
$(\mathbf{e}_1, \mathbf{e}_2, \mathbf{e}_3)$	Unit vectors in the Cartesian coordinate system.
$(\mathbf{e}_r, \mathbf{e}_\psi, \mathbf{e}_\zeta)\,(\mathbf{e}_{r_a,r_b}, \mathbf{e}_\theta, \mathbf{e}_\phi)$	Unit vectors in the spherical coordinate system.
E, E^1, E^M, E^D, E^{EF}	Energy density, where superscript "1" is for single scattering, "M" for multiple scattering of the order larger or equal to two, "D" for the diffusion model, and "EF" for the energy-flux model.
$EI_{1,2,3}$	Integral of energy density used for the MLTW analysis.
f	Frequency.
f_C	Corner frequency.
$F, {}^T F$	Scattering amplitude, where prefix "T" is for that with travel-time correction.
$g, {}^T g$	Scattering coefficient, where prefix "T" is for that with travel-time correction.
g_π, g_π^{SS}	Backscattering coefficient (of S to S scattering).
g_0	Total scattering coefficient (of S to S scattering).
g_F	Turbidity coefficient.
G	Green function. Defined in each section.
I	Intensity of wavefield.
$\hat{I}$	Intensity spectral density.
J	Energy-flux density.
k	Wavenumber of P-wave (or scalar wave).
$\mathbf{k}_\perp$	Wavenumber vector in the transverse plane perpendicular to the mean propagation direction.

k_M, t_M	Characteristic scales of multiple scattering.
K	Characteristic function for the single isotropic scattering model.
K_C	Characteristic function for the single isotropic scattering model with conversion between P- and S-waves.
l	Wavenumber of S-wave.
ℓ	Mean free path.
L	Extension of the inhomogeneities.
L_F	Dimension of the fault.
$M(t)$	Seismic moment time function.
M_0	Seismic moment of an earthquake.
M_L	Local magnitude of an earthquake.
n	Number density of point-like scatterers.
$\mathbf{n}$	Unit vector normal to a fault.
$p, \mathbf{p}$	Slowness and slowness vector.
P	Power spectral density function of the fractional fluctuation of wave propagation velocity.
P_{fw}	Frequency-wavenumber power spectral density in 2-D space.
Q^{-1}, $^I Q^{-1}$, $^{Sc} Q^{-1}$, $^{BSc} Q^{-1}$, $^{TSc} Q^{-1}$	Attenuation parameter, reciprocal of the quality factor, where prefix "I" is for intrinsic attenuation and "Sc" for scattering attenuation. Prefix "BSc" is for scattering attenuation based on the ordinary Born approximation, and "TSc" for scattering attenuation based on the travel-time corrected Born approximation.
Q_P^{-1}, Q_S^{-1}	Attenuation parameter, where subscript "P" and "S" are for P- and S-waves, respectively.
Q_C^{-1}	Coda attenuation factor.
(r, ψ, ζ) $(r_{a,b}, \theta, \phi)$	Spherical coordinates.
r	Radial distance or hypocentral distance.
$\bar{r}$	Normalized radial distance.
R	Autocorrelation function of the fractional fluctuation of wave propagation velocity.
$\mathbf{s}$	Unit slip vector of a fault.
S_i	Power spectral density of the ith component velocity seismogram.
$\bar{t}$	Normalized lapse time.
T_{ij}	Stress tensor.
u	Scalar wavefield.
$\mathbf{u}$, u_i	Vector wavefield.
$^T u$, $^T \mathbf{u}$	Travel-time corrected wavefield.
$\dot{u}_{ij}$	Velocity wavefield for source i at receiver j.
V, V_0	Wave propagation velocity. Subscript zero is for the average.
(w, v, ϕ)	Prolate spheroidal coordinates.
W, W^P, W^S	Radiated energy from the source. Superscript "P" is for P-wave and "S" for S-wave.

$\widehat{W}^{P}, \widehat{W}^{S}$ Spectral density of radiated energy from the source. Superscript "P" is for P-wave and "S" for S-wave.

$\overline{\mathbf{x}}$ Normalized coordinate vector.

$\mathbf{x}_{\perp}$ Coordinate vector in the transverse plane perpendicular to the mean propagation direction.

$\mathbf{x}_c, z_c, k_c, \omega_c$ Center-of-mass coordinates.

$\mathbf{x}_d, z_d, k_d, \omega_d$ Difference coordinates.

$X_\bullet^{\bullet\bullet}(\psi,\zeta), {}^{T}X_\bullet^{\bullet\bullet}(\psi,\zeta)$ Basic scattering pattern of elastic vector waves, where prefix "T" is for that with travel-time correction.

α, α_0 P-wave velocity. Subscript zero is for the average.

β, β_0 S-wave velocity. Subscript zero is for the average.

χ Normalized radius in the transverse plane with respect to $a_{\perp}$.

δ_{Ω} Delta function in a solid angle.

δf_i Equivalent body force.

${}^{c}\delta f_i$ Equivalent body force due to travel-time correction.

δt Travel-time fluctuation due to the long-wavelength component of velocity fluctuation.

$\Delta \ln A_0$ Log-amplitude fluctuation ($\Xi = \Delta \ln A_0 + i\Delta\varphi$).

$\Delta\varphi$ Phase fluctuation ($\Xi = \Delta \ln A_0 + i\Delta\varphi$).

ε Root-mean-square of the fractional fluctuation of velocity.

Φ Scattering strength.

γ_0 Ratio of average P-wave velocity to average S-wave velocity.

Γ_1 Mutual coherence function.

Γ_2 Two-frequency mutual coherence function.

φ Phase function.

λ Lamé coefficient.

$\lambda_w, \lambda_w{}^{P}, \lambda_w{}^{S}$ Wavelength of scalar, P-, and S-waves.

λ_c Cutoff wavelength ($= \lambda_w / \upsilon_C$).

μ Lamé coefficient.

ν Ratio of the density fractional fluctuation to the velocity fractional fluctuation.

ρ, ρ_0 Mass density. Subscript zero is for the average.

σ, σ_0 Scattering cross section of a scatterer. Subscript zero is for the total scattering cross section.

τ Normalized distance in the mean propagation direction.

υ_C Cutoff value for the travel-time correction.

ω Angular frequency.

ξ, ξ^s, ξ^L Fractional fluctuation of velocity, where superscript "s" is for the short-wavelength component and "L" for the long-wavelength component with respect to cutoff wavelength λ_C.

Ξ Perturbation term of the Rytov method.

ψ Scattering angle.

ψ_C Cutoff scattering angle.

Ψ	Radiation pattern of energy from a source.
Ω	Solid angle.
$\Delta_\perp$	Laplacian in the transverse plane perpendicular to the mean propagation direction.
$\nabla_\perp$	Gradient in the transverse plane perpendicular to the mean propagation direction.
$\begin{pmatrix} l_1 & l_2 & l_3 \\ j_1 & j_2 & j_3 \end{pmatrix}$	Wigner 3-j symbol.
$\langle \cdots \rangle$	Ensemble average.
$\langle \cdots \rangle_T$	Moving average in time.
$\langle \cdots \rangle_D$	Moving average in distance.
$\overline{E}$	Overbar denotes the normalized non-dimensional quantity.
$\hat{E}$	Caret means the Laplace transform in time.
$\tilde{E}$	Tilde means the Fourier transform in space.
$\dot{u}$	Over-dot means the partial derivative with respect to time.
$\breve{u}$	2-D Fourier transform of u in the transverse plane perpendicular to the mean propagation direction.

References

Abramowitz, M. and I. A. Stegun, *Handbook of Mathematical Functions with Formulas, Graphs, and Mathematical Tables*, Dover, New York, 1970.

Abubakirov, I. R. and A. A. Gusev, Estimation of scattering properties of lithosphere of Kamchatka based on Monte-Carlo simulation of record envelope of a near earthquake, *Phys. Earth Planet. Inter.* **64**, 52–67, 1990.

Akamatsu, J., Coda attenuation in the Lützow-Holm Bay region, East Antarctica, *Phys. Earth Planet. Inter.* **67**, 65–75, 1991.

Aki, K., Scaling law of the seismic spectrum, *J. Geophys. Res.* **72**, 1217–1231, 1967.

Aki, K., Analysis of seismic coda of local earthquakes as scattered waves, *J. Geophys. Res.* **74**, 615–631, 1969.

Aki, K., Scattering of P waves under the Montana LASA, *J. Geophys. Res.* **78**, 1334–1346, 1973.

Aki, K., Attenuation of shear-waves in the lithosphere for frequencies from 0.05 to 25 Hz, *Phys. Earth Planet. Inter.* **21**, 50–60, 1980a.

Aki, K., Scattering and attenuation of shear waves in the lithosphere, *J. Geophys. Res.* **85**, 6496–6504, 1980b.

Aki, K., Attenuation and scattering of short-period seismic waves in the lithosphere, in *Identification of Seismic Sources - Earthquake or Underground Explosion* (eds. E. S. Husebye and S. Mykkeltveit), pp. 515–541, D. Reidel, Dordrecht, Holland, 1981.

Aki, K., Scattering and attenuation, *Bull. Seismol. Soc. Am.* **72**, S319–S330, 1982.

Aki, K., Theory of earthquake prediction with special reference to monitoring of the quality factor of lithosphere by the coda method, *Earthq. Predict. Res.* **3**, 219–230. 1985.

Aki, K., Scattering conversions P to S versus S to P, *Bull. Seismol. Soc. Am.* **82**, 1969–1972, 1992.

Aki, K., Interrelation between fault zone structures and earthquake processes, *Pure Appl. Geophys.* **145**, 647–676, 1995.

Aki, K. and M. Tsujiura, Correlation study of near earthquake waves, *Bull. Earthq. Inst. Univ. Tokyo* **37**, 207–232, 1959.

Aki, K. and B. Chouet, Origin of coda waves: Source, attenuation and scattering effects, *J. Geophys. Res.* **80**, 3322–3342, 1975.

Aki, K. and W. H. K. Lee, Determination of three-dimensional velocity anomalies under a seismic array using first P arrival times from local earthquakes, Part I. A homogeneous initial model, *J. Geophys. Res.* **81**, 4381–4399, 1976.

Aki, K. A. Christoffersson, and E. S. Husebye, Three-dimensional seismic structure of the lithosphere under Montana LASA, *Bull. Seismol. Soc. Am.* **66**, 501–524, 1976.

Aki, K., A. Christoffersson, and E. S. Husebye, Determination of the three-dimensional seismic structure of the lithosphere, *J. Geophys. Res.* **82**, 277–296, 1977.

Aki, K. and P. Richards, *Quantitative Seismology - Theory and Methods*, Vols. 1 and 2, W. H. Freeman, San Francisco, 1980.

Akinci, A., E. Del Pezzo, and J. M. Ibáñez, Separation of scattering and intrinsic attenuation in southern Spain and western Anatolia (Turkey), *Geophys. J. Int.* **121**, 337–353, 1995.

Alford, R. M., K. Kelly, and D. Boore, Accuracy of finite-difference modeling of the acoustic wave equation, *Geophysics* **39**, 834–842, 1974.

Aminzadeh, F., N. Burkhard, F. Rocca, and K. Wyatt, SEG/EAEG 3-D modeling project: 2nd update, *Leading Edge* 13, 949–952, 1994.

Anderson, D. L., A. Ben-Menahem, and C. B. Archambeau, Attenuation of seismic energy in the upper mantle, *J. Geophys. Res.* 70, 1441–1448, 1965.

Anderson, D. L. and R. S. Hart, Q of the Earth, *J. Geophys. Res.* 83, 5869–5882, 1978.

Antolik, M., R. Nadeau, R. C. Aster, and T. V. McEvilly, Differential analysis of coda Q using similar microearthquakes in seismic gaps. Part 2: Application to seismograms recorded by the Parkfield high resolution seismic network, *Bull. Seismol. Soc. Am.* 86, 890–910, 1996.

Arfken, G. B. and H. J. Weber, *Mathematical Methods for Physicists*, Fourth Edition, Academic Press, San Diego, 1995.

Aster, R, C., G. Slad, J. Henton, and M. Antolik, Differential analysis of coda Q using similar microearthquakes in seismic gaps. Part 1: Techniques and application to seismograms recorded in the Anza Seismic Gap, *Bull. Seismol. Soc. Am.* 86, 868–889, 1996.

Atkinson, G. M., Notes on ground motion parameters for Eastern North America: Duration and H/V ratio, *Bull. Seismol. Soc. Am.* 83, 587–596, 1993.

Bakun, W. H., C. G. Bufe, and R. M. Stewart, Body-wave spectra of Central California earthquakes, *Bull. Seismol. Soc. Am.* 66, 363–384, 1976.

Bakun, W. H. and A. G. Lindh, Local magnitudes, seismic moments, and coda durations for earthquakes near Oroville, California, *Bull. Seismol. Soc. Am.* 67, 615–629, 1977.

Banik, N. C., I. Lerche, and R. T. Shuey, Stratigraphic filtering, Part I: Derivation of the O'Doherty-Anstey formula, *Geophysics*, 50, 2768–2774, 1985.

Bannister, S. C., E. S. Husebye, and B. O. Ruud, Teleseismic P coda analyzed by three-component and array techniques: Deterministic location of topographic P-to-Rg scattering near the NORESS array, *Bull. Seismol. Soc. Am.* 80, 1969–1986, 1990.

Barabanenkov, Yu. N., Yu. A. Kravtsov, S. M. Rytov, and V. I. Tamarskii, Status of the theory of propagation of waves in randomly inhomogeneous medium, *Soviet Phys. Usp.* (Eng. Trans.) 13, 551–680, 1971.

Baskoutas, I., Dependence of coda attenuation on frequency and lapse time in central Greece, *Pure Appl. Geophys.* 147, 483–496, 1996.

Baskoutas, I. and H. Sato, Coda attenuation Q_C^{-1} for 4 to 64 Hz signals in the shallow crust measured at Ashio, Japan, *Boll. Geof. Teor. Appl.* 21, 277–283, 1989.

Beaudet, P. R., Elastic wave propagation in heterogeneous media, *Bull. Seismol. Soc. Am.* 60, 769–784, 1970.

Benites, R., K. Aki, and K. Yomogida, Multiple scattering of SH waves in 2-D media with many cavities, *Pure Appl. Geophys.* 138, 353–390, 1992.

Ben-Menahem A. and A. J. Singh, *Seismic Waves and Sources*, Springer-Verlag, New York, 1981.

Beroza, G. C., A. T. Cole, and W. L. Ellsworth, Stability of coda wave attenuation during the Loma Prieta, California earthquake sequence, *J. Geophys. Res.* 100, 3977–3987, 1995.

Berteussen, K, A., A. Christoffersson, E. S. Husebye, and A. Dahle, Wave scattering theory in analysis of P-wave anomalies observed at NORSAR and LASA, *Geophys. J. R. Astron. Soc.* 42, 403–417, 1975.

Biot, M. A. Theory of propagation of elastic waves in a fluid-saturated porous solid. I. Low frequency range, *J. Acoust. Soc. Am.* 28, 168–178, 1956a.

Biot, M. A. Theory of propagation of elastic waves in a fluid-saturated porous solid. II. Higher frequency range, *J. Acoust. Soc. Am.* 28, 179–191, 1956b.

Birch, F., The velocity of compressional waves in rocks to 10 kilobars, Part 1, *J. Geophys. Res.* 65, 1083–1102, 1960.

Birch, F., The velocity of compressional waves in rocks to 10 kilobars, Part 2, *J. Geophys. Res.* 66, 2199–2224, 1961.

Biswas, N. N. and K. Aki, Characteristics of coda waves: Central and south central Alaska, *Bull. Seismol. Soc. Am.* 74, 493–507, 1984.

Block, L. V., C. H. Cheng, M. C. Fehler, and W. S. Phillips, Seismic imaging using microearthquakes induced by hydraulic fracturing, *Geophysics*, 59, 102–112, 1994.

Boettcher, A. L., The role of amphiboles and water in circum-Pacific volcanism, in *High Pressure Research* (eds., M. H. Manghnani and S. Akimoto), pp. 107–125, Academic Press, New York, 1977.

Braile, L., G. R. Keller, S. Mueller, and C. Prodehl, Seismic Techniques, in *Continental Rifts: Evolution, Structure, Tectonics* (ed. K. Olsen), pp. 61–92, Elsevier, New York, 1995.

Brockman, S. R. and G. A. Bollinger, Q estimates along the Wasatch front in Utah derived from Sg and Lg wave amplitudes, *Bull. Seismol. Soc. Am.* **82**, 135–147, 1992.

Brune, J. N., Tectonic stress and the spectra of seismic shear waves from earthquakes, *J. Geophys. Res.* **75**, 4997–5009, 1970.

Budiansky, B. and R. O'Connell, Elastic moduli of a cracked solid, *Int. J. Solid Struct.* **12**, 81–97, 1976.

Burridge, R., G. S. Papanicolaou, and B. S. White, One-dimensional wave propagation in a highly discontinuous medium, *Wave Motion* **10**, 19–44, 1988.

Burridge, R., G. Papanicolaou, P. Sheng, and B. White, Probing a random medium with a pulse, *SIAM J. Appl. Math.* **49**, 582–607, 1989.

Butler, R., C. S. McCreery, L. N. Frazer, and D. A. Walker, High-frequency seismic attenuation of oceanic P and S waves in the western Pacific, *J. Geophys. Res.* **92**, 1383–1396, 1987.

Campillo, M., Propagation and attenuation characteristics of the crustal phase Lg, *Pure Appl. Geophys.* **132**, 1–19, 1990.

Campillo, M. and J. L. Plantet, Frequency dependence and spatial distribution of seismic attenuation in France: Experimental results and possible interpretations, *Phys. Earth Planet. Inter.* **67**, 48–64, 1991.

Campillo, M. and A. Paul, Influence of the lower crustal structure on the early coda of regional seismograms, *J. Geophys. Res.* **97**, 3405–3416, 1992.

Capon, J., Characterization of crust and upper mantle structure under LASA as a random medium, *Bull. Seismol. Soc. Am.* **64**, 235–266, 1974.

Carpenter, P. J. and A. R. Sanford, Apparent Q for upper crustal rocks of the Central Rio Grande Rift, *J. Geophys. Res.* **90**, 8661–8674, 1985.

Cerveny, V., Ray tracing algorithms in three-dimensional laterally varying layered structures, in *Seismic Tomography* (ed. G. Nolet), pp. 99–133, D. Reidel, Boston, 1987.

Cerveny, V. and R. Ravindra, *Theory of Seismic Head Waves*, Univ. Toronto Press, Toronto, 1971.

Cessaro, R. K. and R. Butler, Observations of transverse energy for P waves recorded on a deep-ocean borehole seismometer located in the northwest Pacific, *Bull. Seismol. Soc. Am.* **77**, 2163–2180, 1987.

Chandrasekhar, S., *Radiative Transfer*, Dover, New York, 1960.

Chapman, C. H., The Radon transform and seismic tomography, in *Seismic Tomography* (ed. G. Nolet), pp. 25–48, D. Reidel, Boston, 1987.

Chen, X. and K. Aki, General coherence functions for amplitude and phase fluctuations in a randomly heterogeneous medium, *Geophys. J. Int.* **105**, 155–162, 1991.

Cheng, C. C. and B. J. Mitchell, Crustal Q structure in the United States from multi-mode surface waves, *Bull. Seismol. Soc. Am.* **71**, 161–181, 1981.

Cheng, N., C. H. Cheng, and M. N. Toksöz, Error analysis of phase screen method in 3-D, *Geophys. Res. Lett.* **23**, 1841–1844, 1996.

Chernov, L. A., *Wave Propagation in a Random Medium* (Engl. trans. by R. A. Silverman), McGraw-Hill, New York, 1960.

Chin, B. H. and K. Aki, Simultaneous study of the source path, and site effects on strong ground motion during the 1989 Loma Prieta earthquake: A preliminary result on pervasive nonlinear site effects, *Bull. Seism. Soc. Am.* **81**, 1859–1884, 1991.

Chouet, B., Temporal variation in the attenuation of earthquake coda near Stone Canyon, California, *Geophys. Res. Lett.* **6**, 143–146, 1979.

Christensen, N. I., Chemical changes associated with upper mantle structure, *Tectonophysics* **6**, 331–342, 1968.

Christensen, N., Poisson's ratio and crustal seismology, *J. Geophys. Res.* **101**, 3139–3156, 1996.

Christensen, N. and W. Mooney, Seismic velocity structure and composition of the continental crust: A global view, *J. Geophys. Res.* **100**, 9761–9788, 1995.

Cichowicz, A. and R. W. E. Green, Changes in the early part of the seismic coda due to localized scatterers: The estimation of Q in a stope environment, *Pure Appl. Geophys.* **129**, 497–511, 1989.

Claerbout, J. F., *Imaging the Earth's Interior*, Blackwell Science Pub., Boston, 1985.

Conrad, V. Laufzeitkurven des Tauernbebens vom 28. November, 1923, *Mitt. Erdb. Komm. Wien. Akad Wiss.* **59**, 1–23, 1925.

Console, R. and A. Rovelli, Attenuation parameters for Friuli region from strong-motion accelerogram spectra, *Bull. Seismol. Soc. Am.* **71**, 1981–1991, 1981.

Craig, M. S., L. T. Long, and A. Tie, Modeling the seismic P coda as the response of a discrete-scatterer medium, *Phys. Earth Planet. Inter.* **67**, 20–35, 1991.

Crosson, R., Crustal structure modeling of earthquake data 1. Simultaneous least squares estimation of hypocenter and velocity parameters, *J. Geophys. Res.* **81**, 3036–3046, 1976.

Dainty, A. M., High-frequency acoustic backscattering and seismic attenuation, *J. Geophys. Res.* **89**, 3172–3176, 1984.

Dainty, A. M. and M. N. Toksöz, Seismic codas on the earth and the moon: A comparison, *Phys. Earth Planet. Inter.* **26**, 250–260, 1981.

Dainty, A. M., R. M. Duckworth, and A. Tie, Attenuation and backscattering from local coda, *Bull. Seismol. Soc. Am.* **77**, 1728–1747, 1987.

Dainty, A. M. and M. N. Toksöz, Array analysis of seismic scattering, *Bull. Seismol. Soc. Am.* **80**, 2242–2260, 1990.

Dainty, A. M. and C. A. Schultz, Crustal reflections and the nature of regional P coda, *Bull. Seismol. Soc. Am.* **85**, 851–858, 1995.

Dashen, R., Path integrals for waves in random media, *J. Math. Phys.* **20**, 894–920, 1979.

Deans, S., *The Radon Transform and Some of Its Applications*, John Wiley & Sons, New York, 1983.

Del Pezzo, E., G. De Natale, G. Scarcella, and A. Zollo, Q_c of three component seismograms of volcanic microearthquakes at Campi Flegrei volcanic area - Southern Italy, *Pure Appl. Geophys.* **123**, 683–696, 1985.

Der, A. Z., M. E. Marshall, A. O'Donnell, and T. W. McElfresh, Spatial coherence structure and attenuation of the Lg phase, site effects, and interpretation of the Lg coda, *Bull. Seismol. Soc. Am.* **74**, 1125–1147, 1984.

Devaney, A. J., A filtered backprojection algorithm for diffraction tomography, *Ultrasonic Imaging* **4**, 336–350, 1982.

De Wolf, D. A., Electromagnetic reflection from an extended turbulent medium: Cumulative forward-scatter single backscatter approximation, *IEEE Trans. Ant. Propag.* **AP-19**, 254–262, 1971.

Dewberry, S. R. and R. S. Crosson, Source scaling and moment estimation for the Pacific Northwest seismograph network using S-coda amplitudes, *Bull. Seismol. Soc. Am.* **85**, 1309–1326, 1995.

Douglas, A, P. D. Marshall, P. G. Gibbs, J. B. Young, and C. Blamey, P signal complexity re-examined, *Geophys. J. R. Astron. Soc.* **33**, 195–221, 1973.

Dubendorff, B. and W. Menke, Time-domain apparent-attenuation operators for compressional and shear waves: Experiment versus single scattering theory, *J. Geophys. Res.* **91**, 14023–14032, 1986.

Dziewonski, A. M., Elastic and anelastic structure of the earth, *Rev. Geophys. Space Phys.* **17**, 303–312, 1979.

Eck, T. V., Attenuation of coda waves in the Dead Sea region, *Bull. Seismol. Soc. Am.* **78**, 770–779, 1988.

Einspruch, N. G., E. J. Witterholt, and R. Truell, Scattering of a plane transverse wave by a spherical obstacle in an elastic medium, *J. Appl. Phys.* **31**, 806–819, 1960.

Ellsworth, W. L., Review for "Temporal change in scattering and attenuation associated with the earthquake occurrence - A review of recent studies on coda waves (H. Sato)", in *Evaluation of Proposed Earthquake Precursors* (ed. M. Wyss), pp. 54–55, AGU, Washington, D. C., 1991.

Fang, Y. and G. Müller, Attenuation operators and complex wave velocities for scattering in random media, *Pure Appl. Geophys.* **148**, 269–285, 1996.

Fedotov, S. A. and S. A. Boldyrev, Frequency dependence of the body-wave absorption in the crust and the upper mantle of the Kuril Island chain, *Izv. Acad. Sci. USSR* (Engl. trans. *Phys. Solid Earth*) **9**, 17–33, 1969.

Fehler, M., Using dual-well seismic measurements to infer the mechanical properties of a hot dry rock geothermal system, *J. Geophys. Res.* **87**, 5485–5494, 1982.

Fehler, M., P. Roberts, and T. Fairbanks, A temporal change in coda wave attenuation observed during an eruption of Mount St. Helens, *J. Geophys. Res.* **93**, 4367–4373, 1988.

Fehler, M. and W. S. Phillips, Simultaneous inversion for Q and source parameters of microearthquakes accompanying hydrofracturing in granitic rock, *Bull. Seismol. Soc. Am.* **81**, 553–575, 1991.

Fehler, M., M. Hoshiba, H. Sato, and K. Obara, Separation of scattering and intrinsic attenuation for the Kanto-Tokai region, Japan, using measurements of S-wave energy versus hypocentral distance, *Geophys. J. Int.* **108**, 787–800, 1992.

Fei, T., M. Fehler, and S. Hildebrand, Finite-difference solutions of the 3-D Eikonal equation, *65th Annual Mtg., Soc. Expl. Geophys., Expanded Abstracts* **95**, 1129–1132, 1995.

Fei, T., M. Fehler, S. Hildebrand, and S. Chilcoat, Depth migration artifacts associated with first-arrival traveltime, *66th Annual Mtg., Soc. Expl. Geophys., Expanded Abstracts* **96**, 489–502, 1996.

Feustel, A. J., C. I. Trifu, and T. I. Urbancic, Rock-mass characterization using intrinsic and scattering attenuation estimates at frequencies from 400 to 1600 Hz, *Pure Appl. Geophys.* **147**, 289–304, 1996.

Fisk, M. D. and G. D. McCartor, The phase screen method for vector elastic waves, *J. Geophys. Res.* **96**, 5985–6010, 1991.

Fisk, M. D., E. E. Charrette, and G. D. McCartor, A comparison of phase screen and finite difference calculations for elastic waves in random media, *J. Geophys. Res.* **97**, 12409–12423, 1992.

Flatté, S. M., R. Dashen, W. H. Munk, K. M. Watson, and F. Zachariasen, *Sound Transmission through a Fluctuating Ocean*, Cambridge Univ. Press, New York, 1979.

Flatté, S. M. and R. S. Wu, Small-scale structure in the lithosphere and asthenosphere deduced from arrival time and amplitude fluctuations at NORSAR, *J. Geophys. Res.* **93**, 6601–6614, 1988.

Foldy, L. L., The multiple scattering of waves- I. General theory of isotropic scattering by randomly distributed scatterers, *Phys. Rev.* **67**, 107–119, 1945.

Fountain, D. M. and N. Christensen, Composition of the continental crust and upper mantle: A review, in *Geophysical Framework of the Continental United States* (eds., Pakiser, L.C. and W. Mooney), *Geol. Soc. Am. Memoir* **172**, 711–742, Geological Society of America, Boulder, Colo., 1989.

Frankel, A., The Effects of attenuation and site response on the spectra of microearthquakes in the northeastern Caribbean, *Bull. Seismol. Soc. Am.*, **72**, 1379–1402, 1982.

Frankel, A., Review for "Observational and physical basis for coda precursor (Jin, A. and K. Aki)" in *Evaluation of Proposed Earthquake Precursors* (ed. M. Wyss), pp. 51–53, AGU, Washington, D. C., 1991.

Frankel, A. and R. W. Clayton, Finite difference simulations of seismic scattering: Implications for the propagation of short-period seismic waves in the crust and models of crustal heterogeneity, *J. Geophys. Res.* **91**, 6465–6489, 1986.

Frankel, A. and L. Wennerberg, Energy-flux model of seismic coda: Separation of scattering and intrinsic attenuation, *Bull. Seismol. Soc. Am.* **77**, 1223–1251, 1987.

Frankel, A., A. McGarr, J. Bicknell, J. Mori, L. Seeber, and E. Cranswick, Attenuation of high-frequency shear waves in the crust: Measurements from New York State, South Africa and Southern California, *J. Geophys. Res.* **95**, 17441–17457, 1990.

Frisch, U., Wave Propagation in random media, in *Probabilistic Method in Applied Mathematics* (Vol. I, ed. A. T. Bharucha-Reid), pp. 76–198, Academic Press, New York, 1968.

Gagnepain–Beyneix, J., Evidence of spatial variations of attenuation in the western Pyrenean range, *Geophys. J. R. Astron. Soc.* **89**, 681–704, 1987.

Gao, L. S., N. N. Biswas, L. C. Lee, and K. Aki, Effects of multiple scattering on coda waves in three-dimensional medium, *Pure Appl. Geophys.* **121**, 3–15, 1983.

Gazdag, J. and P. Sguazzero, Migration of seismic data by phase shift plus interpolation, *Geophysics* **49**, 124–131, 1984.

Gibowicz, S. J. and A. Kijko, *An Introduction to Mining Seismology*, Academic Press, San Diego, 1994.

Gibson, R. L. and A. Ben-Menahem, Elastic wave scattering by anisotropic obstacles: Application to fractured volumes, *J. Geophys. Res.* **96**, 19905–19924, 1991.

Godano, C., E. Der Pezzo, and S. D. Martino, Dependence of the apparent seismic quality factor on epicentral distance: An interpretation in terms of fractal structure of the seismic medium, *Phys. Earth Planet. Inter.* **82**, 271–276, 1994.

Görich, U. and G. Müller, Apparent and intrinsic Q: One-dimensional case, *J. Geophys.* **61**, 46–54, 1987.

Got, J. L., G. Poupinet, and J. Fréchet, Changes in source and site effects compared to coda Q^{-1} temporal variations using microearthquakes doublets in California, *Pure Appl. Geophys.* **134**, 195–228, 1990.

Gradshteyn, I. S. and I. M. Ryzhik, *Table of Integrals, Series, and Products* (5th Ed. in Engl., ed. A. Jeffrey), Academic Press, San Diego, 1994.

Gritto, R., V. A. Korneev, and L. R. Johnson, Low-frequency elastic-wave scattering by an inclusion: Limits of applications, *Geophys. J. Int.* **120**, 677–692, 1995.

Gupta, I. N., C. S. Lynnes, and R. A. Wagner, Broadband *f–k* analysis of array data to identify source of local scattering, *Geophys. Res. Lett.* **17**, 183–186, 1990.

Gusev, A. A., Vertical profile of turbidity and coda Q, *Geophys. J. Int.* **123**, 665–672, 1995a.

Gusev, A. A., Baylike and continuous variations of the relative level of the late coda during 24 years of observation on Kamchatka, *J. Geophys. Res.* **100**, 20311–20319, 1995b.

Gusev, A. A. and V. K. Lemzikov, Properties of scattered elastic waves in the lithosphere of Kamchatka: Parameters and temporal variations, *Tectonophysics* **112**, 137–153, 1985.

Gusev, A. A. and I. R. Abubakirov, Monte-Carlo simulation of record envelope of a near earthquake, *Phys. Earth Planet. Inter.* **49**, 30–36, 1987.

Gusev, A. A. and V. M. Pavlov, Deconvolution of squared velocity waveform as applied to study of non-coherent short-period radiator in earthquake source, *Pure Appl. Geophys.* **136**, 236–244, 1991.

Gusev, A. A. and I. R. Abubakirov, Simulated envelopes of anisotropically scattered waves as compared to observed ones: New evidence for fractal heterogeneity, *Abstracts IUGG XXI General Assembly* **B-408**, Boulder, Colo., 1995.

Gutenberg, B., The energy of earthquakes, *Quart. J. Geol. Soc. London*, **112**, 1–14, 1956.

Haddon, R. A. W. and E. S. Husebye, Joint interpretation of P-wave time and amplitude anomalies in terms of lithospheric heterogeneities, *Geophys. J. R. Astron. Soc.* **55**, 19–43, 1978.

Hadley, K., Comparison of calculated and observed crack densities and seismic velocities in Westerly granite, *J. Geophys. Res.* **81**, 3484–3494, 1976.

Hardy, J., Y. Pomeau, and O. de Pazzis, Time evolution of two-dimensional model system, I: Invariant states and time correlation functions, *J. Math. Phys.* **14**, 1746–1759, 1973.

Hartse, H. E., A. R. Sanford, and J. S. Knapp, Incorporating Socorro magma body reflections into the earthquake location process, *Bull. Seismol. Soc. Am.* **82**, 2511–2532, 1992.

Hartse, H., W. S. Phillips, M. C. Fehler, and L. S. House, Single-station spectral discrimination using coda waves, *Bull. Seismol. Soc. Am.* **85**, 1464–1474, 1995.

Hatzidimitriou, P. M., S-wave attenuation in the crust in northern Greece, *Bull. Seismol. Soc. Am.* **85**, 1381–1387, 1995.

Herraiz, M. and A. F. Espinosa, Coda waves: A review, *Pure Appl. Geophys.* **125**, 499–577, 1987.

Hestholm, S. O., E. S. Husebye, and B. O. Ruud, Seismic wave propagation in complex crust–upper mantle media using 2-D finite-difference synthetics, *Geophys. J. Int.* **118**, 643–670, 1994.

Hiramatsu, Y., M. Ando, and F. Takeuchi, Temporal change in coda Q^{-1} in the Hida region, central Japan, *Geophys. Res. Lett.* **19**, 1403–1406, 1992.

Hirata, T. and S. Imoto, Multifractal analysis of spatial distribution of microearthquakes in the Kanto region, *Geophys. J. Int.* **107**, 155–162, 1991.

Hoang-Trong, P, Some medium properties of the Hohenzollerngraben (Swabian Jura, W. Germany) inferred from Q_P/Q_S analysis, *Phys. Earth Planet. Inter.* **31**, 119–131, 1983.

Holliger, K. and A. Levander, A stochastic view of lower crustal fabric based on evidence from the Ivrea zone, *Geophys. Res. Lett.* **19**, 1153–1156, 1992.

Holliger, K., A. Levander, and J. Goff, Stochastic modeling of the reflective lower crust: Petrophysical and geological evidence from the Ivrea zone (Northern Italy), *J. Geophys. Res.* **98**, 11967–11980, 1993.

Holme, R., and D. Rothman, Lattice-gas and lattice-Boltzman models of miscible fluids, *J. Stat. Phys.* **68**, 409–429, 1992.

Hong, S. T. and A. Ishimaru, Two-frequency mutual coherence function, coherence, bandwidth, and coherence time of millimeter and optical waves in rain, fog, and turbulence, *Radio Sci.* **11**, 551–559, 1976.

Hoshiba, M., Simulation of multiple-scattered coda wave excitation based on the energy conservation law, *Phys. Earth Planet. Inter.* **67**, 123–136, 1991.

Hoshiba M., Separation of scattering attenuation and intrinsic absorption in Japan using the multiple lapse time window analysis of full seismogram envelope, *J. Geophys. Res.* **98**, 15809–15824, 1993.

Hoshiba, M., Simulation of coda wave envelope in depth dependent scattering and absorption structure, *Geophys. Res. Lett.* **21**, 2853–2856, 1994.

Hoshiba, M., Estimation of nonisotropic scattering in western Japan using coda wave envelopes: Application of a multiple nonisotropic scattering model, *J. Geophys. Res.* **100**, 645–657, 1995.

Hoshiba, M., H. Sato, and M. Fehler, Numerical basis of the separation of scattering and intrinsic absorption from full seismogram envelope - A Monte-Carlo simulation of multiple isotropic scattering, *Pa. Meteorol. Geophys., Meteorol. Res. Inst.* **42**, 65–91, 1991.

Hough, S. E., J. G. Anderson, J. Brune, F. Vernon, III., J. Berger, J. Fletcher, L. Haar, T. Hanks, and L. Baker, Attenuation near Anza, California, *Bull. Seismol. Soc. Am.* **78**, 672–691, 1988.

Howe, M. S., Wave propagation in random media, *J. Fluid Mech.* **45**, 769–783, 1971.

Howe, M. S., Conservation of energy in random media, with application to the theory of sound absorption by an inhomogeneous flexible plate, *Proc. R. Soc. Lond. A.* **331**, 479–496, 1973.

Huang, L-J., and P. Mora, The phononic lattice solid by interpolation for modeling P waves in heterogeneous media, *Geophys. J. Int.* **119**, 766–778, 1994.

Huang, L-J., and P. Mora, Numerical simulation of wave propagation in strongly heterogeneous media using a lattice solid approach, *Soc. Photo-Optical Inst. Eng.* **2822**, 170–179, 1996.

Huang, L-J. and R. S. Wu, Prestack depth migration with acoustic screen propagators, *66th Annual Mtg., Soc. Expl. Geophys., Expanded Abstracts* **96**, 415–418, 1996.

Huang, L-J., and M. Fehler, Accuracy analysis of the split-step Fourier propagator: Implications for seismic modeling and migration, *Bull. Seismol. Soc. Am.* in press, 1998.

Huang, L-J., M. Fehler, and R. S. Wu, Generalized local Born Fourier migration for complex structures, Preprint, submitted to *Geophysics*, 1998.

Hudson, J. A. and J. R. Heritage, The use of the Born approximation in seismic scattering problems, *Geophys. J. R. Astron. Soc.* **66**, 221–240, 1981.

Husebye, E. S., ed., Contribution of scattering to the complexity of seismograms, *Phys. Earth Planet. Inter.* **26**, 233–291, 1981.

Husebye, E. S., A. Christoffersson, K. Aki, and C. Powell, Preliminary results on the 3-dimensional seismic structure under the USGS central California seismic array, *Geophys. J. R. Astron. Soc.* **46**, 319–340, 1976.

Ikelle, L. T., S. K. Yung, and F. Daube, 2-D random media with ellipsoidal autocorrelation functions, *Geophysics* **58**, 1359–1372, 1993.

Ishida, M. and A. H. Hasemi, Three-dimensional fine velocity structure and hypocentral distribution of earthquakes beneath the Kanto-Tokai district, Japan, *J. Geophys. Res.* **93**, 2076–2094, 1988.

Ishimaru, A., *Wave Propagation and Scattering in Random Media*, Vols. 1 and 2, Academic Press, New York, 1978.

Jacobson, R. S., An investigation into the fundamental relationships between attenuation, phase dispersion, and frequency using seismic refraction profiles over sedimentary structures, *Geophysics* **52**, 72–87, 1987.

Jackson, D. D. and D. L. Anderson, Physical mechanisms of seismic-wave attenuation, *Rev. Geophys. Space Phys.* **8**, 1–63, 1970.

Jannaud, L. R., P, M. Adler, and C. G. Jacquin, Spectral analysis and inversion of codas, *J. Geophys. Res.* **96**, 18215–18231, 1991a.

Jannaud, L. R., P. M. Adler, and G. G. Jacquin, Frequency dependence of the Q factor in random media, *J. Geophys. Res.* **96**, 18233–18243, 1991b.

Jannaud, L. R., P, M. Adler, and C. G. Jacquin, Wave propagation in random anisotropic media, *J. Geophys. Res.* **97**, 15277–15289, 1992.

Jensen, F., W. Kuperman, M. Porter, and H. Schmidt, *Computational Ocean Acoustics*, American Institute of Physics Press, New York, 1994.

Jin, A., T. Cao, and K. Aki, Regional change of coda Q in the oceanic lithosphere, *J. Geophys. Res.* **90**, 8651–8659, 1985.

Jin, A. and K. Aki, Temporal change in coda Q before the Tangshan earthquake of 1976 and the Haicheng earthquake of 1975, *J. Geophys. Res.* **91**, 665–673, 1986.

Jin, A. and K. Aki, Spatial and temporal correlation between coda Q and seismicity in China, *Bull. Seismol. Soc. Am.* **78**, 741–769, 1988.

Jin, A. and K. Aki, Spatial and temporal correlation between coda Q^{-1} and seismicity and its physical mechanism, *J. Geophys. Res.* **94**, 14041–14059, 1989.

Jin, A. and K. Aki, Observational and physical basis for coda precursor, in *Evaluation of Proposed Earthquake Precursors* (ed. M. Wyss), pp. 33–46, AGU, Washington, D. C., 1991.

Jin, A. and K. Aki, Temporal correlation between coda Q^{-1} and seismicity: Evidence for a structural unit in the brittle-ductile transition zone, *J. Geodyn.* **17**, 95–119, 1993.

Jin, A., K. Mayeda, D. Adams, and K. Aki, Separation of intrinsic and scattering attenuation in southern California using TERRAscope data, *J. Geophys. Res.* **99**, 17835–17848, 1994.

Kagan, Y. Y. and L. Knopoff, Spatial distribution of earthquakes: The two point correlation function, *Geophys. J. R. Astron. Soc.* **62**, 303–320, 1980.

Kakehi, Y. and K. Irikura, Estimation of high-frequency wave radiation areas on the fault plane by the envelope inversion of acceleration seismograms, *Geophys. J. Int.* **125**, 892–900, 1996.

Kanamori, H., The energy release in great earthquakes, *J. Geophys. Res.* **82**, 2981–2987, 1977.

Kanamori, H. and H. Mizutani, Ultrasonic measurements of elastic constants of rocks under high pressures, *Bull. Earthq. Res. Inst. Univ. Tokyo* **43**, 173–194, 1965.

Karal, F. C. and Keller, J. B., Elastic, electromagnetic and other waves in a random medium, *J. Math. Phys.* **5**, 537–547, 1964.

Kato K., K. Aki, and M. Takemura, Site amplification from coda waves: Validation and application to S-wave site response, *Bull. Seismol. Soc. Am.* **85**, 467–477, 1995.

Kawahara, J. and T. Yamashita, Scattering of elastic waves by a fracture zone containing randomly distributed cracks, *Pure Appl. Geophys.* **139**, 121–144, 1992.

Kawahara, J. and K. Yomogida, Scattering of SH waves in 2-D media with circular and elliptic cavities: A stochastic approach, *Abst. 1996 Japan Earth Planet. Sci. Joint Meeting*, 331, 1996.

Keenan, J. H., F. G. Keys, P. G. Hill and J. G. Moore, *Steam Tables - Thermodynamic Properties of Water Including Vapor, Liquid, and Solid Phases*, John Wiley & Sons, New York, 1969.

Kennett, B. L., *Seismic Wave Propagation in Stratified Media*, Cambridge Univ. Press, Cambridge, UK, 1985.

Key, F. A., Signal-generated noise recorded at the Eskdalemuir seismometer array station, *Bull. Seismol. Soc. Am.* **57**, 27–37, 1967.

Kikuchi, M., Dispersion and attenuation of elastic waves due to multiple scattering from cracks, *Phys. Earth Planet. Inter.* **27**, 100–105, 1981.

Kikuchi, M. and M. Ishida, Source retrieval for deep local earthquakes with broadband records, *Bull. Seismol. Soc. Am.* **83**, 1855–1870, 1993.

Kinoshita, S., Frequency-dependent attenuation of shear waves in the crust of the southern Kanto, Japan, *Bull. Seismol. Soc. Am.* **84**, 1387–1396, 1994.

Kishimoto, S. and S. Kinoshita, On the representation of the whole path S wave attenuation (in Japanese), *Abstract 45th Ann. Meeting Japan Soc. Civil. Eng.* 1032–1033, 1990.

Knopoff, L., Q, *Rev. Geophys.* **2**, 625–660, 1964.

Knopoff, L. and J. A. Hudson, Scattering of elastic waves by small inhomogeneities, *J. Acoust. Soc. Am.* **36**, 338–343, 1964.

Kopnichev, Y. F., A model of generation of the tail of the seismogram, *Dok. Akad. Nauk, SSSR* (Engl. trans.) **222**, 333–335, 1975.

Kopnichev, Y. F., The role of multiple scattering in the formulation of a seismogram's tail, *Izv. Acad. Nauk USSR* (Engl. trans. *Phys. Solid Earth*) **13**, 394–398, 1977.

Kopnichev, Yu. F., *Short-Period Seismic Wave Fields* (in Russian), Nauka, Moscow, 1985.

Korn, M., A modified energy flux model for lithospheric scattering of teleseismic body waves, *Geophys. J. Int.* **102**, 165–175, 1990.

Korn, M., Determination of site-dependent scattering Q from P-wave coda analysis with an energy-flux model, *Geophys. J. Int.* **113**, 54–72, 1993.

Korn, M., H. Sato, and F. Scherbaum, eds., Stochastic seismic wave fields and realistic media, *Phys. Earth Planet. Inter.* in press, 1997.

Korneev, V. A. and L. R. Johnson, Scattering of elastic wave by a spherical inclusion -I. Theory and numerical results, *Geophys. J. Int.* **115**, 230–250, 1993a.

Korneev, V. A. and L. R. Johnson, Scattering of elastic wave by a spherical inclusion -II. Limitations of asymptotic solutions, *Geophys. J. Int.* **115**, 251–263, 1993b.

Korneev, V. A. and L. R. Johnson, Scattering of P and S waves by a spherically symmetric inclusion, *Pure Appl. Geophys.* **147**, 675–718, 1996.

Kostrov, V., An inversion problem for seismic coda, *Proc. 17th Assembly E. S. C.* (eds. E. Bisztrisany and G. Y. Szeidovitz), 205–216, 1980.

Kostrov, B. V. and S. Das, *Principles of Earthquake Source Mechanics*, Cambridge University Press, Cambridge, UK, 1988.

Kosuga, M., Dependence of coda Q on frequency and lapse time in the western Nagano region, Japan, *J. Phys. Earth* **40**, 421–445, 1992.

Kurita, T., Attenuation of shear waves along the San Andreas fault zone in central California, *Bull. Seismol. Soc. Am.* **65**, 277–292, 1975.

Kuwahara, Y., H. Ito, M. Shinohara, and H. Kawakatsu, Small-array observation of seismic coda waves in Izu-Ohshima, *Zisin* (in Japanese) **43**, 359–371, 1990.

Kuwahara, Y., H. Ito, T. Ohminato, H. Kawakatsu, T. Kiguchi, and T. Miyazaki, Analysis of short period seismic coda waves observed with a small-span array (in Japanese), *Prog. Abs. Seism. Soc. Japan* **2**, 244, 1991.

Kuwahara, Y., H. Ito, H. Kawakatsu, T. Ohminato and T. Kiguchi, Crustal heterogeneity as inferred from seismic coda wave: Decomposition by small-aperture array observations, *Phys. Earth. Planet. Inter.* in press, 1997.

Kvamme, L. B. and J. Havskov, Q in southern Norway, *Bull. Seismol. Soc. Am.* **79**, 1575–1588, 1989.

Lacoss, T., E. J. Kelly, and M. N. Toksöz, Estimation of seismic noise structure using arrays, *Geophysics* **34**, 21–38, 1969.

Landau, L. D. and E. M. Lifshitz, *Quantum Mechanics* (3rd Ed., Engl. trans. by J. B. Sykes and J. S. Bell), Pergamon, Oxford, UK, 1989.

Langston, C. A., Scattering of teleseismic body waves under Pasadena, California, *J. Geophys. Res.* **94**, 1935–1951, 1989.

Lay, T., Analysis of near-source contributions to early P-wave coda for underground explosions. II. Frequency dependence, *Bull. Seismol. Soc. Am.* **77**, 1252–1273, 1987.

Lay, T. and T. C. Wallace, *Modern Global Seismology*, Academic Press, San Diego, 1995.

Leary, P. and R. Abercrombie, Frequency dependent crustal scattering and absorption at 5–160 Hz from coda decay observed at 2.5 km depth, *Geophys. Res. Lett.* **21**, 971–974, 1994.

Lee, L. C. and J. R. Jokipii, Strong scintillations in astrophysics. I. The Markov approximation, its validity and application to angular broadening, *Astrophys. J.* **196**, 695–707, 1975a.

Lee. L. C. and J. R. Jokipii, Strong scintillations in astrophysics. II. A theory of temporal broadening of pulses, *Astrophys. J.* **201**, 532–543, 1975b.

Lee, W. H. K. and S. Stewart, *Principles and Applications of Microearthquake Networks*, Academic Press, New York, 1981.

Lerche, I., Scintillations in astrophysics. I. An analytic solution of the second-order moment equation, *Astrophys. J.* **234**, 262–274, 1979.

Lerche, I. and W. Menke, An inversion method for separating apparent and intrinsic attenuation in layered media, *Geophys. J. R. Astron. Soc.* **87**, 333–347, 1986.

Levander, A. R. and K. Holliger, Small-scale heterogeneity and large-scale velocity structure of the continental crust, *J. Geophys. Res.* **97**, 8797–8804, 1992.

Lin, J. C. and A. Ishimaru, Multiple scattering of waves by a uniform random distribution of discrete isotropic scatterers, *J. Acoust. Soc. Am.* **56**, 1695–1700, 1974.

Liu, H., P., D. L. Anderson, and H. Kanamori, Velocity dispersion due to anelasticity; Implications for seismology and mantle composition, *Geophys. J. R. Astron. Soc.* **47**, 41–58, 1976.

Liu, Y. B. and R. S. Wu, A comparison between phase screen, finite difference, and eigen function expansion calculations for scalar waves in inhomogeneous media, *Bull. Seismol. Soc. Am.* **84**, 1154–1168, 1994.

Lundquist, G. M. and V. C. Cormier, Constraints on the absorption band model of Q, *J. Geophys. Res.* **85**, 5244–5256, 1980.

Madariaga, R., High-frequency radiation from crack (stress drop) models of earthquake faulting, *Geophys. J. Roy. Astron. Soc.* **51**, 625–651, 1977.

Main, I. G., S. Peacock, and P. G. Meredith, Scattering, attenuation and the fractal geometry of fracture systems, *Pure Appl. Geophys.* **133**, 283–304, 1990.

Malamud, A. S., On a possible precursor of strong earthquake, *Dok. Akad. Nauk Tadjik SSR* (in Russian) **17**, 31–34, 1974.

Malin, P. E., A first order scattering solution for modeling elastic wave codas - I. The acoustic case, *Geophys. J. Roy. Astron. Soc.* **63**, 361–380, 1980.

Malin, P. E. and R. A. Phinney, On the relative scattering of P- and S-waves, *Geophys. J. R. Astron. Soc.* **80**, 603–618, 1985.

Mandelbrot, B. B., *The Fractal Geometry of Nature*, W. H. Freeman, San Francisco, 1983.

Manghnani, M. H., R. Ramananantoandro, and S. P. Clark, Jr., Compressional and shear wave velocities in granulite faces rocks and eclogites to 10 kb, *J. Geophys. Res.* **79**, 5427–5446, 1974.

Marquardt, D. W., An algorithm for least squares estimation of nonlinear parameters, *J. Soc. Ind. Appl. Math.* **11**, 431–441. 1963.

Mason, W. P., Internal friction mechanism that produces an attenuation in the earth's crust proportional to the frequency, *J. Geophys. Res.* **74**, 4963–4966, 1969.

Mason, W. P., K. J. Marfurt, D. N. Beshers, and J. T. Kuo, Internal friction in rocks, *J. Acoust. Soc. Am.* **63**, 1596–1603, 1978.

Matsumoto, S., Characteristics of coda waves and inhomogeneity of the earth, *J. Phys. Earth* **43**, 279–299, 1995.

Matsumoto, S. and A. Hasegawa, Two-dimensional coda Q structure beneath Tohoku, NE Japan, *Geophys. J. Int.* **99**, 101–108, 1989.

Matsumoto, S. and A. Hasegawa, Estimation of relative site effect, coda Q and scattering strength from records obtained by a large airgun experiment, *Phys. Earth Planet. Inter.* **67**, 95–103, 1991.

Matsumoto, S. and A. Hasegawa, Distinct S wave reflector in the midcrust beneath Nikko-Shirane volcano in the northeastern Japan arc, *J. Geophys. Res.* **101**, 3067–3083, 1996.

Matsumura, S., Three-dimensional expression of seismic particle motions by the trajectory ellipsoid and its application to the seismic data observed in the Kanto district, Japan, *J. Phys. Earth* **29**, 221–239, 1981.

Matsunami, K., Laboratory measurements of spatial fluctuation and attenuation of elastic waves by scattering due to random heterogeneities, *Pure Appl. Geophys.* **132**,197–220, 1990.

Matsunami, K., Laboratory tests of excitation and attenuation of coda waves using 2-D models of scattering media, *Phys. Earth Planet. Inter.* **67**, 36–47, 1991.

Mavko, G. M. and A. Nur, Wave attenuation in partially saturated rocks, *Geophysics* **44**, 161–178, 1979.

Mavko, G., E. Kjartansson, and K. Winkler, Seismic wave attenuation in rocks, *Rev. Geophys. Space Phys.* **17**, 1155–1164, 1979.

Mayeda, K., High frequency scattered S-waves in the lithosphere: Application of the coda method to the study of source, site, and path effects, *Ph. D. Thesis*, Univ. Southern California, Los Angeles, 1991.

Mayeda, K., S. Koyanagi, and K. Aki, Site amplification from S-wave coda in the Long Valley Caldera region, California, *Bull. Seismol. Soc. Am.* **81**, 2194–2213, 1991.

Mayeda, K., S. Koyanagi, M. Hoshiba, K. Aki and Y. Zeng, A comparative study of scattering, intrinsic, and coda Q^{-1} for Hawaii, Long Valley and Central California between 1.5 and 15 Hz, *J. Geophys. Res.* **97**, 6643–6659, 1992.

Mayeda, K. and W. R. Walter, Moment, energy, stress drop, and source spectra of western United States earthquakes from regional coda envelopes, *J. Geophys. Res.* **101**, 11195–11208, 1996.

McCall, K. and R. Guyer, Equation of state and wave propagation in hysteretic nonlinear elastic materials, *J. Geophys. Res.* **99**, 23887–23897, 1994.

McLaughlin, K. L. and L. M. Anderson, Stochastic dispersion of short-period P-waves due to scattering and multipathing, *Geophys. J. R. Astron. Soc.* **89**, 933–963, 1987.

Menke, W. *Geophysical Data Analysis: Discrete Inverse Theory*, Academic Press, New York, 1984a.

Menke, W., Asymptotic formulas for the apparent Q of weakly scattering three-dimensional media, *Bull. Seismol. Soc. Am.* **74**, 1079–1081, 1984b.

Menke, W. and R. Chen, Numerical studies of the coda falloff rate of multiply scattered waves in randomly layered media, *Bull. Seismol. Soc. Am.* **74**, 1605–1614, 1984.

Mereu, R. F. and S. B. Ojo, The scattering of seismic waves through a crust and upper mantle with random lateral and vertical inhomogeneities, *Phys. Earth Planet. Inter.* **26**, 233–240, 1981.

Miles, J. W., Scattering of elastic waves by small inhomogeneities, *Geophysics*, **15**, 642–648, 1960.

Modiano, T. and D. Hatzfeld, Experimental study of the spectral content for shallow earthquakes, *Bull. Seismol. Soc. Am.* **72**, 1739–1758, 1982.

Mohorovicic, A., Das Beben vom 8. X, *Jahrb. Meterol. Obs. Zagreb* 9, 1–63, 1909.

Mooney, W., Seismic methods for determining earthquake source parameters and lithospheric structure, in *Geophysical Framework of the Continental United States* (eds., L. Pakiser and W. Mooney), *Geol. Soc. Am. Memoir* 172, 11–34, Geological Society of America, Boulder, Colo., 1989.

Mora, P., The lattice Boltzman phononic lattice solid, *J. Stat. Phys.* **68**, 591–609, 1992.

Mori J. and K. Shimazaki, Inversion of intermediate-period Rayleigh waves for source characteristics of the 1968 Tokachi-Oki earthquake, *J. Geophys. Res.* **90**, 11374–11382, 1985.

Mori, J. and A. Frankel, Correlation of P wave amplitudes and travel time residuals for teleseisms recorded on the southern California seismic network, *J. Geophys. Res.* **97**, 6661–6674, 1992.

Morse, P. M. and H. Feshbach, *Methods of Theoretical Physics*, Vols. I and II, McGraw-Hill, New York, 1953.

Müller, G., M. Roth, and M. Korn, Seismic wave traveltimes in random media, *Geophys. J. Int.* **110**, 29–41, 1992.

Murai, Y., J. Kawahara, and T. Yamashita, Multiple scattering of SH waves in 2-D elastic media with distributed cracks, *Geophys. J. Int.* **122**, 925–937, 1995.

Nakahara, H., Seismogram envelope inversion for the high frequency energy radiation from the 1994 far east off Sanriku earthquake (in Japanese), *M. Sc. Thesis*, Tohoku Univ., Sendai, Japan, 1996.

Nakahara, H., T. Nishimura, H. Sato, and M. Ohtake, Seismogram envelope inversion for the spatial distribution of high-frequency energy radiation on the earthquake fault: Application to the 1994 far east off Sanriku earthquake (Mw=7.7), *J. Geophys. Res.* in press, 1997.

Nakamura, Y., Seismic energy transmission in an intensively scattering environment, *J. Geophys.* **43**, 389–399, 1977.

Neidell, N. S. and M. T. Taner, Semblance and other coherency measures for multi-channel data, *Geophysics* **36**, 482–497, 1971.

New Energy and Industrial Technology Development Organization, *1990 Annual Report of Geothermal Exploration Technology by Using Seismic Method for the Identification of Fractured Reservoirs (Summary)*, (in Japanese) 1992a.

New Energy and Industrial Technology Development Organization, *1991 Annual Report of Geothermal Exploration Technology by Using Seismic Method for the Identification of Fractured Reservoirs (Summary)*, (in Japanese), 1992b.

Nikolaev, A. V., *The Seismics of Heterogeneous and Turbid Media* (Engl. trans. by R. Hardin), Israel Program for Science translations, Jerusalem, 1975.

Nikolajev, A. V. and F. S. Tregub, A statistical model of the earth's crust: Method and results, *Tectonophysics* **10**, 573–578, 1970.

Nishigami, K., A new inversion method of coda waveforms to determine spatial distribution of coda scatterers in the crust and uppermost mantle, *Geophys. Res. Lett.* **18**, 2225–2228, 1991.

Nishimura, T., H. Nakahara, H. Sato, and M. Ohtake, Source process of the 1994 far east off Sanriku earthquake, Japan, as inferred from a broad-band seismogram, *Sci. Rep. Tohoku Univ.* **34**, 121–134, 1996.

Nishimura, T., M. Fehler, W. Baldridge, P. Roberts, and L. Steck, Heterogeneous structure around the Jemez Volcanic field, New Mexico, USA, as inferred from envelope inversion of active experiment seismic data, *Geophys. J. Int.* in press, 1997.

Nishizawa, O., C. Pearson, and J. Albright, Properties of seismic wave scattering around water injection well at Fenton Hill hot dry rock geothermal site, *Geophys. Res. Lett.* **10**, 101–104, 1983.

Nishizawa, O., T. Satoh, X. Lei, and Y. Kuwahara, Laboratory studies of seismic wave propagation in inhomogeneous media using a laser Doppler vibrometer, *Bull. Seismol. Soc. Am.* **87**, 809–823, 1997.

Noguchi, S. Regional difference in maximum velocity amplitude decay with distance and earthquake magnitude (in Japanese), *Res. Notes Nat. Res. Ctr. Disast. Prev.* **86**, 1–40, 1990.

Nolet, G. Seismic wave propagation and seismic tomography, in *Seismic Tomography* (ed. G. Nolet), pp. 11–23, D. Reidel, Boston, 1987.

Novelo–Casanova, D. A., E. Berg, V. Hsu, and E. Helsley, Time-space variations of seismic S-wave coda attenuation Q_c^{-1} and magnitude distribution (b-value) for the Petatlan earthquake. *Geophys. Res. Lett.* **12**, 789–792, 1985.

Nur, A. Viscous phase in rocks and the low-velocity zone, *J. Geophys. Res.* **76**, 1270–1277, 1971.

Nur, A. and G. Simmons, The effect of saturation on velocity in low porosity rocks, *Earth Planet. Sci. Lett.* **7**, 183–193, 1969.

Obara, K., Regional extent of the S wave reflector beneath the Kanto district, Japan, *Geophys. Res. Lett.* **16**, 839–842, 1989.

Obara, K. and H. Sato, Existence of an S wave reflector near the upper plane of the double seismic zone beneath the southern Kanto district, Japan, *J. Geophys. Res.* **93**, 15037–15045, 1988.

Obara, K. and H. Sato, Regional differences of random inhomogeneities around the volcanic front in the Kanto-Tokai area, Japan, revealed from the broadening of S wave seismogram envelopes, *J. Geophys. Res.* **100**, 2103–2121, 1995.

O'Connell, R. J. and B. Budiansky, Viscoelastic properties of fluid-saturated cracked solids, *J. Geophys. Res.* **82**, 5719–5735, 1977.

O'Doherty, R. F. and N. A. Anstey, Reflections on amplitudes, *Geophys. Prospecting* **19**, 430–458, 1971.

Olson, A. H. and J. G. Anderson, Implications of frequency-domain inversion of earthquake ground motions for resolving the space-time dependence of slip on an extended fault, *Geophys. J. Int.* **94**, 443–455, 1988.

Ordaz, M. and S. K. Singh, Source spectra and spectral attenuation of seismic waves from Mexican earthquakes, and evidence of amplification in the hill zone of Mexico city, *Bull. Seismol. Soc. Am.* **82**, 24–43, 1992.

Papanicolaou, G. C., L. V. Ryzhik, and J. B. Keller, Stability of the P to S wave energy ratio in the diffusive regime, *Bull. Seismol. Soc. Am.* **86**, 1107–1115, 1996.

Peng, J. Y., Spatial and temporal variation of coda Q_c^{-1} in California, *Ph. D. Thesis*, Univ. Southern California, Los Angeles, 1989.

Perry, F., W. S. Baldridge, D. J. DePaolo, and M. Shafiqullah, Evolution of a magmatic system during continental extension: The Mount Taylor volcanic field, New Mexico, *J. Geophys. Res.* **95**, 19327–19348, 1990.

Phillips, W. S. and K. Aki, Site amplification of coda waves from local earthquakes in central California, *Bull. Seismol. Soc. Am.* **76**, 627–648, 1986.

Phillips, W. S., S. Kinoshita, and H. Fujiwara, Basin-induced Love waves observed using the Strong-motion array at Fuchu, Japan, *Bull. Seismol. Soc. Am.* **83**, 65–84, 1993.

Powell, C. A. and A. S. Meltzer, Scattering of P-waves beneath SCARLET in southern California, *Geophys. Res. Lett.* **11**, 481–484, 1984.

Press, F., Seismic Velocities, in *Handbook of Physical Constants* (ed., S. P. Clark), *Geol. Soc. Am. Memoir* **97**, 195–218, Geological Society of America, New York, 1966.

Pujol, J., An integrated 3D velocity inversion – joint hypocentral determination relocation analysis of events in the Northridge area, *Bull. Seismol. Soc. Am.* **86**, s138–s155, 1996.

Pulli, J. J., Attenuation of coda waves in New England, *Bull. Seismol. Soc. Am.* **74**, 1149–1166, 1984.

Radon, J., Über die bestimmung von funktionen durch ihre integralwerte längs gewisser mannigfaltigkeiten, *Ber. Verh. Sachs. Akad. Wiss. Leipzig, Math. Phys. Kl.* **69**, 262–267, 1917.

Rastogi, P. K. and K. F. Scheucher, Range dependence of volume scattering from a fractal medium: Simulation results, *Radio Sci.* **25**, 1057–1063, 1990.

Ratcliffe, J. A., Some aspects of diffraction theory and their application to the ionosphere, *Rep. Progr. Phys.* **19**, 188–267, 1956.

Rautian, T. G. and V. I. Khalturin, The use of the coda for determination of the earthquake source spectrum, *Bull. Seismol. Soc. Am.* **68**, 923–948, 1978.

Rautian, T. G. and V. I. Khalturin, V. G. Martinov, and P. Molnar, Preliminary analysis of the spectral content of P and S waves from local earthquakes in the Garm, Tadjikistan region, *Bull. Seismol. Soc. Am.* **68**, 949–971, 1978.

Rautian, T. G., V. I. Khalturin, M. S. Zakirov, A. G. Zemchova, A. P. Proskurin, B. G. Pustovitenko, A. N. Pustovitenko, L. G. Sinelinikova, A. G. Filina, and I. S. Tchengelia, *Experimental Studies of Seismic Coda* (in Russian), Nauka, Moscow, 1981.

Reha, S., *Q* determined from local earthquakes in the South Carolina coastal plain, *Bull. Seismol. Soc. Am.* **74**, 2257–2268, 1984.

Richards, P. G., Weakly coupled potentials for high-frequency elastic waves in continuously stratified media, *Bull. Seismol. Soc. Am.* **64**, 1575–1588, 1974.

Richards, P. G. and W. Menke, The apparent attenuation of a scattering medium, *Bull. Seismol. Soc. Am.* **73**, 1005–1021, 1983.

Richter, C. F., An instrumental earthquake magnitude scale, *Bull. Seismol. Soc. Am.* **25**, 1–32, 1935.

Ritter, J. R. R., P. M. Mai, G. Stoll, and K. Fuchs, Scattering of teleseismic waves in the lower crust observations in the Massif Central, France, *Phys. Earth Planet. Inter.* in press, 1997.

Rodriguez, M., J. Havskov, and S. K. Singh, Q from coda waves near Petatlan, Gurrero, Mexico, *Bull. Seismol. Soc. Am.* 73, 321–326, 1983.

Roecker, S. W., B. Tucker, J. King, and D. Hatzfeld, Estimation of Q in central Asia as a function of frequency and depth using the coda of locally recorded earthquakes, *Bull. Seismol. Soc. Am.* 72, 129–149, 1982.

Roth, M., Statistical interpretation of traveltime fluctuations, *Phys. Earth Planet. Inter.* in press, 1997.

Roth, M. and M. Korn, Single scattering theory versus numerical modelling in 2-D random media, *Geophys. J. Int.* 112, 124–140, 1993.

Roth, M., G. Müller, and R. Snieder, Velocity shift in random media, *Geophys. J. Int.* 115, 552–563, 1993.

Rothman, D., Cellular-automaton fluids: A model for flow in porous media, *Geophysics* 53, 509–518, 1988.

Rovelli, A., Frequency relationship for seismic Q_β of central southern Italy from accelerograms for the Irpinia earthquake (1980), *Phys. Earth Planet. Inter.* 32, 209–217, 1983.

Rovelli, A., Seismic Q for the lithosphere of the Montenegro region (Yugoslavia): Frequency, depth and time windowing effect, *Phys. Earth Planet. Inter.* 34, 159–172, 1984.

Rytov, S. M., Yu. A. Kravstov, and V. I. Tatarskii, *Principles of Statistical Radiophysics (Vol. 4) Wave Propagation Through Random Media*, Springer-Verlag, Berlin, 1989.

Ryzhik, L. V., G. C. Papanicolaou, and J. B. Keller, Transport equations for elastic and other waves in random media, *Wave Motion* 24, 327–370, 1996.

Sadovskiy, M. A., T. V. Golbeva, V. F. Pisarenko, and M. G. Shnirman, Characteristics dimensions of rock and hierarchical properties of seismicity, *Izv. Acad. Sci. USSR* (Engl. trans. *Phys. Solid Earth*), 20, 87–96, 1984.

Sakamoto, Y., M. Ishiguro, and G. Kitagawa, *Akaike Information Criterion Statistics*, D. Reidel, Dordrecht, Holland, 1986.

Sanford, A. and L. T. Long, Microearthquake crustal reflections, Socorro, New Mexico, *Bull. Seismol. Soc. Am.* 55, 579–586, 1965.

Sato, H., Energy propagation including scattering effect: Single isotropic scattering approximation, *J. Phys. Earth* 25, 27–41, 1977a.

Sato, H., Single isotropic scattering model including wave conversions: Simple theoretical model of the short period body wave propagation, *J. Phys. Earth* 25, 163–176, 1977b.

Sato, H., Mean free path of S-waves under the Kanto district of Japan, *J. Phys. Earth* 26, 185–198, 1978.

Sato, H., Wave propagation in one dimensional inhomogeneous elastic media, *J. Phys. Earth* 27, 455–466, 1979.

Sato, H., Amplitude attenuation of impulsive waves in random media based on travel time corrected mean wave formalism, *J. Acoust. Soc. Am.* 71, 559–564, 1982a.

Sato, H., Attenuation of S waves in the lithosphere due to scattering by its random velocity structure, *J. Geophys. Res.* 87, 7779–7785, 1982b.

Sato, H., Coda wave excitation due to nonisotropic scattering and nonspherical source radiation, *J. Geophys. Res.* 87, 8665–8674, 1982c.

Sato, H., Attenuation and envelope formation of three-component seismograms of small local earthquakes in randomly inhomogeneous lithosphere, *J. Geophys. Res.* 89, 1221–1241, 1984a.

Sato, H., Scattering and attenuation of seismic waves in the lithosphere: Single scattering theory in a randomly inhomogeneous lithosphere (in Japanese), *Rep. Nat. Res. Ctr. Disast. Prev.* 33, 101–186, 1984b.

Sato, H., Temporal change in attenuation intensity before and after the eastern Yamanashi earthquake of 1983 in central Japan, *J. Geophys. Res.* 91, 2049–2061, 1986.

Sato, H., A precursorlike change in coda excitation before the western Nagano earthquake (Ms=6.8) of 1984 in central Japan, *J. Geophys. Res.* **92**, 1356–1360, 1987.

Sato, H., Is the single scattering model invalid for the coda excitation at long lapse time?, *Pure Appl. Geophys.* **128**, 43–47, 1988a.

Sato, H., Fractal interpretation of the linear relation between logarithms of maximum amplitude and hypocentral distance, *Geophys. Res. Lett.* **15**, 373–375, 1988b.

Sato, H., Temporal change in scattering and attenuation associated with the earthquake occurrence - A review of recent studies on coda waves, *Pure Appl. Geophys.* **126**, 465–497, 1988c.

Sato, H., Broadening of seismogram envelopes in the randomly inhomogeneous lithosphere based on the parabolic approximation: Southeastern Honshu, Japan, *J. Geophys. Res.* **94**, 17735–7747, 1989.

Sato, H., Unified approach to amplitude attenuation and coda excitation in the randomly inhomogeneous lithosphere, *Pure. Appl. Geophys.* **132**, 93–121, 1990.

Sato, H., Study of seismogram envelopes based on scattering by random inhomogeneities in the lithosphere: A review, *Phys. Earth Planet. Inter.* **67**, 4–19, 1991a.

Sato, H., ed., Scattering and attenuation of seismic waves, *Phys. Earth Planet. Inter.* **67**, 1–210, 1991b.

Sato, H., Energy transportation in one- and two-dimensional scattering media: Analytic solutions of the multiple isotropic scattering model, *Geophys. J. Int.* **112**, 141–146, 1993.

Sato, H., Multiple isotropic scattering model including P–S conversions for the seismogram envelope formation, *Geophys. J. Int.* **117**, 487–494, 1994a.

Sato, H., Formulation of the multiple non-isotropic scattering process in 2-D space on the basis of energy transport theory, *Geophys. J. Int.* **117**, 727–732, 1994b.

Sato, H., Formulation of the multiple non-isotropic scattering process in 3-D space on the basis of energy transport theory, *Geophys. J. Int.* **121**, 523–531, 1995a.

Sato, H., Fractal model for the power law decay of seismic wave amplitudes and the uniform distribution of coda energy, *Abstracts IUGG XXI General Assembly* B-399, Boulder, Colo., 1995b.

Sato, H. and S. Matsumura, Q^{-1} value for S-waves (2–32 Hz) under the Kanto district in Japan, *Zisin* (in Japanese) **33**, 541–543, 1980.

Sato, H., A. M. Shomahmadov, V. I. Khalturin, and T. G. Rautian, Temporal change in spectral coda attenuation Q^{-1} associated with the K=13.3 earthquake of 1983 near Garm, Tadjikistan region in Soviet central Asia, *Zisin* (in Japanese), **41**, 39–46, 1988.

Sato, H., H. Nakahara, and M. Ohtake, Synthesis of scattered energy density for non-spherical radiation from a point shear dislocation source based on the radiative transfer theory, *Phys. Earth Planet. Inter.* in press, 1997.

Savage, J. C., Attenuation of elastic waves in granular medium, *J. Geophys. Res.* **70**, 3935–3942, 1966a.

Savage, J. C., Thermoelastic attenuation of elastic waves by cracks, *J. Geophys. Res.* **71**, 3929–3938, 1966b.

Scherbaum, F. and H. Sato, Inversion of full seismogram envelopes based on the parabolic approximation: Estimation of randomness and attenuation in southeast Honshu, Japan, *J. Geophys. Res.* **96**, 2223–2232, 1991.

Scherbaum, F., D. Gillard, and N. Deichmann, Slowness power spectrum analysis of the coda composition of two microearthquake clusters in northern Switzerland, *Phys. Earth Planet. Inter.* **67**, 137–161, 1991.

Schilt, S., J. Oliver, L. Brown, S. Kaufman, D. Albauch, J. Brewer, F. Cook, L. Jensen, P. Krumhansl, G. Long, and D. Steiner, The heterogeneity of the continental crust: Results from deep crustal reflection profiling using the VIBROSEIS technique, *Rev. Geophys. Space Phys.* **17**, 354–368, 1979.

Schneider, W. A., Integral formulation for migration in two and three dimensions, *Geophysics* **43**, 49–76, 1978.

Shang, T. and L. Gao, Transportation theory of multiple scattering and its application to seismic coda waves of impulsive source, *Scientia Sinica* (series B, China) **31**, 1503–1514, 1988.

Shapiro, S. A., Elastic wave scattering and radiation by fractal inhomogeneity of a medium, *Geophys. J. Int.* **110**, 591–600, 1992.

Shapiro, S. A. and G. Kneib, Seismic attenuation by scattering: Theory and numerical results, *Geophys. J. Int.* **114**, 373–391, 1993.

Shapiro, S. A. and Zien, H, The O'Doherty–Anstey formula and localization of seismic waves, *Geophysics* **58**, 736–740, 1993.

Shiomi, K., H. Sato, and M. Ohtake, Broad power spectra of well-log data obtained in Kyushu and Kanto, Japan, *Proc. VIIth Int. Symp. Obs. Continental Crust through Drilling*, Tsukuba, Japan, pp. 253–258, 1996.

Shiomi, K., H. Sato and M. Ohtake, Broad-band power-law spectra of well-log data in Japan, *Geophys. J. Int.* **130**, 57–64, 1997.

Shriner, J. F. and W. J. Thompson, Angular momentum coupling coefficients: New and improved algorithm, *Comp. Phys.* **7**, 144–148, 1993.

Simmons, G, and A. Nur, Granites - Relation of properties in situ to laboratory measurements, *Science*, **162**, 789–791, 1968.

Simmons, G. and H. Wang, *Single Crystal Elastic Constants and Calculated Aggregate Properties: A Handbook*, MIT Press, Cambridge, Mass. 1971.

Singh, S. and R. B. Herrmann, Regionalization of crustal coda Q in the continental United States, *J. Geophys. Res.* **88**, 527–538, 1983.

Singh, S. K., R. J. Apsel, J. Fried, and J. N. Brune, Spectral attenuation of SH waves along the Imperial fault, *Bull. Seismol. Soc., Am.* **72**, 2003–2016, 1982.

Solov'ev, S. L., Seismicity of Sakhalin, *Bull. Earthq. Res. Inst.* **43**, 95–102, 1965.

Spencer, Jr. J. W., Stress relaxation at low frequencies in fluid-saturated rocks: Attenuation and modulus dispersion, *J. Geophys. Res.* **86**, 12803–1812, 1981.

Spivack, M. and B. Uscinski, The split-step solution in random wave propagation, *J. Comp. Appl. Math.* **27**, 349-361, 1989.

Spudich, P. and T. Bostwick, Studies of the seismic coda using an earthquake cluster as a deeply buried seismograph array, *J. Geophys. Res.* **92**, 10526–10546, 1987.

Sreenivasiah, I., A. Ishimaru, and S. T. Hong, Two-frequency mutual coherence function and pulse propagation in a random medium: An analytic solution to the plane wave case, *Radio Sci.* **11**, 775–778, 1976.

Stoffa, P. L., J. T. Fokkema, R. M. L. Freire, and W. P. Kessinger, Split-step Fourier migration, *Geophysics* **55**, 410–421, 1990.

Su, F. and K. Aki, Temporal and spatial variation on coda Q^{-1} associated with the North Palm Springs earthquake of July 8, 1986, *Pure Appl. Geophys.* **133**, 23–52, 1990.

Su, F., K. Aki, and N. Biswas, Discriminating quarry blasts from earthquakes using coda waves, *Bull. Seismol. Soc. Am.* **81**, 162–178, 1991.

Sugimura, A., Zonal arrangement of some geophysical and petrological features in Japan and its environs, *J. Fac. Sci. Univ. Tokyo* (Sect. 2) **12**, 133–153, 1960.

Suzuki, H., R. Ikeda, T. Mikoshiba, S. Kinoshita, H. Sato, and H. Takahashi, Deep well logs in the Kanto-Tokai area (in Japanese), *Rev. Nat. Res. Ctr. Disast. Prev.* **65**, 1–162, 1981.

Takahara, M. and K. Yomogida, Estimation of coda Q using the maximum likelihood method, *Pure Appl. Geophys.* **139**, 255–268, 1992.

Takemura, M., K. Kato, T. Ikeura, and E. Shima, Site amplification of S-waves from strong motion records in special relation to surface geology, *J. Phys. Earth* **39**, 537–552, 1991.

Tarantola, A., *Inverse Problem Theory*, Elsevier Science, Amsterdam, 1987.

Tatarskii, V. I., *The Effects of the Turbulent Atmosphere on Wave Propagation*, Israel Program for Science translations, Jerusalem, 1971.

Tatsumi, Y., Formation of the volcanic front in subduction zones, *Geophys. Res. Lett.* **13**, 717–720, 1986.

Tatsumi, Y., Migration of fluid phases and genesis of basalt magmas in subduction zones, *J. Geophys. Res.* **94**, 4697–4707, 1989.

Taylor, S. R., B. P. Bonner, and G. Zandt, Attenuation and scattering of broadband P and S waves across North America, *J. Geophys. Res.* **91**, 7309–7325, 1986.

Telford, W., L. Geldart, R. Sheriff, and D. Keys, *Applied Geophysics*, Cambridge University Press, Cambridge, UK. 1976.

Thatcher, W. and T. C. Hanks, Source parameters of southern California earthquakes, *J. Geophys. Res.* **78**, 8547–8576, 1973.

Thomson, D. J. and N. R. Chapman, A wide-angle split-step algorithm for the parabolic equation, *J. Acoust. Soc. Am.* **74**, 1848–1854, 1983.

Thurber, C., Local earthquake tomography: Velocities and V_p / V_s - theory, in *Seismic Tomography: Theory and Practice* (eds. H. M. Iyer and K. Hirahara), pp. 563–583, Chapmann & Hall, London, 1993.

Tittmann, B. R., Internal friction measurements and their implications in seismic Q structure model of the crust, in *The Earth's Crust*, pp. 197–213, AGU monograph, 1977.

Tittmann, B. R., L. Ahlberg, and J. Curnow, Internal friction and velocity measurements, *Proc. 7th Lunar Sci. Conf.* 3123–3132, 1976.

Tittmann, B. R., V. A. Clark, and J. M. Richardson, Possible mechanism for seismic attenuation in rocks containing small amount of volatiles, *J. Geophys. Res.* **85**, 5199–5208, 1980.

Toksöz, N. and D. H. Johnston, eds., *Seismic Wave Attenuation*, Soc. Expl. Geophys., Tulsa, Okla., 1981.

Toksöz, M. N., A. M. Dainty, and E. E., Charrette, Coherency of ground motion at regional distances and scattering, *Phys. Earth Planet. Inter.* **67**, 162–179, 1991.

Tsuboi, C., Determination of the Gutenberg–Richter's magnitude of earthquakes occurring in and near Japan, *Zisin* (in Japanese), **7**, 185–193, 1954.

Tsujiura, M., Spectral analysis of the coda waves from local earthquakes, *Bull. Earthq. Inst. Univ. Tokyo* **53**, 1–48, 1978.

Tsujiura, M., Characteristic Seismograms, *Bull. Earthq. Res. Inst. Univ. Tokyo* Sup. **5**, 1–212, 1988.

Tsukuda, T., Coda Q before and after the 1983 Misasa earthquake of M 6.2, Tottori Prefecture, Japan, *Pure Appl. Geophys.* **128**, 261–280, 1988.

Tsumura, K., Determination of earthquake magnitude from duration of oscillation, *Zisin* (in Japanese) **20**, 30–40, 1967a.

Tsumura, K., Determination of earthquake magnitude from total duration of oscillation, *Bull. Earthq. Res. Inst. Univ. Tokyo* **45**, 7–18, 1967b.

Tucker, B. and J. Brune, Source mechanism and $m_b - M_s$ analysis of aftershocks of the San Fernando earthquake, *Geophys. J. R. Astron. Soc.* **49**, 371–426, 1977.

Tucker, B. and J. King, Dependence of sediment-filled valley response on input amplitude and valley properties, *Bull. Seismol. Soc. Am.* **74**, 153–165, 1984.

Um, J. and C. Thurber, A fast algorithm for two-point seismic ray tracing, *Bull. Seismol. Soc. Am.* **77**, 972–986, 1987.

Uscinski, J., *The Elements of Wave Propagation in Random Media*, McGraw–Hill, New York, 1977.

Uyeda, S. and K. Horai, Terrestrial heat flow in Japan, *J. Geophys. Res.* **69**, 2121–2141, 1964.

Varadan, V. K., V. V. Varadan, and Y. H. Pao, Multiple scattering of elastic waves by cylinders of arbitrary cross section. I. SH waves, *J. Acoust. Soc. Am.* **63**, 1310–1319, 1978.

Vasco, D. W., L. R. Johnson, R. J. Pulliam, and P. Earle, Robust inversion of IASP91 travel time residuals for mantle P and S velocity structure, earthquake mislocations, and station corrections, *J. Geophys. Res.* **99**, 13727–13755, 1994.

Vidale, J., Finite-difference calculation of travel times, *Bull. Seismol. Soc. Am.* **78**, 2062–2076, 1988.

Vinogradov, S. D., P. A. Troitskiy, and M. S. Solov'yeva, Study of propagation of elastic waves in medium with oriented cracks, *Izv. Acad. Sci. USSR* (Engl. trans. *Phys. Solid Earth*) **28**, 367–384, 1992.

Wagner, G. S., Regional wave propagation in southern California and Nevada: Observations from a three-component seismic array, *J. Geophys. Res.* **102**, 8285-8311, 1997.

Wagner, G. S. and C. A. Langston, A numerical investigation of scattering effects for teleseismic plane wave propagation in a heterogeneous layer over a homogeneous half-space, *Geophys. J. Int.* **110**, 486–500, 1992a.

Wagner, G. S. and C. A. Langston, Body-to-surface-wave scattered energy in teleseismic coda observed at the NORESS seismic array, *Bull. Seismol. Soc. Am.* **82**, 2126–2138, 1992b.

Wagner, G. S. and T. J. Owens, Broadband bearing-time records of three-component seismic array data and their application to the study of local earthquake coda, *Geophys. Res. Lett.* **20**, 1823–1826, 1993.

Walsh, J. B., The effects of cracks on the compressibility of rocks, *J. Geophys. Res.* **70**, 381–389, 1965.

Walsh, J. B., Seismic wave attenuation in rock due to friction, *J. Geophys. Res.* **71**, 2591–2599, 1966.

Walsh, J. B., New analysis of attenuation in partially melted rock, *J. Geophys. Res.* **74**, 4333–4337, 1969.

Warner, M., Absolute reflection coefficients from deep seismic reflections, *Tectonophysics* **173**, 15–23, 1990a.

Warner, M., Basalts, water, or shear zones in the lower continental crust?, *Tectonophysics* **173**, 163–174, 1990b.

Watanabe, H., Determination of earthquake magnitude at regional distance in and near Japan, *Zisin* (in Japanese) **24**, 189–200, 1971.

Watanabe, K., M. Ohtake, and H. Sato, Model of high-frequency seismogram envelope on the basis of energy transport theory: Formulation of multiple non-isotropic scattering and non-spherical source radiation (in Japanese), *Prog. Abs. Seism. Soc. Japan* **2**, B23, 1996.

Weiland, C. M., L. K. Steck, P. B. Dawson, and V. Korneev, Nonlinear teleseismic tomography at Long Valley Caldera, using three-dimensional minimum travel time ray tracing. *J. Geophys. Res.* **100**, 20379–20390, 1995.

Wennerberg, L., Multiple-scattering interpretation of coda-Q measurements, *Bull. Seismol. Soc. Am.* **83**, 279–290, 1993.

Wenzel, A. R., Radiation and attenuation of waves in a random medium, *J. Acoust. Soc. Am.* **71**, 26–35, 1982.

Wesley, J. P., Diffusion of seismic energy in the near range, *J. Geophys. Res.* **70**, 5099–5106, 1965.

White, J. E., *Seismic Waves; Radiation, Transmission, and Attenuation*, McGraw–Hill, New York, 1965.

Williamson, I. P., Pulse broadening due to multiple scattering in the interstellar medium, *Mon. Not. R. Astron. Soc.* **157**, 55–71, 1972.

Williamson, I. P., The broadening of pulses due to multi-path propagation of radiation, *Proc. R. Soc. London. A.* **342**, 131–147, 1975.

Williamson, P. R., A guide to the limits of resolution imposed by scattering in ray tomography, *Geophysics* **56**, 202–207, 1991.

Wolfram, S., *Mathematica: A System for Doing Mathematics by Computer* (2nd Ed.), Addison-Wesley, Redwood, Calif., 1991.

Wu, R. S. Attenuation of short period seismic waves due to scattering, *Geophys. Res. Lett.* **9**, 9–12, 1982a.

Wu, R. S., Mean field attenuation and amplitude attenuation due to wave scattering, *Wave Motion* **4**, 305–316, 1982b.

Wu, R. S., Multiple scattering and energy transfer of seismic waves - separation of scattering effect from intrinsic attenuation - I. Theoretical modeling, *Geophys. J. R. Astron. Soc.* **82**, 57–80, 1985.

Wu, R. S., The perturbation method in elastic wave scattering, *Pure Appl. Geophys.* **131**, 605–638; 131, 1989.

Wu, R. S., Separation of scattering and absorption in 1-D random media from spatio-temporal distribution of seismic energy, *63rd Annual Internat. Mtg., Soc. Expl. Geophys., Expanded Abstracts* **93**, 1014–1017, 1993.

Wu, R. S., Wide-angle elastic wave one-way propagation in heterogeneous media and elastic wave complex-screen method, *J. Geophys. Res.* **99**, 751–766, 1994.

Wu, R. S. and K. Aki, The fractal nature of the inhomogeneities in the lithosphere evidenced from seismic wave scattering, *Pure Appl. Geophys.* **123**, 805–818, 1985a.

Wu, R. S. and K. Aki, Elastic wave scattering by a random medium and the small-scale inhomogeneities in the lithosphere, *J. Geophys. Res.* **90**, 10261–10273, 1985b.

Wu, R. S. and M. N. Toksöz, Diffraction tomography and multisource holography applied to seismic imaging, *Geophysics* **52**, 11–25, 1987.

Wu, R. S. and K. Aki, Multiple scattering and energy transfer of seismic waves - Separation of scattering effect from intrinsic attenuation. II. Application of the theory to Hindu-Kush region, *Pure Appl Geophys.* **128**, 49–80, 1988a.

Wu, R. S. and K. Aki, Introduction: Seismic wave scattering in three-dimensionally heterogeneous earth, *Pure Appl. Geophys.* **128**, 1–6, 1988b.

Wu, R. S. and K. Aki, eds., Scattering and attenuation of seismic waves, *Pure Appl. Geophys.* **128**, 1–447, 1988c; **131**, 551–739, 1989; **132**, 1–437, 1990.

Wu, R. S. and S. M. Flatté, Transmission fluctuations across an array and heterogeneities in the crust and upper mantle, *Pure Appl. Geophys.* **132**, 175–196, 1990.

Wu, R. S. and X-B. Xie, Separation of scattering and absorption in 1-D random media: II. Numerical experiments on stationary problems, *64th Annual Mtg., Soc. Expl. Geophys., Expanded Abstracts* **94**, 1302–1305, 1994.

Wu, R. S., Z. Xu, and X. P. Li, Heterogeneity spectrum and scale-anisotropy in the upper crust revealed by the German Continental Deep-Drilling (KTB) holes, *Geophys. Res. Lett.* **21**, 911–914, 1994.

Wu, R. S., L. J. Huang, and X. B. Xie, Backscattered wave calculation using the De Wolf approximation and a phase-screen propagator, *65th Annual Mtg., Soc. Expl. Geophys., Expanded Abstracts* **95**, 1923–1926, 1995.

Yamakawa, N., Scattering and attenuation of elastic waves, *Geophys. Mag.* **31**, 63–103, 1962.

Yamashita, T., Attenuation and dispersion of SH waves due to scattering by randomly distributed cracks, *Pure Appl. Geophys.* **132**, 545–568, 1990.

Yan, F. and H. Mo, Coda of the 1982 Jianchuan earthquake, *J. Seismol. Res.* (in Chinese) **7**, 505–510, 1984.

Ying, C. F. and R. Truell, Scattering of a plane longitudinal wave by a spherical obstacle in an isotropically elastic solid, *J. Appl. Phys.* **27**, 1086–1097, 1956.

Yomogida, K. and R. Benites, Relation between direct wave Q and coda Q: A numerical approach, *Geophys. J. Int.* **123**, 471–483, 1995.

Yomogida, K., K. Aki, and R. Benites, Coda Q in two-layer random media, *Geophys. J. Int.* **128**, 425–433, 1996.

Yoshimoto, K., Formation of seismogram envelopes due to short wavelength components of the crustal heterogeneity (in Japanese), *Ph. D. thesis*, Science, Tohoku Univ., Sendai, Japan, 1995.

Yoshimoto, K., H. Sato, and M. Ohtake, Frequency-dependent attenuation of P and S waves in the Kanto area, Japan, based on the coda-normalization method, *Geophys. J. Int.* **114**, 165–174, 1993.

Yoshimoto, K., H. Sato, H. Ito, T. Ohminato and M. Ohtake, Attenuation of high-frequency P and S waves in shallow crustal rocks obtained from the extended coda normalization method, *IASPEI Abstracts of 27th General Assembly* S1.1, Wellington, 1994.

Yoshimoto, K., H. Sato, and M. Ohtake, Three-component seismogram envelope synthesis in randomly inhomogeneous semi-infinite media based on the single scattering approximation, *Phys. Earth Planet. Inter.* in press, 1997a.

Yoshimoto, K., H. Sato, and M. Ohtake, Short-wavelength crustal inhomogeneities in the Nikko area, central Japan, revealed from the three-component seismogram envelope analysis, *Phys. Earth Planet. Inter.* in press, 1997b.

Yuhara, K., Effect of the hydrothermal systems of the terrestrial heat flow (in Japanese), *Bull. Volcanol. Soc. Japan* **18**, 129–142, 1973.

Zelt, C. A. and R. B. Smith, Seismic traveltime inversion for 2-D crustal velocity structure, *Geophys. J. Int.* **108**, 16–34, 1992.

Zener, G., *Elasticity and Anelasticity of Metals*, University of Chicago Press, Chicago, 1948.

Zeng, Y. Compact solutions for multiple scattered wave energy in time domain, *Bull. Seismol. Soc. Am.* **81**, 1022–1029, 1991.

Zeng, Y. Theory of scattered P- and S-wave energy in a random isotropic scattering medium, *Bull. Seismol. Soc. Am.* **83**, 1264–1276, 1993.

Zeng, Y., F. Su, and K. Aki, Scattering wave energy propagation in a random isotropic scattering medium 1. Theory, *J. Geophys. Res.* **96**, 607–619, 1991.

Zeng, Y., K. Aki, and T. L. Teng, Mapping of the high-frequency source radiation for the Loma Prieta earthquake, California, *J. Geophys. Res.* **98**, 11981–11993, 1993.

Zhao, D., A. Hasegawa, and S. Horiuchi, Tomographic imaging of P and S wave velocity structure beneath Northeastern Japan, *J. Geophys. Res.* **97**, 19909–19928, 1992.

Zhu, T., K. Chun, and G. F. West, Geometrical spreading and Q of Pn waves: An investigative study in eastern Canada, *Bull. Seismol. Soc. Am.* **81**, 882–896, 1991.

Subject Index